Climate to a Fish Sandwich: Why We Study the Ocean's Circulation

Robert Weisberg

Climate to a Fish Sandwich: Why We Study the Ocean's Circulation

 Springer

Robert Weisberg
College of Marine Science
University of South Florida
St. Petersburg, FL, USA

ISBN 978-3-031-77591-8   ISBN 978-3-031-77592-5 (eBook)
https://doi.org/10.1007/978-3-031-77592-5

© The Editor(s) (if applicable) and The Author(s), under exclusive license to Springer Nature Switzerland AG 2025

This work is subject to copyright. All rights are solely and exclusively licensed by the Publisher, whether the whole or part of the material is concerned, specifically the rights of translation, reprinting, reuse of illustrations, recitation, broadcasting, reproduction on microfilms or in any other physical way, and transmission or information storage and retrieval, electronic adaptation, computer software, or by similar or dissimilar methodology now known or hereafter developed.
The use of general descriptive names, registered names, trademarks, service marks, etc. in this publication does not imply, even in the absence of a specific statement, that such names are exempt from the relevant protective laws and regulations and therefore free for general use.
The publisher, the authors and the editors are safe to assume that the advice and information in this book are believed to be true and accurate at the date of publication. Neither the publisher nor the authors or the editors give a warranty, expressed or implied, with respect to the material contained herein or for any errors or omissions that may have been made. The publisher remains neutral with regard to jurisdictional claims in published maps and institutional affiliations.

This Springer imprint is published by the registered company Springer Nature Switzerland AG
The registered company address is: Gewerbestrasse 11, 6330 Cham, Switzerland

If disposing of this product, please recycle the paper.

When growing up, my family had a rental property at Rockaway Beach, Queens, New York, where we spent our summers. I would accompany my father on many spring weekends to help with cleaning, painting, repairing and whatever other chores were necessary to prepare the property for rental showings. I recall as a child commenting that with this work being difficult, it would be expensive to hire someone to do it, to which my dad said just the opposite. He advised me that unskilled, manual labor pays very little so I should take my studies seriously and attend college to avoid such a future. Education was a priority within my family, and both my mother and father were quite pleased when I was accepted at Cornell University. It is an understatement to say just how important parental advice and encouragement is. Achievements generally begin at home so I am dedicating this book in memory to my parents, Lillian and Leonard Weisberg.

Foreword

The answer to why I became an oceanographer is multifaceted. In its simplest sense, I enjoy walking along the beach and doing puzzles. The vastness of the oceans and the mysteries that it exudes come to mind as one strolls the strand, and solving these mysteries is akin to doing puzzles. The answer, of course, is more complex, and in that lies important advice for both our youth and their mentors.

I was always interested in science, physical science in particular. Reading was enjoyable, but playing with things was more so. Grammar was of no interest, whereas mathematics was. In the eighth grade, my teacher told me something to the effect: Bob, your affinity for science and mathematics exceeds that for English and the arts so I recommend that you take the entrance examination for Brooklyn Technical High School and attend there should you qualify. I followed her advice and becoming an engineer was my adopted goal. Metallurgy caught my attention, specifically new materials that could facilitate space exploration. Upon graduation from Brooklyn Tech, I even had a few credits from the nearby Brooklyn Polytechnical Institute, and off I went to Cornell University to pursue my career in Materials Science and Engineering.

Two things intervened: the Vietnam War and the Apollo Moon Mission. My initial patriotic fervor gave way to a more reasoned, albeit still patriotic, opposition, once it appeared that political gain more so than national interest was steering the war. Nonetheless, I engaged in ROTC and committed myself to a United States Army Reserve commission, when, in that year, graduate school deferments were no longer automatic—I wanted to attend graduate school.

At some point, my interest in engineering waned. Working for an engineering company during the summer after my sophomore year, I took every opportunity to seek advice from those already in the field. I was disappointed to learn that the quickest way to advance was to eschew engineering science and, instead, enter engineering management. Had management been my goal, I would have pursued a business curriculum and not studied nearly as much. Later in time came the first Apollo-crewed mission. Watching it on television led to the realization that I knew neither how the television worked nor the rocketry to send someone to the Moon; hence my disillusion with an engineering career.

No longer interested in engineering, what was the sensible thing to do? My scientific interest remained keen, but for larger, versus molecular scale applications, and, of course, I still liked taking walks along the beach (the summers of my youth were spent on Rockaway Beach) so why not oceanography? Upon admission to the University of Rhode Island's Graduate School of Oceanography, I proposed to Cindy, my partner for now over 55 years, and off we went.

The lessons are several. First, despite a youthful exuberance of knowing exactly what you want to do, the reality is that you really do not. Getting a well-rounded education is important because that opens doors to anything should your initial interest change. Regardless of my penchant for science, my favorite courses at Cornell were Art History and British Literature. The university that you attend will offer both meaning and opportunity—take advantage of that, attend a good one, and don't waste time. I was fortunate to attend a great one. Second, effort pays off. Cornell engineering was a difficult curriculum, but it taught me how to work, and while I did not use all of my specific engineering courses, the science and mathematics translated directly to physical oceanography (the study of the ocean's circulation and how the ocean and atmosphere interact)—nothing of my undergraduate studies was wasted. Third, my involvement in the United States Army Reserve was a terrific augmentation to my other studies, and what I took from that experience served me well over the years. I credit the basic training mantra: Stay alert, stay alive, for why no one on my many sea-going research expeditions, even while working the deck in heavy seas, ever got hurt. It also taught me how to prioritize and appreciate those who wear the uniform in service to our nation.

My career has been one of solving puzzles, trying to explain how things work. My goal in writing this book is to transfer some of my excitement of the oceans to the reader. The more that we understand the workings of nature, the more we can appreciate the world around us and how to critically assess how science at times may be misused for political purposes.

The topics to be covered begin with climate, specifically the role of the oceans in determining the Earth's climate, and they end with a Grouper (fish) sandwich. The bookends are therefore existential, the environment in which we live and the food that sustains us, and in between are individual topics that I hope will inspire more of you to stroll along the strand and ponder what lies beyond the horizon.

Preface

This is intended to be a lay-person-oriented book. It attempts to explain topics of Earth science without mathematical assistance. In other words, it is meant to inspire the learning of ocean circulation and ocean-atmosphere interaction-related topics, verses explaining them rigorously. I recall a quote from my Mathematics Methods of Physics professor to the effect that: "The ultimate state of rigor is rigor mortis." Science requires rigor, but an inspiration to study science does not. Needed for that is a desire to learn more, plus the fortitude to think through concepts that may not be entirely intuitive.

Three attributes are required to be successful in science. First, is being bright, not brilliant, just bright and with a sense of curiosity. Second, one must be creative enough to think of hypotheses that may explain observed phenomena and/or to design experiments necessary to test such hypotheses, either yours or someone else's. Third, is the tenacity to follow through and finish what you may begin. Lacking any of these will result in disappointment. Brilliant people, lacking either creativity or tenacity, may fail to contribute to the growth of knowledge. Similarly, there may be brilliant, tenacious workers lacking creativity who may still make excellent contributions; but, instead of by themselves independently, through the colleagues with whom they interact. Being an independent scientist requires the whole package. That does not mean that one should work alone; collaborations are enriching, especially when those collaborations are with others who may have attributes exceeding your own. As in sports, advancements come when you interact with those who challenge you. Learning from others is fulfilling. I always measured my professional interactions by what I learned during such an interaction. Walking away with a new understanding is in itself enriching. Fear of appearing ignorant is limiting. Have the self-confidence to push your own boundaries.

Once I realized that I would not pursue an engineering career, I was fortunate to find two oceanography courses at Cornell, one of a very general nature, and the other about geological oceanography, both of which I took as electives in my senior year. Thus, along with being suitable for a general audience, I can also envision this book being used at the undergraduate level, either to encourage students to study science (such as a survey course at the freshman level) or to help steer a science-oriented student into the field of ocean physics (as I was during my senior year).

One last thing: Cindy and I arrived in Rhode Island with a few bucks in our pocket, a $2,300 per year graduate student assistantship for me and a $6,400 elementary teaching position for her. To make this livable, I minored in edible oceanography, although I admit that this was more for fun than necessity. Narragansett Rhode Island was a terrific place for seafood foraging. Striped bass and bluefish were seasonally abundant and easily caught while casting from rocky coastline sites like Hazard Avenue; there were little necks, quahogs, surf clams, oysters, mussels, and periwinkles, each with a favorite location for capture and, of course, there were lobsters.

I have no idea where these critters may be distributed today, but back then, a quick free-dive with mask and snorkel just to the north of the Graduate School of Oceanography pier would yield all of the little necks and larger quahogs that one might want. The largest mussels were in Newport offshore of the Hammersmith Farm estate. Oysters were abundant in the Pettaquanscutt (Narrow) River just across the street from our rental house on Bridgetown Road. I could walk over with a mask and snorkel to retrieve these. Oyster beds were also abundant farther north in the river, as I discovered one day after falling into one while water skiing. Others were located south along the river near where we settled in Mettatuxet. Another spot was in Jamestown north of the causeway leading to Beavertail. Beavertail also provided an unending source of periwinkles, but those were hardly worth the effort. For surf clams, the beach at Point Judith had an abundance, although these were not good for much except chowder, and quahogs for that purpose are much better. Surf clams can be cut into strips and fried, but you could get the same effect using rubber bands. Thus, I recall the surf clam experience being one and done.

The lobsters at Point Judith were the real food value. The Point Judith Harbor of Refuge is enclosed by a series of long rock jetties, each providing excellent lobster habitat. A small group of us would swim the length of one of these jetties along the bottom at night with SCUBA, carrying an underwater light and a dive bag. Lobsters come out of the rocks at night to forage. When lit up, they stop just long enough to be grabbed gingerly from behind. By the end of the jetty swim, we would each have a bag full and, despite wearing gloves, a few cut fingers.

My favorite edible oceanography gambit was a class project for Introduction to Biological Oceanography in which I chose to study the molting of lobsters. This required dozens of lobsters to be housed in tanks for observation over a period of several weeks to describe their molting behavior. To keep them under similar conditions, I chose to feed them flounder, further requiring the acquisition of plenty of flounder, achieved aboard the Billie II, a fishing vessel used for local Narragansett Bay research. The experiment was fascinating, it resulted in a report that helped me get an A in the course, and all the lobster and flounder that Cindy and I could eat.

The foregoing edible oceanography discussion demonstrates that graduate school, albeit difficult, can also be quite enjoyable. I tell my students that as hard as it may seem at the time, graduate school is the easiest part of one's career. Your primary, if only, job is to learn. That learning part never ends, but added to it after the degree conferral, are all of the other responsibilities associated with being an independent scientist. Not everyone is suited for this, and outlets for enjoyment and relaxation are necessary. Along with water sports, cooking, for me, is one of those. This started as an undergraduate, when away from the freshman meal plan and sophomore fraternity living, it became quite evident that I could either eat junk food or purchase good ingredients when on sale and learn how to cook them. This approach continued into graduate school and through the present time. Consequently, along with Cindy, both of our children are excellent home cooks. Cindy and I would split up the cooking chores. On days when it was my turn, I would leave the office, and during the 15 min drive to the market, I would think about what to make for dinner and hence what to purchase. This required that I set my oceanography mind at rest and think about other matters, something that I found to be quite relaxing. Dinner usually turned out to be edible as well.

With this in mind, I am adding a recipe at the end of each chapter; well, not really a recipe but a suggested preparation guide. Recipes are work, especially having to follow directions. For relaxation, I always found it better to improvise. Hence, my suggestions will include ingredients, but without detailed measurements. That is something requiring a personal touch. I learned this from my grandmother while growing up. She did not measure; she just had a certain feel, and I recall everything turning out great.

St. Petersburg, USA   Robert Weisberg

Acknowledgments

Every profession requires sacrifice, much of which falls upon the immediate family, which in my case are Cindy, my partner in marriage for 55 years, and our two children, Seth and Ari. Seth was born after my Ph.D. comprehensive exams and fieldwork were completed so I had opportunity to work at home with him more so than usual. But at about nine months of age, I departed on my first cruise as chief scientist, which had me away at sea for two months. I had to start over again reacquainting myself upon return. Travel, especially extended deployments, is tough on a family. Never again was I away for two full months at one time, but thereafter, nearly annual cruises lasting about a month, plus countless professional workshops, conferences or just collaborative meetings must have added up to two months a year. At our thirtieth wedding anniversary, when it was difficult to find more than two other couples who had remained married for that long, I was able to joke that we were still together because by being away about a quarter of the time, we were really only married for 23 years. I could not have had such a career had my family not sacrificed for it.

It may sound strange, but I must next thank Harry Chapin, in my opinion the best balladeer of my time. His song: "Cat's in the Cradle" had a profound effect upon my interactions with Seth and Ari. Frightened by the lyrics, I spent as much time as possible with them, from weekend hikes on my back and rides on my bicycle to coaching their youth soccer teams, taking them skiing or torturing them as crew in sailing regattas. We enjoyed our family time together and it seemed to work—they both live close by in St. Petersburg, and we see them often.

In the text, you will find mention to my three major professors: Tony Sturges, Kern Kenyon and John Knauss. How they each treated me as a student became reflected in how I treated my students. I also learned some negative lessons from others (not mentioned) that taught me how not to treat students. Positive and negative lessons are equally important.

Early on, after giving presentations at national meetings, two senior scientists, John Allen and Chris Mooers showed interest in my work, which greatly increased my confidence that I could function in an independent role. It cannot be overemphasized just how important it is to encourage a young person. My first international, GATE Program experience introduced me to several senior and junior scientists, which led to further professional interactions. I learned just how important it is to hang out with smart people and to learn from others instead of being fearful of doing so. My advice to people new to the field, or to any field, is that the last thing you want to be is the smartest person in the room. You are better off being on the other end of the spectrum because you will be inspired to learn more. Just as in sports, one improves when challenged. I can name many such professional interactions, but one that stood out consistently from GATE through the present time is with George Philander, Professor Emeritus, Princeton University.

Perhaps more so than any other interaction after family, it is the students and the research associates that impart joy and enable success. My first Ph.D. student was Allen Horigan. I must have made a great first impression when I picked him up at the Recife, Brazil airport, mid-way between two back-to-back GATE research cruises. Somehow, I found myself going the wrong way down a busy, one-way avenue. My only recourse was to jump the curb and drive down the sidewalk until I could bail out from the oncoming traffic. It would have made good footage for a movie getaway scene, but no one was hurt, and Allen took it in stride. His work on identifying equatorially trapped waves was excellent, but he took ill prior to completion and was awarded the Ph.D. posthumously.

I subsequently had a number of students, but none of distinction until David Tang. I learned just what a difference exists when a student arrives well-enough prepared. Together, using forced equatorial wave theory, we explained why the region of maximum upwelling along the equator in the Atlantic Ocean is distinct from the eastern boundary and why this upwelling occurs abruptly as the Southeast Tradewinds begin to intensify with a westward expanding fetch, which also resulted in observed westward moving momentum and temperature pulses as near resonant Rossby wave responses. We then used similar, near resonance ideas, but with the wind fetch moving eastward, to explain why the 1982/83 El Nino response in the Pacific Ocean was so disproportionately large. This case entailed near resonant Kelvin waves. David remained with me for several years post-Ph.D. helping to explain the observed evolution of the Atlantic equatorial thermocline to wind forcing on the basis of forced and reflected equatorial trapped waves before returning to Taiwan, where he had an excellent career at the National Taiwan University before also meeting with an untimely death.

Tom Weingartner overlapped with David. He also worked with us on observations from the Seasonal Equatorial Atlantic (SEQUAL) experiment, which ensued subsequent to my GATE and Gulf of Guinea projects. Tom contributed to our understanding of Tropical Instability Wave energetics, their region of generation by barotropic instability just to the north of the equator, the duration of their active season and why this differs between the Atlantic and the Pacific Oceans owing to the zonal pressure gradient force adjustment time after the tradewind's onset each year. He then moved to Alaska and enjoyed an excellent and productive academic career as professor at the University of Alaska, Fairbanks, where he concentrated on Arctic Oceanography. Tom was also among the most gifted of the writers who came through my group. He kindly read through the first draft of this book and made many helpful suggestions for improvements.

With strong backgrounds in mathematics and physics, my next three students were from China: Chunzai Wang, Lin Qiao and Zhenjiang Li, who were all quite gifted and hard-working. Chunzai Wang studied air-sea interaction topics in the tropical Pacific Ocean with applications to the El Nino - Southern Oscillation (ENSO), and together we added another mechanism affecting ENSO (a Western Pacific Oscillator, which Dennis Mayer and Jyotika Virmani also contributed to). He went on to a career with NOAA before moving back to China, where he is a member of the Chinese Academy of Science and a Distinguished University Professor.

Lin Qiao did observational work with me on Tropical Instability Waves in the Pacific. We found results similar to what Tom Weingartner found in the Atlantic, but with more data, we were able to produce a more complete description of the wave kinematics and energetics, as well as adding some fundamental descriptions of the equatorial undercurrent and the vertical distribution of equatorial upwelling as related to the dynamical balances of the equatorial undercurrent. Upon completion of the Ph.D., he decided to find work in private industry outside of oceanography. He wanted more stability. Unbeknownst to most is that academic careers can be chancy. Many institutional salaries only cover an academic (nine-month) year; some even cover lesser times. Thus, salary for the full year depends on grant support, which may be difficult to secure. Additionally, the support of a research team requires much more grant support than for one's own salary. My lawyer friends, who refer to eating what they kill (they personally benefit from the money that they bring in) would comment that as a professor, I had to kill much more than I could eat. Not everyone is willing to do this.

Zhenjiang Li was my first graduate student during my transition from equatorial to coastal oceanography. He added a numerical circulation modeling component, which allowed us to describe both the kinematics and dynamics of the west Florida continental shelf responses to wind forcing. Upon completion of the Ph.D., he also decided to seek employment elsewhere. Like many quantitatively oriented Chinese students in the early 2000s, he went to Wall Street in an attempt to prosper more so than he thought that he could in oceanography. This drain of qualitied students for the study of ocean physics continues through the present time because earning potential and stability is often thought to be better elsewhere.

On an invited trip to China in 1997 (arranged by Chunzai Wang), I was introduced to two potential students, Ruoying He and Yonggang Liu, both of whom eventually came to USF and continued successfully in academia. Ruoying, being several years ahead of Yonggang is Distinguished University Professor at NCSU, coincidentally where I was prior to being recruited at USF. Yonggang returned to USF after a post-doctoral stint and is now an Associate Professor heading my former research group. Academic opportunities remain quite desirable for some of us. Both Ruoying and Yonggang added substantially to the program of study on the west Florida continental shelf, Ruoying primarily through numerical ocean circulation modeling and Yonggang primarily through observational analyses, but together, we all arrived at an appreciation for why the two facets, modeling and observations are best done in combination, especially when looking at the influences of deep ocean, shelf interactions and local forcing, plus how their joint effects impact topics of societal concern.

Four more Ph.D. students rounded out my career prior to retirement in May, 2022. Bob Helber worked on equatorial Pacific and equatorial Atlantic circulation and ended up with a research scientist position at the Naval Research Laboratory at Stennis, MS. He described a perhaps non-intuitive finding of upwelling beneath the western Pacific warm pool and expanded upon how divergence causes the region of coldest water on the equator in the Atlantic to be positioned where it is. Jyotika Virmani tackled ocean-atmosphere interaction issues and is presently the Executive Director of the Schmidt Oceanographic Institute, a private concern. She may have most interesting career of all of my students. Upon my retirement (and along with her husband Ben Alpi), she generously initiated a named graduate student endowment in my honor (the Robert H. Weisberg Fellowship in Physical Oceanography) that we hope will continue to grow in support of the next generation of students.

Next came Jun Zhu who made some fundamental applications to the workings of the Tampa Bay estuary prior to returning to China for an academic position there. Jing Chen followed up on Jun Zhu's work with an automated nowcast/forecast model of Tampa Bay, with several important environmental applications, before becoming a research scientist at the NOAA Geophysical Fluid Dynamics Lab, Princeton, NJ.

As emeritus professor, I still have two Ph.D. students finishing up, Luis Sorinas on ocean-atmosphere interactions and Alex Nickerson with applications to how the Gulf of Mexico Loop Current transitions through its various penetrative, eddy shedding and retracted stages, plus secular rises observed both locally and globally in SST and a detailed analysis of why Hurricane Ian rapidly intensified on the west Florida continental shelf.

There were also a suite of Master of Science students who were all successful with their career goals, either in private industry, federal agencies, the US Navy, the US Coast Guard, and, in one case, a subsequent Ph.D. and an academic career in computer science. Working with so many young people was, in itself, rewarding, but it must also be recognized that the academic enterprise is not something performed alone. Each student contributed in ways that greatly added to my ability to learn about the oceans. What may begin as a flow of information from professor to a student quickly becomes an equal exchange of ideas and accomplishments.

Along with students, the other form of mentoring that we engage in is the support of post-doctoral associates and visiting scientists. This is also a two-way street. Along with my own students, who extended their stay with me prior to going elsewhere, I was honored to have in residence Alex Barth and Aida Alvera Azcarate (both from Belgium and now married with children), who added both modeling and analysis techniques prior to returning to Belgium for research positions there; Lianyuan Zheng, who developed our present numerical circulation modeling capabilities and added to our knowledge of both estuarine circulation, hurricane storm surge inundation and destruction by waves prior to securing a permanent position with NOAA; Yong Huang, who added wave modeling before transitioning to private industry; Yang Yang, who visited from China and made some fundamental contributions to Gulf of Mexico Loop Current behavior, before returning to China, plus a few others. Post-doctoral positions are transitional, and they provide different levels of engagement and mentoring, often as rewarding as graduate student mentoring.

Running a coordinated, sea-going experimentation and circulation modeling program also requires technical support. Personnel stability is critical. Computationally, I was fortunate to have two long-serving computer system/data managers, Jack Hickman at NCSU, who relocated with me to USF, followed by Jeff Donovan. Instrument record downloading, data preconditioning and quality control are an ever-ongoing activity that was greatly assisted by Dennis Mayer, a former science collaborator who came to work for me part time after his retirement from NOAA. For sea-going technical support, after first utilizing the Graduate School of Oceanography at URI technical group, I then relied upon Paul Blankinship while at NCSU and Rick Cole, followed by Jay Law, at USF. It is not possible to mount sea-going operations without dedicated, competent support personnel familiar with the equipment and the deployment platforms and are able to run sea-going operations in conditions that are often unfavorable, if not outright dangerous. Along with sea-going operations were shore-based ones. Cliff Merz was responsible for these, maintaining an array of high frequency radar for mapping surface currents on the west Florida shelf.

By now it should be evident that science is a team sport. Sure, leadership and creativity are of critical importance, but as is clear from the author's bibliography, the student, post-doctoral and other associate's contributions over time take on an ever-increasing role in productivity. I am greatly indebted to all of my students and associates for the career that I was able to engage in, and it is for that reason that I appended my bibliography to serve as a tribute to them.

Finally, I have to thank Peter Betzer, who wrested me away from NCSU to help build a program in physical oceanography at USF. Peter then became the founding Dean of the College of Marine Science and was quite successful at bringing new resources to the college, which all of the faculty and students benefitted from. The best thing that a College Dean can do is help with resources and then get out of the way so that the faculty can perform in teaching and research with the least of burdens. Peter was excellent at this, and his successors continued along that vein. Upon retirement, Tom Frazer, our present Dean, was as accommodating as he could be in helping me to sustain my research group through an orderly transition, without which I would not have felt comfortable retiring. No one lost their employment, all of the students were able to continue, and my former Ocean Circulation Lab, now directed by Prof. Yonggang Liu, never missed a beat.

Bringing this full circle, however, harkens back to my immediate family, for without their support throughout my journey as an academic scientist, none of this would have been possible. My love and thanks go to Cindy, Seth and Ari!

Introduction

The basic premise of this book is that just about everything that we experience on Earth depends upon the ocean circulation, or the movement of the ocean water. The text begins on a global scale, addressing the rudiments of climate and ecology before working down to applications on a more local level. Recognizing that the Earth's surface consists of land, water and air, each with their own density and heat capacity, we begin by exploring how the movement of the ocean water, initiated by the non-uniform input of solar energy, determines how the ocean and atmosphere interact, thereby yielding the global distribution of properties and hence climate, as discussed in Chap. 1.

Temperature is one such property, and by virtue of the Earth's rotation and its spherical shape, there tends to be a zonal (i.e., east-west) asymmetry in the distribution of temperature at any given latitude, which is caused by a western intensification in the ocean's currents, as explained in Chap. 2. Along with these properties, we also find that just like land-based plants tend to thrive in certain places, the microscopic ocean plants, the phytoplankton, also exhibit lush, versus desert-like regions. By looking at what lies beneath the ocean surface and how the circulation distributes these water properties in Chap. 3, we can account for the global distribution of plant life, and hence how the ocean circulation provides the basic underpinning for global ecology.

While global distributions are of fundamental ecological importance, it is the coastal ocean where society literally meets the sea. How the shallower coastal ocean works, in contrast to the deeper ocean, is the topic of Chap. 4. Specific applications are made to the west Florida continental shelf, a region of gently sloping bottom bathymetry, extending from the coast to the shelf break, which is about as wide as the State of Florida's peninsula landmass. Given such width, the west Florida shelf offers a natural laboratory in which to study the continental shelf circulation because there we can distinguish between the inner-shelf, where friction is a dominant factor, and the outer-shelf, where deeper ocean influences are more prevalent.

Even closer to major population centers are the estuaries, the regions where river and coastal ocean waters blend together. There, the circulation, and hence the distribution of water properties, becomes even more complex and of ever more concern due to the many competing societal and natural habitat utilizations of estuarine waters. Whereas the Tampa Bay, Florida estuary is the focal point of Chap. 5, the resulting understandings derived therein apply to most estuary systems found worldwide.

Perhaps the first of the ocean observations that kindled human interest was the rhythmic rise and fall of sea level and how these vary daily, seasonally, interannually and with the passage of each weather system. Recently heightened concerns about global warming now lead to fears of where sea level may be heading. Thus, the effects of tides, winds, seasonal heating and cooling and larger, global scale phenomenon on local sea level form the topics of Chap. 6, again with application to Tampa Bay, Florida, but only as a regional example that applies equally elsewhere.

Physical oceanography or the physics of the oceans and ocean-atmosphere interactions also includes the topic of waves. Most of us are familiar with surface gravity waves, as readily observed on the ocean surface when the wind is blowing locally (referred to as seas) or propagating in from afar (referred to as swell). Rarely is the ocean surface flat because such waves can travel great distances from their region of generation to where they may eventually break upon a distant beach. Visible surface gravity waves are only one manifestation of a much broader suite of ocean waves, some even occurring sight unseen beneath the surface. As nature's way of adjusting a medium (such as water) to a disturbance, waves may take on many different forms and behaviors owing to the Earth's rotation, its spherical shape and how fluid density varies internally within the ocean and the atmosphere. These concepts are introduced in Chap. 7.

Surface gravity waves may grow to great heights if the transmission of their energy is large enough. This brings us to the concept of tsunamis, dangerously large waves that may break along the shore subsequent to an earthquake, a landslide or any disturbance to the sea surface of large enough magnitude. Warranting their own discussion, tsunamis are the topic of Chap. 8.

Of equal, if not a greater, threat to those living near the shoreline are the sea level extremes that may accompany a hurricane near its point of landfall. Not only is the rise in sea level by such a storm surge of concern, even more destructive are the waves that can ride atop the storm surge, thereby attacking structures that might otherwise be flooded, but not completely destroyed. Misconceptions abound regarding hurricane storm surge and waves, which Chap. 9 attempts to clarify. Used as a primary example is Hurricane Katrina, which devastated the entire coast of Mississippi in 2005. The reader may recall all of the attention given to the flooding of New Orleans where several levees there failed. Not to downplay that catastrophe, the larger story, that went relatively untold, was the destruction along the entire adjacent Mississippi coast. Chapter 9 explains not only how storm surge evolves, but by making use of specific case study examples it also explains how damage may ensure on the basis of wind, storm surge and waves.

Introduction xxi

With the Katrina example explored in detail, attention then shifts to the Tampa Bay, Florida region, which is similarly vulnerable to such disaster in the event that a hurricane may make landfall just north of Tampa Bay. This is particularly important given that three recent storms [Michael (2018), Ian (2022) and Idalia (2023)] hit the west coast of Florida, all within about 70 to 200 miles of Tampa Bay. Detailed discussions on the Tampa Bay vulnerability are provided, and such discussions are equally instructive for readers from other coastal regions.

With these frightening aspects of physical oceanography now aside, explanations are given to the seasonal march of ocean temperature, again with a specific west coast of Florida application. Ocean water temperature does follow the sun angle and the length of the day, but the actual energy balances that control this are much more complex, entailing not only incoming and outgoing radiation, but also turbulent processes. How these all add up to cause the water temperature to vary seasonally are the topics of Chap. 10. If the local air-sea interactions themselves are not complex enough, there are years when the effects of the adjacent deeper ocean also come into play, making summer months either warmer or cooler than the perceived norm.

The west coast of Florida is subjected almost annually to blooms of the harmful algae, *Karenia brevis*, a microscopic, toxin-producing, phytoplankton species that discolors the water, causes respiratory distress and kills copious quantities of fish, whose rotting carcasses then adversely impact tourism, resulting in several billion dollars in losses. Intuition may suggest that this is a biological problem, but we now understand that the ocean circulation is as important to red tide ecology as is the organism biology. What was learned in prior chapters may now be employed in Chap. 11 to explain not only what may cause a red tide, but also what may ultimately lead to a bloom's termination.

Whereas it is evident that the average temperature of the Earth is rising, superimposed upon such secular rise (of slightly less than one degree centigrade per century) are variations of larger magnitude that occur quite naturally. Chapter 12 discusses a few of these natural variations that we are just beginning to observe and understand. This is important because by conflating human-induced with natural phenomenon, we may overestimate human impacts on climate resulting in policy decisions that may be more detrimental to humankind than the perceived climate change itself.

One way to mitigate climate change is to switch the way by which we generate power for human consumption, i.e., the movement toward a net zero carbon emission environment by using alternative means of electric power generation. What may or may not be feasible and why are discussed in Chap. 13, again with specific application to the State of Florida by considering solar inputs, winds, ocean currents and ocean waves. The results, which are also applicable elsewhere, may be surprising to some readers.

Finally, Chap. 14 closes with a discussion on gag grouper recruitment. Gag, the most abundant of the commercial and recreational grouper species landings on the west coast of Florida, is one reason why a grouper (fish) sandwich is such an important component of any Florida seafood restaurant menu. For years, it was a mystery of how gag larvae could get from the region where adults spawn to the region where juveniles settle. The ocean circulation provided the answer to this conundrum, as will be demonstrated. Thus, from climate to a fish sandwich, this book endeavors to demonstrate how the ocean circulation is of existential human importance and, hence, why we bother to study it.

Contents

1	**The Global Ocean Circulation and Climate**	1
	Addendum: Salmon with Mango Salsa	9
2	**Additional Aspects of the Global Ocean Circulation and Climate**	11
	Addendum: Barbeque Salmon	19
3	**The Global Ocean Circulation and Ecology**	21
	Addendum: Baked Cod	29
4	**The Coastal Ocean: How It Is Driven**	31
	4.1 The Wind-Driven Coastal Ocean Circulation	32
	4.2 Deeper Ocean Forcing	40
	Addendum: Pesce Delicioso	51
5	**Estuarine Circulation**	53
	5.1 A Narragansett Bay, Graduate School Digression	55
	5.2 Back to Estuaries	58
	Addendum: Pasta with Scallops and Calamari	74
6	**Sea Level: Why It Goes Up, Down and Where It Might Be Heading**	75
	Addendum: A Simple Bouillabaisse	92
7	**Waves of All Sizes**	95
	Addendum: Ahi Poke	107
8	**Sea Level Extremes by Tsunamis**	109
	Addendum: Fish Cakes	115

9	**Sea Level Extremes by Hurricane Storm Surge**		117
	9.1	Hurricane Storm Surge Overview	117
	9.2	The Accompanying Waves	121
	9.3	A Conceptual Overview on the Relative Effects of Winds and Water	122
	9.4	A Hurricane Katrina Case Study	123
		9.4.1 An Overview of the Storm	123
		9.4.2 The Various Damage Mechanisms by Hurricane Katrina	123
		9.4.3 A Specific Case Study of How Katrina Destroyed a Particular House	136
		9.4.4 The Winds, Surge and Waves at the Subject Location	141
		9.4.5 The Relative Forces by Wind and Water	148
		9.4.6 Results	150
	9.5	Hurricane Inundation and Damage Potential for Tampa Bay	151
	Addendum: Grilled Halibut		161
10	**The Air–Sea Interactions that Determine Water Temperature**		163
	Addendum: Flounder Meuniere		172
11	**Florida Red Tides**		175
	Addendum: Grilled Lamb		191
12	**Natural Climate Variability: What We Are Just Beginning to Learn**		193
	Addendum: Braised Short Ribs		199
13	**Alternative Energy Generation from the Ocean: What May or May Not Be Feasible and Why**		201
	13.1	Wind Observations	202
	13.2	The Conversion of Wind Speed Observations to Electrical Power Generation Potential	202
	13.3	Ocean Currents	204
	13.4	Ocean Waves	209
	13.5	Solar	211
	13.6	Discussion	211
	Addendum: Meatloaf		213
14	**Why Grouper Sandwiches Are Popular on Florida's West Coast**		215
	Addendum: A Grouper Sandwich		225
Epilog			227
Author Bibliography			231

Chapter 1
The Global Ocean Circulation and Climate

The present configurations of the Earth's oceans and its land masses have been in place for many millions of years, and it is over this span of time that human beings evolved to where we are today. Hardly steady, in view of the ice-age and interglacial cycles that have occurred over the past million years, the Earth's climate has at least settled into a certain rhythm, allowing for the civilizations as we know them to develop, especially during the present interglacial period that began some 8000 years ago subsequent to the last glacial maximum that peaked around 20,000 years ago. Previous Earth configurations experienced vastly different climate scenarios so it is reasonable to ask: Given the present, relatively steady ocean and land configurations, what determines climate?

The starting point is the energy that the Earth receives from the Sun. In a relative sense (to less than a tenth of one percent), we may neglect the heat that flows to the surface from the Earth's interior and the work performed on the Earth by its tidal forces. Thus, the Sun is our essential energy source, and it remains relatively constant when averaged globally around the Earth's surface. Ultimately, this energy source is what determines the various material properties that are observed all around the Earth. One such property is land vegetation, a depiction of which may be viewed in Fig. 1.1 for September 2000.

Observed are forests and other vegetated regions, along with deserts and ice. These are not randomly placed. They all have a certain spatial coherence that begs explanation. It is relatively easy to explain why ice should be at and near to the poles because these regions, by receiving less solar energy, tend to be colder than either the tropics, or the mid-latitudes, but why do some tropical areas appear barren (as distinguished by their beige color) while others are lushly verdant?

To begin answering this question, we must consider not only the land, but also the oceans that border the land, plus the atmosphere that overlays both the land and the oceans. Thus, a glance at the cloudiness for September 2000, as depicted in Fig. 1.2, is instructive. Excepting the desert regions, most of the Earth's surface is covered by clouds. Evident from a comparison of Figs. 1.1 and 1.2 is that the properties observed

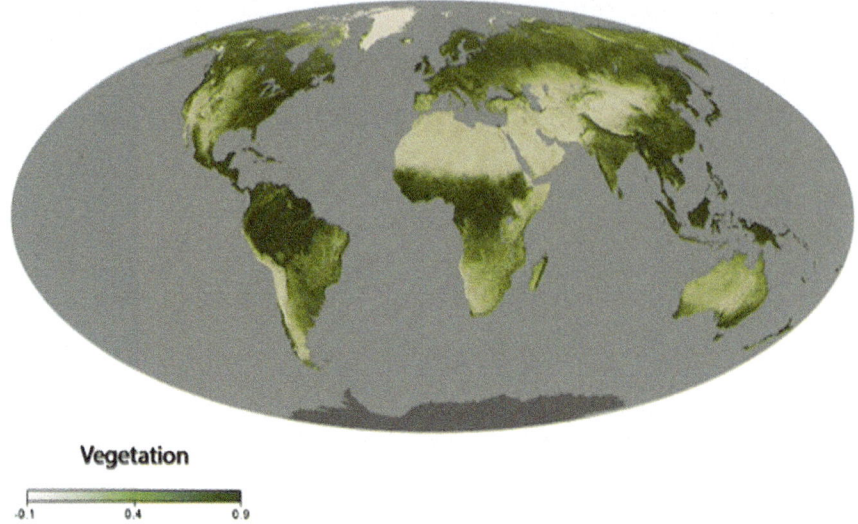

Fig. 1.1 Global distribution of vegetation for September 2000 (courtesy of NASA Earth Observatory, https://earthobservatory.nasa.gov/global-maps)

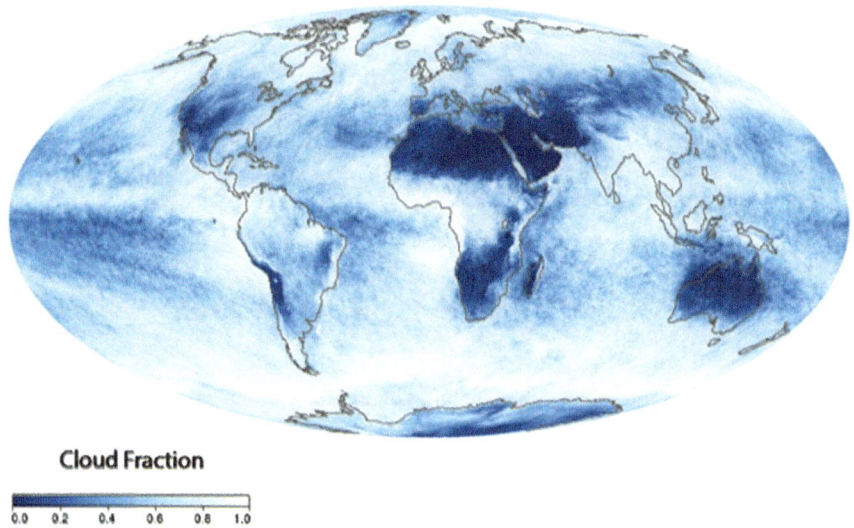

Fig. 1.2 Global distribution of clouds for September 2000 (courtesy of NASA Earth Observatory, https://earthobservatory.nasa.gov/global-maps)

on land must be viewed in the context of the overlying atmosphere. The attendant question then becomes: From whence do the clouds originate? The answer is the oceans.

1 The Global Ocean Circulation and Climate

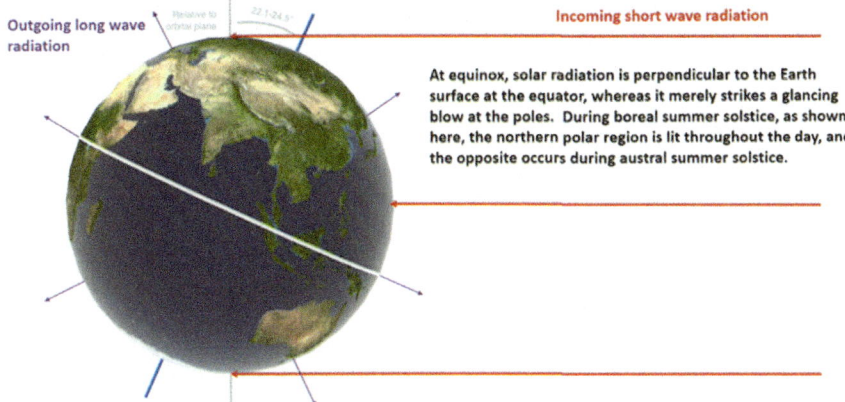

Fig. 1.3 A schematic of how the Earth both receives solar energy in the form of shortwave radiation and emits energy back into space in the form of longwave radiation (Earth image courtesy of NASA)

Whereas land areas consist essentially of solids, the oceans and the atmosphere both consist of fluids, liquid for the ocean and gas for the atmosphere. Solids are stationary, whereas the fluids may move, and by moving, these fluids transport heat, water vapor and other materials that determine what may prosper on land. The Earth's climatic zones, in their most fundamental sense, are therefore owing to the transports of heat and water vapor by the fluid ocean and the fluid atmosphere. It is the coupling between these two fluid ocean and atmosphere media, as set in motion by the heat from the Sun's radiation, that determines the Earth's present-day configuration climate and hence our existence on this planet.

While the Sun fuels climate, the next most important ingredient is water. The physical chemistry of the water molecule yields three important properties: (1) large density, (2) high heat capacity and (3) the ability to undergo state transitions between solid (ice), liquid (water) and gas (water vapor), with each transition requiring either the input or release of heat (energy). These properties, along with the solar energy input, provide the basis for discussing climate.

Let's begin with a thought experiment by asking: How does the ocean and atmosphere respond to solar heating? Given the distance between the Sun and the Earth, the Sun's rays approach the Earth in parallel with one another. Solar radiation near the tropics is therefore directed nearly perpendicular to the Earth's surface, whereas solar radiation near the poles merely strikes a glancing blow (see Fig. 1.3). As a result of the Earth's spherical geometry, the solar energy input to the Earth is much larger at tropical latitudes than at polar latitudes.

The energy input to the Earth must be balanced by an equal energy output; otherwise, the Earth would continually warm. All known matter radiates energy in proportion to the fourth power of its absolute temperature and the emitting body's cross-sectional area. The proportionality constant is the Stefan–Boltzmann constant, and the absolute temperature, measured in degrees Kelvin, is degrees Celsius, plus

273.15°. Absolute zero (in Kelvin degrees) is the temperature at which atoms cease to move, which in Celsius units is $-273.15°$. Given that radiation is calculated using Kelvin degrees, and because the Sun is both hot and large, it radiates an enormous amount of energy. In contrast to the Sun, the Earth is relatively cool, and while the difference in temperature between the tropics and the poles is about 30 °C, when converted to Kelvin degrees this relative difference does not appear to be nearly as large. Thus, while the Earth receives much more heat in the tropics than at the poles, the difference in the radiation of energy back into space from these regions is relatively small. As a result, when added together, and averaged annually, the Earth would appear to have an energy surplus in the tropics and an energy deficit at the poles. Yet, the annually averaged temperature of these regions does not change appreciably over time so something must ameliorate these differences. That is where the coupling between the two fluid media, the oceans and the atmosphere comes into play.

The relatively large density (or mass per unit volume) and heat capacity (or the amount of heat energy needed to raise the temperature of a given mass) of water, when compared either to the air or to the land, means that a given volume of ocean water at a certain temperature contains much more internal energy through heating than an equal volume of either the atmosphere or the land. Moreover, the atmosphere, without clouds or water vapor, is nearly transparent to incoming solar radiation so the ocean is what absorbs the bulk of the Sun's radiant energy. Warm surface temperatures will result in an exchange of heat to the overlying atmosphere through both sensible and latent heat fluxes [where the heat flux is the amount of heat transferred per unit area and time (i.e., in units of Joules/m^2/s or Watts/m^2)].

Sensible heat flux is just that. If two media are in contact and one is warmer than the other, then heat will diffuse from the warmer medium to the cooler one. Latent heat flux is a little more involved. In this instance, water undergoes a phase transformation from liquid to vapor (through evaporation), which requires a certain amount of heat to accomplish such a phase change. This heat of transformation is called latent heat because it is not released to the atmosphere until the vapor transforms back to water in the form of rain. It is only then that the atmosphere itself is heated.

Together, the sensible and latent heat flux from the ocean to the atmosphere lowers the density of the overlying atmosphere because the density of air decreases with both increasing temperature and increasing water vapor. This decrease in density decreases the local atmosphere pressure (force per unit area owing to the weight of the overlying atmosphere), the consequences of which are twofold. First, the warmed, vapor-laden air of decreased density rises, and second, the lowered pressure allows for the surrounding higher-pressure air to flow in to fill the void left by the rising air. This explains the origin of winds, or the movement of atmosphere mass owing to spatial differences (or gradients) in atmosphere pressure. The winds, in turn, by doing work on the ocean surface through friction (the rubbing against the ocean surface), cause the ocean water to move, resulting in ocean currents. The ocean currents then redistribute the internal energy of the ocean (derived from the incoming solar radiation), which changes the distribution of the heat flux (by the sensible and latent heat transfers) between the ocean and the atmosphere. It is in this manner that the

1 The Global Ocean Circulation and Climate

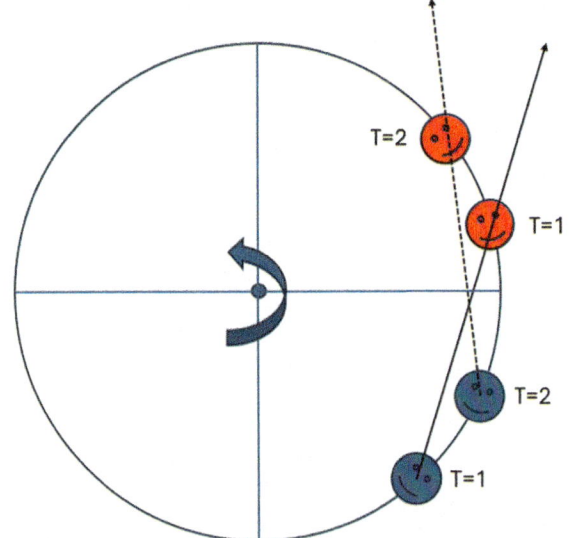

Fig. 1.4 An anticlockwise rotating merry-go-round with a blue rider attempting to throw a ball to a red rider at time $T = 1$. As the ball travels along the straight, solid line, the red rider rotates away from it to the left at $T = 2$. Thus, to any observer on the merry-go-round, it will appear as if the ball turned to the right, e.g., the difference between the dashed, line of sight at $T = 2$ and the direction in which the ball was thrown at $T = 1$

movement of the two fluid media, the ocean and the atmosphere, must be considered to be an intimately coupled system, each driving the other in response to the spatially and temporally varying radiative heat flux originating from the Sun.

Now that the two fluid media are set in motion, one additional concept must be considered, the fact that the Earth rotates about its own axis. While reading this text may appear to be a sedentary activity, we are actually traveling at a very high rate of speed. With the Earth's circumference being 24,901 miles at the equator, the 24-h rotation interval means that objects on Earth are traveling at a speed of over 1038 miles/hour at the equator and at a somewhat lesser speed at mid-latitudes. We do not sense this because we evolved under such conditions, and, with our surroundings moving at the same speed, there is no relative motion to sense. However, for any object moving relative to the rotating Earth, there is an important effect called the Coriolis effect that we must account for by introducing a Coriolis force.

A simple way of appreciating the Coriolis force and how it arises is by considering two individuals attempting to play catch on an anticlockwise rotating merry-go-round (Fig. 1.4).

Envision the person in blue tossing a ball toward the other person in red. According to Newton's second law of motion, the ball will travel along a straight line (the solid line in Fig. 1.4), as observed by someone who is not on the merry-go-round, i.e., someone fixed in space for which Newton's laws of motion are defined. In principle, the observer would also have to be positioned somewhere off of the rotating Earth, but for this illustration we can neglect the Earth's rotation because the merry-go-round is rotating so much faster than the Earth (one rotation in a few minutes, versus in 24 h). In contrast to what is observed by someone not on the merry-go-round, relative to the person in blue, who tossed the ball toward the person in red, the ball will appear to be deflected to the right (the difference between the dashed and solid lines) because both persons are rotating (given the merry-go-round's anticlockwise spin) away from the straight flight path of the ball.

Returning now to the Earth, to make allowance for this phenomenon by an observer fixed to and rotating with the Earth (in analogy with the merry-go-round), we must add a force of magnitude equal to twice the rate of rotation times the speed of the fluid relative to the Earth, or $2\rho\omega v$, where ρ is the fluid density, ω is the Earth's angular rate of rotation, or $2\pi/24$ h, and v is the speed of the fluid relative to the Earth. Moreover, this Coriolis force (per unit volume), by being directed perpendicular to the velocity vector (to the right in the northern hemisphere and to the left in the southern hemisphere), does not perform any work, and hence it is referred to as a fictitious force. Despite the added Coriolis force being fictitious with regard to a Newtonian frame of reference fixed in space, it is quite real to an observer in a rotating reference frame, as all of us are when residing on the Earth.

Armed with this knowledge, let's return to our thought experiment. Beginning at the equator, where the sun's rays at equinox (that is when the sun is over the equator in December and June) are directly normal to the ocean surface, the sun's rays will initially heat the ocean surface uniformly all along the equator. The heated ocean will warm the atmosphere, causing air to rise and be replaced by air converging on the equator from higher northern and southern latitudes. In view of the Coriolis force, these converging air masses must also turn westward (becoming the trade winds), thereby driving a westward ocean current. Also, by virtue of the Coriolis force, the westward moving ocean water will be deflected northward, just to the north of the equator and southward, just to the south of the equator. Together, these poleward deflections will cause a divergence of water at the equator, requiring the upward movement (or upwelling) of deeper, colder water to fill what would otherwise be a void.

Over time, the transport of warmer surface waters to the west by the trade wind-driven westward surface current and the appearance of cooler, upwelled water in the east, will result in an east to west atmosphere pressure difference because the cooler water in the east will result in cooler, heavier air in the east, and the warmer, lighter water in the west will result in warmer, lighter air in the west. These changes in atmosphere pressure and their consequent east to west directed pressure gradient force will further increase the east to west directed trade winds.

Over the same time interval, the wind-induced transports of water and the density difference between cooler and warmer water will result in sea level rising in the west and falling in the east until such time as the resulting oceanic pressure gradient force (due to the sloping sea surface) is sufficient to balance the frictional stress of the wind acting upon the sea surface.

As this redistribution of heat and mass evolves in the ocean, the atmosphere also adjusts, with the rising air along the equator diverging aloft and flowing poleward to replace the surface air that is converging on the equator. It is in this manner that both the zonal (east–west) circulation (called the Walker circulation) and the meridional (north–south) circulation (called the Hadley circulation) sets up. The Hadley circulation can only extend a certain distance away from the equator before

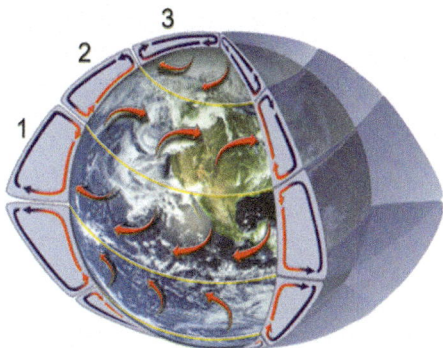

Fig. 1.5 A depiction of the Hadley (1), Ferrel (2) and Polar (3) cells on a rotating Earth. The regions of ascending and descending air have low and high atmospheric pressure, respectively. The surface winds tend to be easterly at the Hadley cell convergence, westerly at the Hadley–Ferrel cell divergence and easterly again at the Ferrel–Polar cell convergence (Courtesy of NOAA, https://www.noaa.gov/jetstream/global/global-atmospheric-circulations)

it must converge and sink back toward the ocean surface, resulting in a region of high (atmospheric) pressure at mid-latitudes. That convergence and sinking connects with other circulation cells at higher latitudes (the Ferrel and Polar cells) until the atmosphere and the ocean over the entire globe are set in motion, as depicted in Fig. 1.5.

The distribution of surface winds, in turn, drives a set of ocean circulation gyres, as depicted in Fig. 1.6. Mid-latitude regions (between the surface trade winds and the Ferrel cell westerlies) exhibit anticyclonic, subtropical gyres in both hemispheres. Poleward of these in the northern hemisphere are cyclonic subpolar gyres, and poleward of these in the southern hemisphere is the cyclonic Southern Ocean, Antarctic Circumpolar Current. Less well defined are the gyres in the vicinity of the equator, the equatorial gyre and the tropical gyre that wax and wane seasonally, plus the circulation of the Arctic Ocean.

Just as the sign of the Coriolis parameter changes between the northern and southern hemispheres, so does the sense of rotation for the cyclonic and anticyclonic gyres. In the northern hemisphere, cyclonic and anticyclonic refer to anticlockwise and clockwise rotation respectively, whereas the opposite applies within the southern hemisphere. The reason for this confusion may be appreciated by considering a walk taken from the north pole to the south pole. At the north pole, your head would point up in the same direction as the Earth's rotation vector. At the equator, your head would be perpendicular to the rotation vector (so the Coriolis force there is zero). Once you cross the equator and proceed to the south pole, your head would be pointing downward relative to the rotation vector and hence the sense of rotation, at least to you, would appear to be opposite from how it appears in the northern hemisphere. In other words, people in the southern hemisphere are basically standing upside down relative to the Earth's rotation vector. It is a good thing that gravity keeps our feet grounded in both hemispheres.

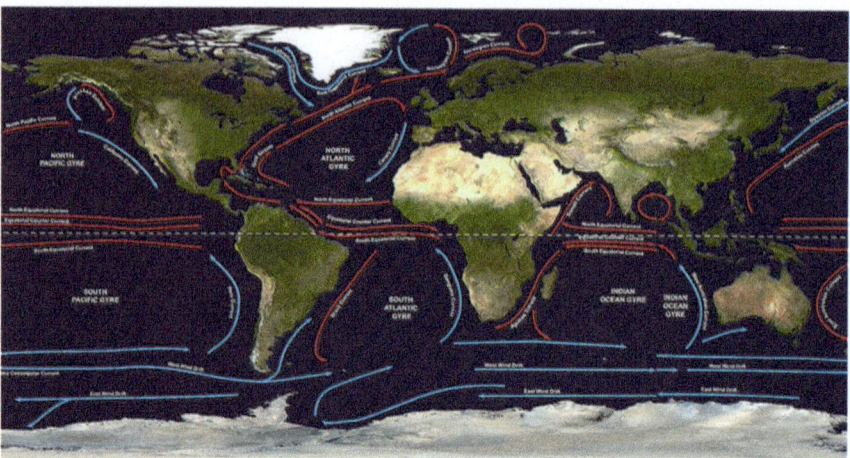

Fig. 1.6 A depiction of the organization of the large scale, ocean circulation gyres as driven by the coupling with the atmosphere and influenced by land masses and the Earth's sphericity (courtesy of NOAA, https://oceanservice.noaa.gov/facts/gyre.html)

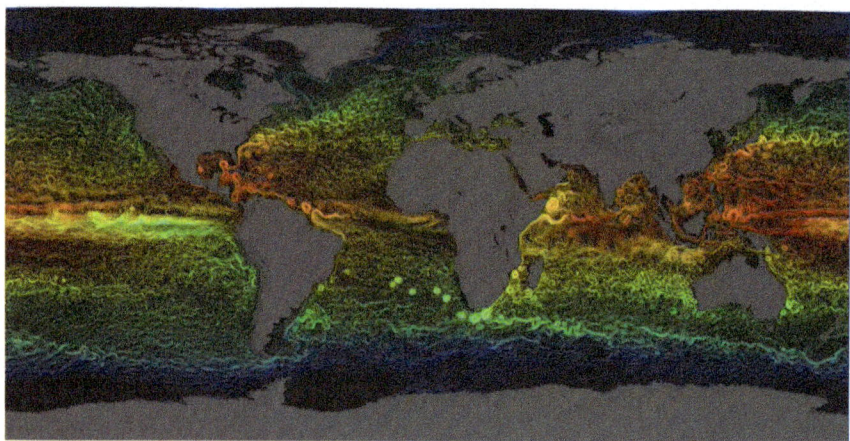

Fig. 1.7 A numerical circulation model rendition of surface currents over the global oceans showing westward intensification and eddies of multiple spatial scales (courtesy NASA/Goddard Space Flight Center Scientific Visualization Studio and a joint JPL/MIT project, Estimating the circulation and climate of the ocean, phase 2 ECCO2, https://svs.gsfc.nasa.gov/)

In actuality, the ocean circulation is much more complex than the average depiction of Fig. 1.6 would lead one to believe. Only recently, with satellites providing views of the entire Earth, have we come to realize that the ocean circulation is highly turbulent, abounding in eddies with multiple scales (sizes), as depicted in the numerical circulation model simulation of Fig. 1.7.

The end result of these coupled, ocean and atmosphere circulations are the transports of internal energy (the flux of heat) from the equator to the poles by both the oceans and the atmosphere. In the simplest sense then, both the cause of the ocean's currents and the atmosphere's winds and their effect is to maintain the Earth's heat balance.

We may now appreciate why the Earth's oceans (in response to the Sun) are the first arbiters of climate. The Sun provides the energy, but there exists an imbalance between the sum of the incoming shortwave and the outgoing longwave radiation between the tropics and the poles. Ocean water, by virtue of its high density and heat capacity (as contrasted with the atmosphere and land), plus its ability to undergo (liquid–vapor–solid) phase transitions, receives the bulk of the solar energy, but differentially between the equator and the poles. The oceans then transfer some of this heat to the atmosphere (by sensible and latent heat fluxes) causing winds to blow, currents to be driven and a systematic set of both ocean and atmosphere circulation cells to be established in such a way that these fluid media together transfer heat between the tropics and poles to maintain a global balance. These are the factors that determine the Earth's climatic zones.

Climate, however, is not steady. It varies annually as the Earth revolves around the Sun; it also varies interannually, decadally and at longer, but still within human lifetime, time scales owing to various coupled, ocean–atmosphere oscillations that we are just beginning to understand. Beyond human lifetimes, the Earth's climate undergoes glacial (ice-age) and interglacial (warm) cycles. These are collectively due to the astronomical changes [the eccentricity of the Earth's orbit around the Sun and its nutation (change in rotation axis tilt) and precession (the rotation axis orientation with respect to the Sun)] referred to as Milankovitch cycles, plus the many internal, positive and negative ocean–atmosphere–cryosphere (ice) feedbacks accompanying these that are also yet to be fully understood. Much remains to be learned about how the coupled ocean–atmosphere system works, leaving plenty of room for new discoveries by the next generation of scientists.

Addendum: Salmon with Mango Salsa

This is my favorite preparation for salmon. I usually purchase Norwegian salmon, which, from what I've read, seems to be the healthiest of the farm-raised salmon. Ora King rates higher, but is also more expensive. Wild caught would be nicer, but not readily available locally.

The next most important ingredient is the mango. These generally require time to ripen so purchase a few several days in advance and wait until they begin to soften and offer a fragrant mango odor of ripeness. When you know that these are ready, purchase the other needed ingredients. I use red and yellow bell peppers, cherry tomatoes, corn, cilantro, red onion, salt, black pepper, lemon juice and olive oil. Peel and cut the mango into ½ inch cubes. Similarly, cut the bell peppers into ½ inch pieces, and cut the cherry tomatoes in half lengthwise. Wash and chop the cilantro.

For corn, I use frozen, packed sweet corn kernels. Dice a thin slice or two of red onion. For two people I use one mango, half of each bell pepper, ¾ cup of corn, one thin slice of red onion, and a fistful of cilantro. Double for four people. Salt and pepper are to taste (add a little and taste—you can always adjust with more). The acid is lemon juice—don't try anything else! Squeeze a whole lemon for two, maybe two lemons for four. Sprinkle on some good olive oil. Don't use cheap cooking oil. The olive oil itself adds wonderful flavor. Toss these ingredients together, taste, adjust and let sit while you prepare the fish. If you don't mind a little heat, dice up a quarter to a half of jalapeno and add. I usually do this separately for myself. If you cannot find mangos, then other fruits like peaches or strawberries will work, although mango is best by far.

The fish is simple. Rinse, dry, season with salt and pepper and bake for about 25 min at 375°. An alternative is to roll the fish (with no skin) in cornmeal and pan saute. This provides a nice crunchy texture, if you don't mind the extra calories. The right portion is about ½ lb of fish per person.

To serve, I generally just put everything atop a green salad of half arugula and half baby spinach. Fill a plate with these greens, then put the fish on top and add lots of salsa over the fish and the greens for a lovely, healthy meal.

A mineral-driven chardonnay is a great accompaniment.

Chapter 2
Additional Aspects of the Global Ocean Circulation and Climate

Evening illumination, as seen from space (Fig. 2.1), is indicative of human population centers. Have you ever wondered why Europe, as far north as Norway, is quite habitable, whereas Labrador, in a relative sense, is not? To understand this, we must add to the prior discussion of the Coriolis force, how gravity and the Earth's spherical shape affect the fluid motions of our planet.

Physical oceanography, the study of the ocean circulation, and meteorology, the study of the atmosphere circulation, both involve the subject of Geophysical Fluid Dynamics (GFD), the physics of fluid motions on a gravitating and rotating Earth. GFD also pertains to other celestial orbs, but here we will only concern ourselves with the Earth. Rotation gives rise to the Coriolis force when motions are viewed relative to the Earth. Gravity gives rise to a force that is directed essentially toward the Earth's center (although there is a small deviation due to the centripetal acceleration, again by rotation), which determines the orientation of the local vertical coordinate. Additionally, being that the gravitational force is of a much larger magnitude than the Coriolis force, it tends to limit motions to being primarily horizontal, i.e., in a plane normal to the local vertical direction. For this reason, we generally neglect the Coriolis force acting on vertical motions because these tend to be much smaller than the horizontal motions, and we consider only the components of the Coriolis force acting in the local horizontal plane, i.e., only acting on the local east–west and north–south components of velocity.

These simplifications allow us to define a Coriolis parameter, $f = 2\omega\sin\varphi$, where φ is the latitude and $\omega\sin\varphi$ is the local vertical component of the Earth's rotation vector. Given the Coriolis parameter, the Coriolis forces for northward (v) and eastward (u) directed motions are fv (directed to the east) and $-fu$ (directed to the south), respectively.

Relative to the rotation of the spherical Earth, whose orientation is constant from the south to the north poles, the orientation of the local horizontal plane is parallel to the rotation axis on the equator, whereas it is perpendicular to the rotation axis at the poles. This sets the Coriolis parameter to being zero on the equator and 2ω, or twice

© The Author(s), under exclusive license to Springer Nature Switzerland AG 2025
R. Weisberg, *Climate to a Fish Sandwich: Why We Study the Ocean's Circulation*,
https://doi.org/10.1007/978-3-031-77592-5_2

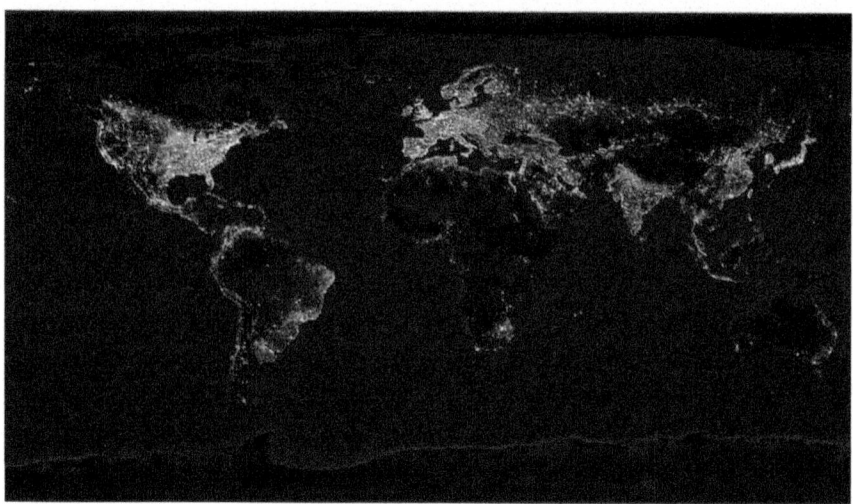

Fig. 2.1 North Atlantic distribution of evening illumination (as seen from space), as an indication of human habitation (courtesy of NASA/NOAA, https://www.nasa.gov/image-article/earth-night/)

the Earth's angular rotational speed, at the poles. It is this equator to pole variation in the magnitude of the Coriolis parameter and hence the Coriolis force (because the Earth is a sphere) that has a profound effect on both ocean currents and atmosphere winds, and hence why we tend to live where we do.

Prior to our physical understanding of these concepts (which only occurred recently), humans discovered these empirically. Once venturing beyond the Pillars of Hercules, propelled by oars and sails, the early explorers navigated over large distances through unknown waters and with unknown winds. Initially their ships sailed best downwind, explaining why Columbus in the fifteenth century landed in the Caribbean, although Phoenician designed lateen sails did allow for some upwind passage. By the eighteenth century and after many explorations, the European trade routes with the new world made use of the prevailing trade winds for westward passage and the prevailing westerlies found farther to the north for eastward passage, with each of these passages assisted by the associated ocean currents. It was only then that the discovery of the Bernoulli principle provided an understanding of why sails could provide lift, leading to more fore and aft sail designs and more effective upwind navigation. Improving ship designs, plus the mounting observational evidence on how ocean currents were organized, led Benjamin Franklin to publish his map of the Gulf Stream, a version (Fig. 2.2) of which appeared in the Transactions of the American Philosophical Society in 1786, that remains reasonably accurate through today.

Revealed in Franklin's map is the phenomenon of western intensification, wherein the currents on the western side of the ocean basins tend to be much narrower and swifter than the currents on the eastern side of the ocean basins. Not only does this apply to the North Atlantic; it is a universal finding for all of the ocean basins. Western

Fig. 2.2 Benjamin Franklin's depiction of the Gulf Stream, as published in 1786 (courtesy, Library of Congress, Digital ID, http://hdl.loc.gov/loc.gmd/g9112g.ct000136)

intensification, while known empirically for two centuries as a fundamental attribute of the global ocean circulation, lacked any physical understanding until recently.

To explain this phenomenon, let's begin with the concept of ocean gyres as depicted in Fig. 1.5. Given the distribution of land that provides east–west boundaries for the Atlantic, Pacific and Indian Oceans, and the prevailing winds, we see that ocean currents approaching a north–south oriented boundary must turn either to the north or to the south. To be more specific, consider the North Atlantic subtropical gyre. In the south is the North Equatorial Current flowing from east to west, as driven by the trade winds, on the west is the northward flowing Gulf Stream that feeds an eastward flowing North Atlantic Drift, as driven by westerly winds, and this circulating gyre is closed on the east by an eastern boundary current sometimes referred to as the Canary Current. The subtropical gyres of the South Atlantic and the North and South Pacific have similar counterparts, as does the Indian Ocean. In all cases, the currents on the western sides of the ocean basins tend to be much swifter and more narrowly confined than on the eastern sides, e.g., the fundamental GFD finding of western intensification.

Whereas early sea-going explorers came to realize that such an organization existed for ocean currents, and Benjamin Franklin had the foresight to draw his

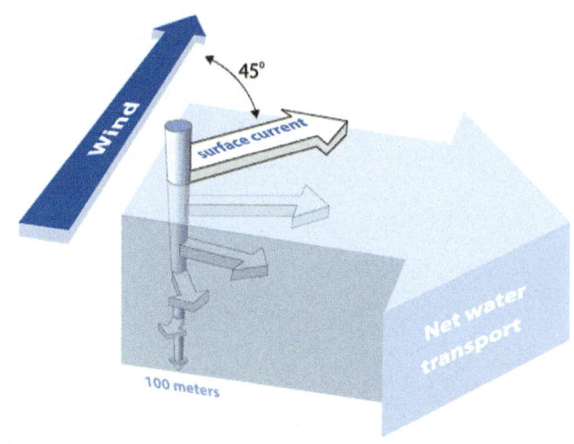

Fig. 2.3 An Ekman spiral schematic showing a surface current directed at 45° angle to the right of the wind direction, a spiral-like decrease in velocity with depth with further right-hand turning and with an overall transport that is directed normal and to the right of the wind (courtesy of NOAA/NOS, https://oceanservice.noaa.gov/education/tutorial_currents)

Gulf Stream map, first in 1769, we had to wait until 1948 for a GFD explanation of this east–west asymmetry in the ocean's gyre scale circulation, underscoring just how young the field of physical oceanography is. Two initial steps had to first occur, one in 1905, the other in 1942.

The first of these gave us what is referred to as Ekman theory, named after the Norwegian scientist Wilfred Ekman who was given the task of explaining why Arctic icebergs tended to drift at an angle to the right of the wind. Using Newtonian physics, as expressed in a set of conservation equations for mass and momentum on a gravitating and rotating Earth, Ekman showed that because of the Coriolis force, a surface current set in motion by frictional coupling with the wind would flow at a 45° angle to the right of the wind, verses directly downwind. Moreover, through successive frictional coupling with depth (i.e., one layer of water rubbing against another deeper layer of water and so on), the wind-driven current would continue turning farther to the right in a spiral of ever decreasing speed, a phenomenon known as the Ekman spiral, as depicted in Fig. 2.3. In light of such spiral-like turning, when averaging over the penetration depth of this Ekman spiral, whose amplitude diminishes rapidly with depth, the total water transport (volume per unit width and time), instead of being downwind, as intuition might suggest, was found to be at a right angle to the wind (to the right in the northern hemisphere and to the left in the southern hemisphere). Note that the 100 m depth is arbitrary. While the Ekman depth is quite shallow, it is the both latitude and friction dependent.

The second step, leading up to our understanding of westward intensification, is also attributed to a Norwegian scientist, Harald Sverdrup, who explained why ocean gyres form in response to winds. For this explanation, we must introduce the concept of angular momentum, another conservation principle used for deductive reasoning, along with mass and momentum conservation. Angular momentum is particularly useful in GFD-related problems because of the Earth's rotation. Just like a change in momentum requires an applied force (a push or a pull), a change in angular momentum requires a torque (or a twist). If the winds operating over the

North Atlantic subtropical gyre are easterly in the south and westerly in the north, then this change in wind from south to north imparts a clockwise torque upon the surface of the North Atlantic Ocean's subtropical gyre. This would tend to impart a clockwise spin on a particle of ocean water. However, recall that the ocean water is already spinning in an anticlockwise sense due to the Earth's rotation and that spin in the local horizontal plane decreases from the north pole to the equator. Hence, if the water moves from north to south in response to the torque applied by the wind, then the decrease in anticlockwise spin can balance the clockwise tendency imparted by the wind-induced torque. In other words, the tendency imparted by the wind stress torque is compatible with the change in fluid parcel spin as the water translates southward. That, in essence, is the Sverdrup relationship, the balance between the change in planetary angular momentum (the fluid particle's spin owing to the Earth's rotation) as ocean water moves south on the eastern side of the ocean basin in response to the clockwise torque imparted by the wind stress. Given that no additional spin relative to the Earth is necessary, a slow, broad southward current on the eastern side of the ocean gyre is sufficient to balance the torque by the wind. While this relatively simple Sverdrup balance can account for the averaged motion over most of the subtropical gyre, something else is required along the western boundary, where the flow direction must be northward to close the gyre and conserve its mass.

Before adding that other ingredient, let's expand upon the Ekman and Sverdrup balances. Ekman theory tells us that the net transport for the thin Ekman layer is at a right angle to the wind direction. Thus, over the northern portion of the North Atlantic subtropical gyre, the net transport for the Ekman layer is southward, whereas for the southern portion the net transport is northward. With the near-surface (Ekman layer) water of the gyre flowing southward from the north and northward from the south, there must be a convergence upon the center of the gyre, where the winds switch direction from westerly to easterly. This convergence results in a bulge in the middle of the gyre and hence a region of high pressure relative to the depressions on both the northern and southern sides of the gyre from where the near-surface water diverges. This wind-related distribution of mass results in a pressure gradient force directed from north to south over the southern portion of the gyre and a pressure gradient force directed from south to north over the northern portion of the gyre. These pressure gradient forces must be balanced by other forces, or else the bulge in the gyre center would collapse. The Coriolis forces associated with North Equatorial Current in the south and the North Atlantic Drift in the north provide for these balances. Such balances between the pressure gradient (the rate and direction at which the pressure forces vary spatially) and the Coriolis forces are referred to as geostrophic balances, and the overall Sverdrup relationship arises as a combination of these Ekman and geostrophic balances, one operating over a relatively thin, near-surface (Ekman) layer, the other operating over the larger, vertical extent of the water column. The Ekman transports cause the convergences and divergences of the ocean water that result in the pressure gradient forces, which, in turn, require Coriolis forces for balance, and the Coriolis forces arise due to the ocean currents being accelerated by the pressure gradient forces until they reach a point where a geostrophic balance is achieved.

The ocean currents are always flowing and the winds are always blowing, thereby rubbing against the surface and continually doing work through frictional coupling. If this were to go unchecked, then the ocean currents would flow ever faster. Another form of dissipation is therefore required to keep the time-averaged ocean circulation in nearly a steady state. An analogy is the stirring of cream into a cup of coffee. As long as you keep stirring, the coffee circles the cup, but once the stirring stops, the coffee quickly settles into a state of no motion because of the friction imparted by the coffee rubbing along both the bottom and the sides of the cup. Just like the cup, the ocean also has a bottom and sides.

In two insightful works, one by Henry Stommel in 1948 and another Walter Munk in 1952, these physical oceanographers explained the phenomenon of western intensification by virtue of the friction that is required on the western side of the basin in order to conserve angular momentum there. Stommel showed how bottom friction could do this, and Munk followed up on the basis of horizontal, side wall friction. Both employed reductive reasoning, employing simplifying assumptions that could facilitate the solution to a closed set of mathematical equations expressing two conservation principles one for mass and the other for momentum.

As an aside, physics provides the understanding necessary to deduce such equations and boundary conditions that together form a closed set, which may be amenable to a mathematical solution, and applied mathematics offers the tools to solve these equations in accordance with their boundary conditions. Reductive reasoning is necessary because these equations and boundary conditions may be quite complex, thereby negating a solution without (reductive) simplifications.

Despite the simplifying assumptions that were necessary for either Stommel or Munk to formulate and solve their respective sets of equations, the essential physics governing western boundary currents was demonstrated. In either case (bottom friction, or horizontal side wall friction), the gain in anticlockwise angular momentum (by virtue of the Earth's rotation as the ocean water moves northward) cannot be balanced by the wind stress torque alone because it is equally clockwise on both sides of the gyre and hence of opposite sign to the change in angular momentum on the western side. Thus, an additional frictional torque other than that provided by the wind stress acting on the surface is necessary on the western side to provide for an angular momentum balance. This frictional torque and its associated dissipation are provided by the generation of clockwise angular momentum relative to the Earth as the currents approach and move along the western boundary. Since frictional dissipation is largest where the currents are both swift and varying over a short distance, we deduce that the western boundary current must be swift and narrow, just as Franklin depicted the Gulf Stream.

Another way to think about this is to consider how an ocean current would vary as it moved northward along the western boundary in the absence of friction. By virtue of the Earth's rotation, each parcel of water would have to gain clockwise spin relative to the Earth to offset the gain in anticlockwise spin by the Earth. We call these two different spins relative vorticity and planetary vorticity, where vorticity is defined as the angular momentum of a fluid parcel spinning about its own axis of rotation. Each fluid parcel has both relative and planetary vorticity, and it is the sum of these two combined that must be conserved if the fluid is not acted upon by an external torque. Without a dissipating frictional torque, a northward flowing western boundary current would grow ever stronger and narrower as it proceeded northward because only in that manner would there be a continual increase in clockwise relative vorticity to offset the increase in anticlockwise planetary vorticity. Under such a scenario, the current would never be able to turn to the right and head offshore, thus closing the North Atlantic subtropical gyre, as is observed. However, by dissipating vorticity via a frictional torque (along the bottom ala Stommel or along the sides ala Munk), the current is able to head offshore and close the gyre circulation. Of course, the situation in nature is more complex; nevertheless, the simplified assumptions by Stommel and Munk, a mere 75 years ago, provided the insights on how the large-scale ocean circulation works, much like either bottom friction or side wall friction acts on a stirred cup of coffee.

As a second aside, both Stommel and Munk assumed a steady state in order to solve their respective sets of equations, thereby negating a description of how the flow field actually sets up. While a necessary assumption leading to an insightful end-result as the previous paragraph attempts to explain, the flow field must evolve to include the clockwise relative vorticity that ultimately must be dissipated through friction. This again shows that physics and mathematics are closely linked to one another. Physical insights are necessary to formulate a problem, but mathematics offers the tools to find a solution to a formulated problem. Students who are interested in a career in science are well advised to gain as much mathematical expertise as possible.

There was a time in my youth when physicians made house calls. While they themselves could readily perform a diagnosis of the illness, without the materials contained in the small black bag that they once all carried, there would be little that they could do as palliative care. Mathematics is the scientist's black bag. It's too bad physicians no longer carry these black bags. Now, if you have a fever you must drive to their office, shivering the entire way while trying not to get into a collision. Then you must travel to a pharmacy to pick up your prescribed medication, and despite modern advances in medicine, you do not recover any more quickly than you did when physicians made house calls.

The ramifications with regard to Earth habitability are profound. Recall that the basis for ocean currents and winds is to transport heat from the tropics to the poles to maintain the Earth's radiant energy balance. Swift western boundary currents provide a primary vehicle for such heat transport. Loosely speaking, the ocean is the primary vehicle for transport from the tropics to mid-latitudes, whereas the atmosphere is the primary vehicle from mid-latitudes to the poles. For the Atlantic Ocean, the primary ocean conveyance is the Gulf Stream, which transports warm water to Cape Hatteras before turning offshore and continuing as the North Atlantic drift. For the Pacific Ocean, the primary ocean conveyance is the Kuroshio, which transports warm water to Japan before turning offshore and continuing as the North Pacific Current.

An example of the resulting distribution of sea surface temperature (SST) for fall/winter months is given in Fig. 2.4. Note that while SST on the Labrador side of the Atlantic is at about 0 °C, it is much warmer at about 14 °C on the England side.

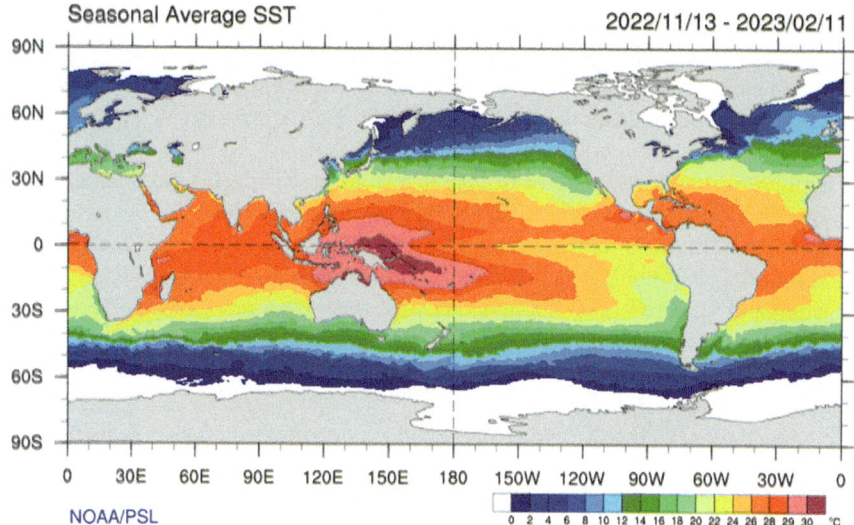

Fig. 2.4 Global distribution of sea surface temperature (SST) averaged from 11/13/22 to 2/11/23. For both the Atlantic and Pacific Oceans note how much warmer SST is on the eastern side than on the western side, a consequence of westward intensification (courtesy of NOAA Physical Science Laboratory, https://psl.noaa.gov/map/clim/sst.shtml)

Whereas SST continues to decrease farther north, the fact that the ocean temperatures are relatively warmer on the eastern sides of the ocean basins impacts the adjacent land temperatures as shown for January 2022 in Fig. 2.5. Thus, the high northern latitude regions on the eastern sides of the ocean basins tend to be habitable, whereas such is not the case on the western sides (e.g., Norway shows 10 °C whereas Greenland shows 2 °C).

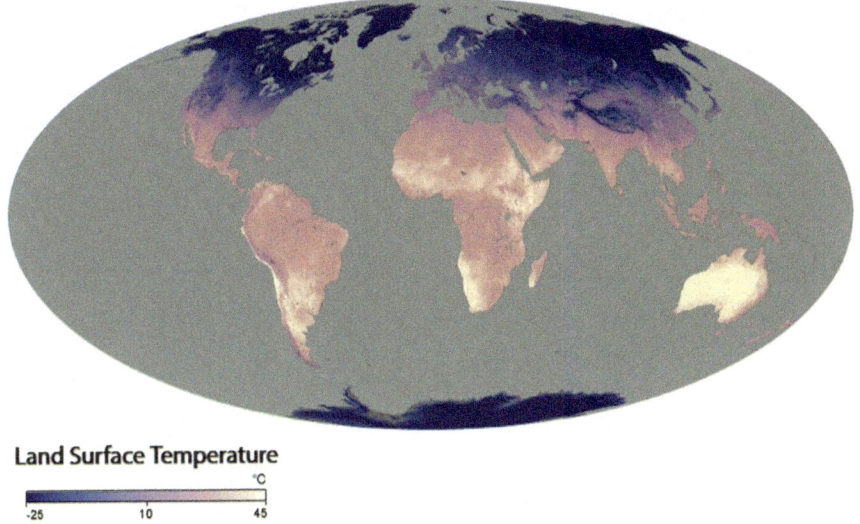

Fig. 2.5 Global distribution of Land Temperature for January 2022. Note the large temperature differences on the Canadian, versus the European sides of the Atlantic Ocean (courtesy of NASA Earth Observatory, https://earthobservatory.nasa.gov/global-maps)

Addendum: Barbeque Salmon

Don't worry, you can still cook this quickly in the oven. Smoked paprika, and lots of it, is the main ingredient besides the salmon. With about a ½ lb of salmon per person, rinse and dry, add salt and black pepper to taste, and lay flat on a baking tray. I usually skin the filet and put the flat side down. Next sprinkle the smoked paprika quite liberally until the entire flesh is covered (don't be stingy). After the paprika, drizzle some honey as little strings about an inch or so apart followed by a little olive oil (less than the honey). Bake in the oven at 375° for about 25 min.

This goes very nicely with a baked sweet potato and a green salad.

You can try variations on this theme, such as ground cumin, and ground coriander, but with lesser amounts than the smoked paprika. The smoked paprika is my favorite.

Another alternative is to use a coarse granular mustard mixed with some honey, maybe 2/3 mustard and 1/3 honey. Put this liberally over the salmon, omitting the olive oil because this mixture is already quite moist.

A pinot noir is a suggested accompaniment.

Chapter 3
The Global Ocean Circulation and Ecology

Having considered the rudiments of climate and human habitability, the next question to ask is how does the ocean circulation impact the most fundamental aspect of Earth ecology, i.e., the distribution of the ocean's primary producers, the phytoplankton? These microscopic plants, which utilize CO_2 and produce O_2 as a product of photosynthesis are the foundation for just about everything that lives in the ocean. Whereas it is estimated that the ocean's phytoplankton comprise only about 1% of the Earth's photosynthetic biomass, these phytoplankton account for nearly half of the world's primary productivity. They do this by reproducing at much more rapid rates than terrestrial plants, and, during the early formation of the Earth's atmosphere (i.e., during the first 2 billion years), it was the phytoplankton that oxygenated the atmosphere, allowing for the subsequent development of terrestrial beings. Whereas the phytoplankton are not as prominent as terrestrial plant life, they play an equal role in moderating atmospheric gases and maintaining the Earth's carbon cycle, plus they are the source (albeit many millions of years removed) of the oil and gas reserves that our economy depends upon.

Varying seasonally, the distribution of phytoplankton within the upper, lighted regions of the world's oceans is far from uniform. Figure 3.1 provides examples of the distribution of Chlorophyll-A, indicative of the surface phytoplankton, as viewed from space. The upper and lower panels are the monthly averages for June and December 2010, respectively. Recognizing that photosynthesis requires both light and nutrients, it is easy to appreciate why the higher latitudes exhibit blooms during summer months (boreal summer in the north and austral summer in the south). Less obvious are the regions that tend not to exhibit blooms in any season and the curious lines along the equator and adjacent to certain continental regions. For explanations, we must return to the ocean circulation.

The previous two chapters were primarily concerned with the near-surface ocean circulation. However, just like the atmosphere, the ocean circulation is fully three-dimensional. In the most general sense, the ocean may be thought of as having two regimes, one being an upper, relatively warm regime of maybe 100–1000 m in depth

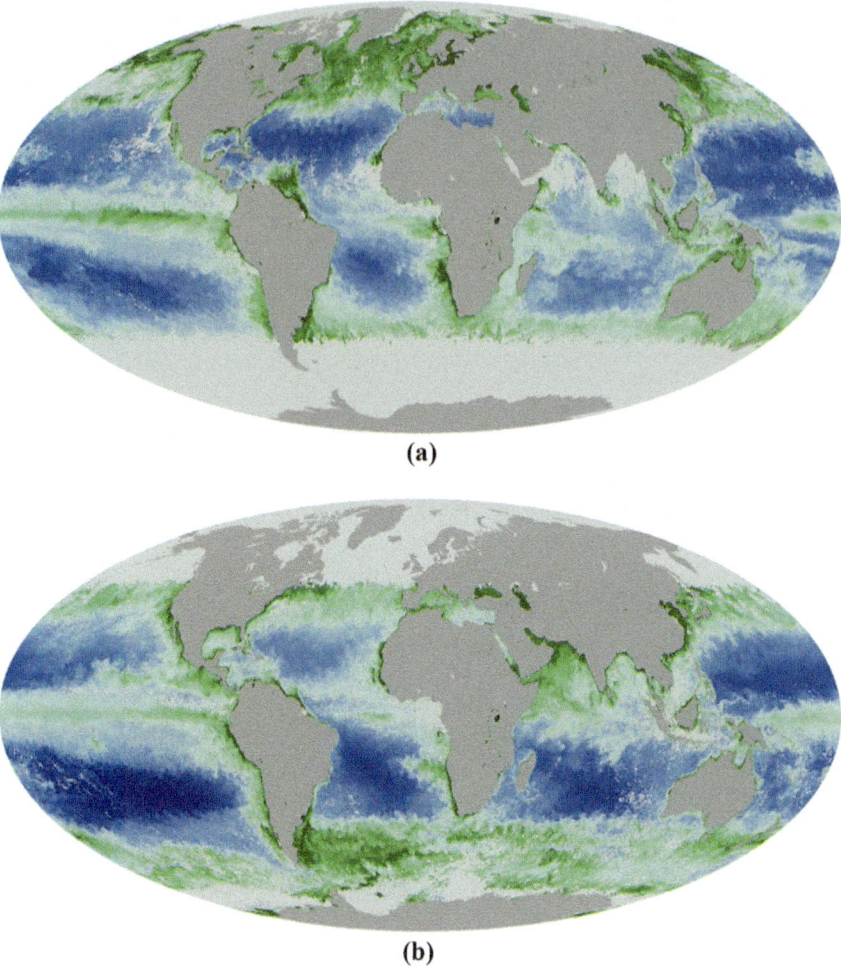

Fig. 3.1 Upper panel (**a**) is the global distribution of Chlorophyll-A as seen from space and averaged over the month of June 2010. The lower panel (**b**) is the global distribution of Chlorophyll-A as seen from space and averaged over the month of December 2010 (courtesy of NASA Earth Observatory, https://earthobservatory.nasa.gov/global-maps)

and the other being a deeper, colder regime extending down to the bottom, which, for the deep ocean basins, is at about 5000 m in depth. Thanks to popular television shows focusing on coral reefs and pretty fishes, we tend to have a slanted view of what comprises the oceans. With reefs generally existing down to depths of less than 100 m, and with SCUBA allowing for dives down to about 50 m, only a miniscule portion of the ocean can be explored by SCUBA diving on coral reefs.

Instead, our understanding of the ocean circulation, and how it impacts everything on Earth, requires much more global scale expeditions, technologically advanced

3 The Global Ocean Circulation and Ecology

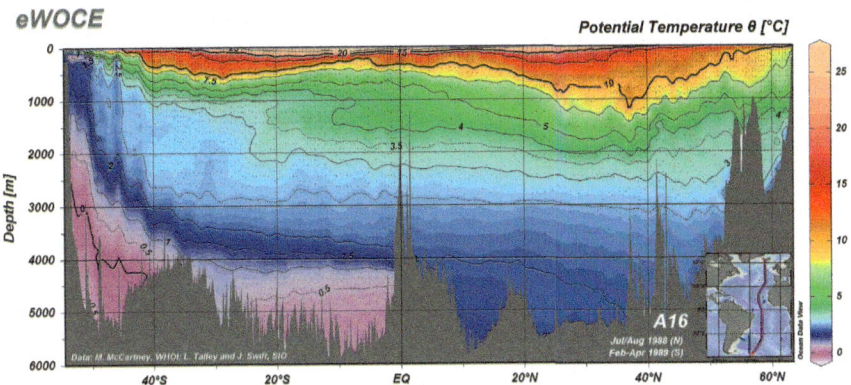

Fig. 3.2 Distribution of potential temperature sampled from Iceland to the Antarctic along the WOCE section A16 (see inset) (courtesy of Schlitzer, R., Electronic Atlas of WOCE Hydrographic and Tracer Data Now Available, *Eos Trans. AGU*, 81(5), 45, 2000)

instrumentation deployed in a variety of ways, and, beginning in the 1970s, various sensors deployed on both orbiting and geostationary satellites. Beginning with the earliest explorations using thermometers lowered on cables, it was found that the temperature of the deeper waters was near freezing, from which it was realized that these waters had to come from high latitudes. Then, as more data began to emerge, including an ensemble of various water properties such as temperature, salinity, nutrients, and oxygen, patterns indicative of potential regions of water mass formation started to become evident.

Consider the distributions of potential temperature (temperature with the effect of pressure removed) and salinity for a mid-Atlantic meridional, trans-ocean section sampled from Iceland to the Antarctic, as shown in Figs. 3.2 and 3.3, respectively. These data were systematically collected during the World Ocean Circulation Experiment (WOCE) conducted from 1990 to 1998. The demarcation between the relatively warm, near-surface waters and the colder waters below is called the thermocline, the region where temperature changes most rapidly with depth. Inspection suggests that the coldest water is formed in the Antarctic, and we refer to this water as Antarctic Bottom Water (ABW).

Adding salinity to the description identifies another water mass of Antarctic origin referred to as Antarctic Intermediate Water (AIW). Its lower salinity allows it to ride atop water of somewhat higher salinity that is traceable to the North Atlantic and referred to as North Atlantic Deep Water (NADW). Thus, we see that the bulk of the ocean's water originates at very high latitudes. Curiously, we also see that the thermocline comes very close to the surface at the equator, whereas it is deeper at mid-latitudes, where relatively high salinity values also penetrate deeper than near the equator.

The waters within and above the thermocline tend to be formed each winter when the surface is cooled, making these waters heavier so that they sink, only to be paved

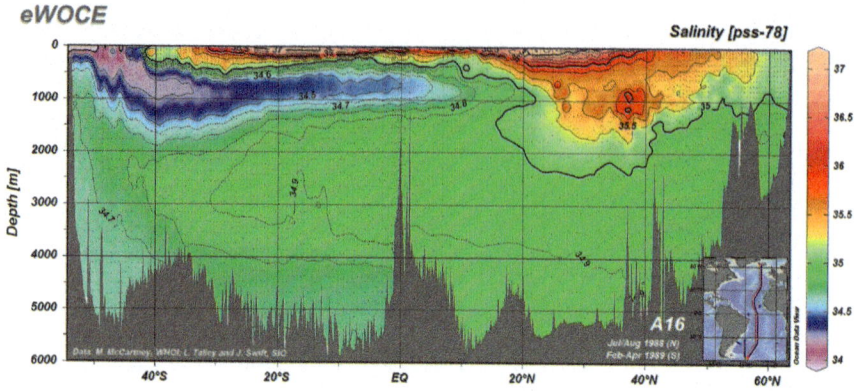

Fig. 3.3 Distribution of salinity sampled from Iceland to the Antarctic along the WOCE section A16 (see inset) (courtesy of Schlitzer, R., Electronic Atlas of WOCE Hydrographic and Tracer Data Now Available, *Eos Trans. AGU*, 81(5), 45, 2000)

over by newly warmed water in the summer months, with this process repeating year upon year. Similarly, but more dramatically, the deeper waters are formed in winter at high latitudes, but since the thermocline is less developed there, those newly formed waters can sink to greater depth; all the way to the bottom in the case of ABW.

We refer to the circulation of the waters above and within the thermocline as the wind-driven circulation and the circulation below the thermocline as the thermohaline circulation. In actuality, both of these circulations are driven by a combination of winds and buoyancy (density) differences, but to varying degrees. Recall that we already discussed how convergences and divergences occurring within the near-surface, wind-driven Ekman layer result in pressure forces that extend to great depth, and ultimately the pressure forces are what drives the deep circulation.

The nutrient distribution is also quite important and interesting. As one might guess, given that the existence of plants depends on both light and nutrients, any nutrients within this euphotic zone (the zone that includes both light and nutrients) are rapidly depleted by photosynthesis, leaving the near-surface waters relatively nutrient depleted. This occurrence is readily seen in Figs. 3.4 and 3.5, the phosphate and nitrate counterparts, respectively, of Figs. 3.2 and 3.3. Both the AAIW and the ABW have high values of phosphate and nitrate.

Lower values are found in the NADW, and, of course, the lowest values are near the surface. Nutrient values at depth are acquired at the surface when and where the water sinks in winter. With the absence of light at depths below about 100 m, photosynthesis shuts down, and the nutrient values remain fairly steady below the euphotic zone except for additions by remineralization from organic matter, as decaying biomass rains down from above.

Given these water property (temperature, salinity and nutrients) distributions, the next question to ask is: why are some regions more photosynthetically active than others? Let's consider these various regions one at a time. First, consider the blue regions of Fig. 3.1. These are located at the centers of the subtropical gyres, where,

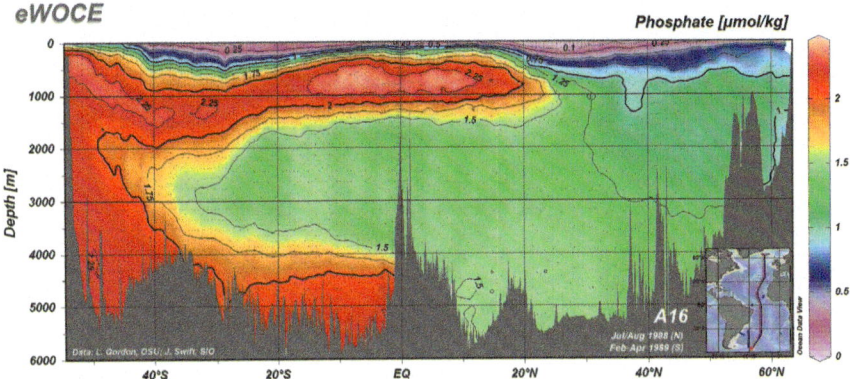

Fig. 3.4 Distribution of phosphate sampled from Iceland to the Antarctic along the WOCE section A16 (see inset) (courtesy of Schlitzer, R., Electronic Atlas of WOCE Hydrographic and Tracer Data Now Available, *Eos Trans. AGU*, 81(5), 45, 2000)

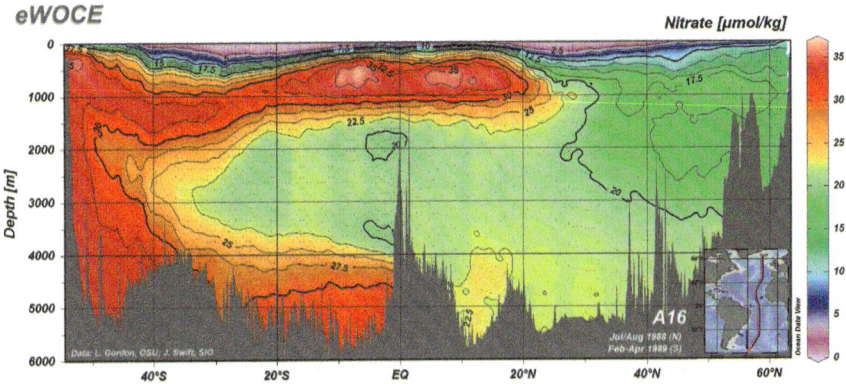

Fig. 3.5 Distribution of nitrate sampled from Iceland to the Antarctic along the WOCE section A16 (see inset) (courtesy of Schlitzer, R., Electronic Atlas of WOCE Hydrographic and Tracer Data Now Available, *Eos Trans. AGU*, 81(5), 45, 2000)

by virtue of the surface Ekman layer transport, there is a convergence of fluid and hence a deepening of the region above and within the thermocline. Recall that the Ekman transport is at a right angle to the wind direction. So the westerly winds to the north of the North Atlantic subtropical gyre transport near-surface water to the south and the easterly winds to the south of the North Atlantic subtropical gyre transport near-surface water to the north. The convergence of these waters in the middle results not only in a bulge at the surface as discussed in the last chapter, but also in a downwelling and hence a deepening of the thermocline. A schematic of the zonal component of the winds and the related distribution of the gyres is shown in Fig. 3.6. As seen in Figs. 3.1, 3.2, 3.3, 3.4 and 3.5, along with the temperature and salinity isopleths, the phosphate and nitrate isopleths are also deepened, leaving very

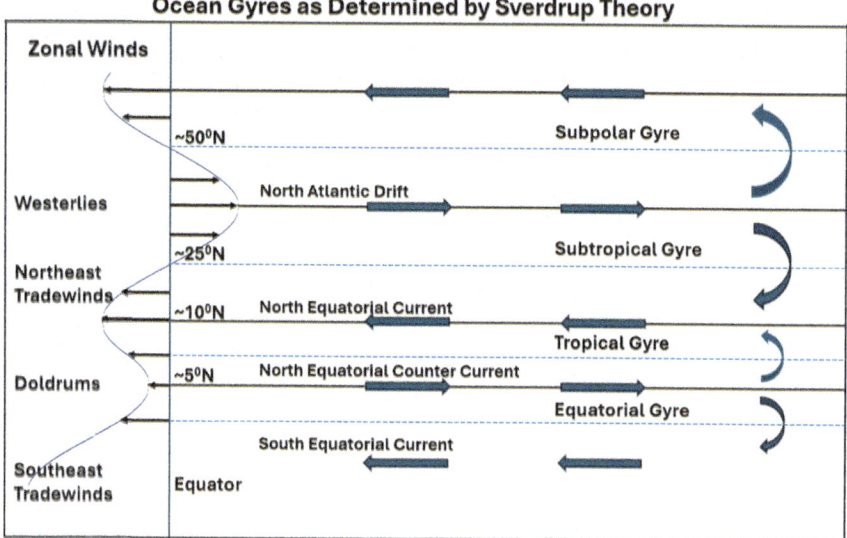

Fig. 3.6 A schematic of the northern hemisphere gyres showing how the south to north distribution of the winds (by virtue of Sverdrup theory, with closure due to westward intensification) give rise to Equatorial, Tropical, Subtropical and Subpolar gyres. These gyres all vary seasonally with the seasonally varying winds. However, due to the change in the Coriolis force with latitude, the adjustment times to wind variations are fast near the equator, versus being much slower toward the poles. Thus, the subtropical and subpolar gyres are robust features year-round, whereas the equatorial and tropical gyres tend to wax and wane annually

little nutrients to support photosynthesis within the subtropical gyre centers; hence, the subtropical gyres are the oceans equivalents of the land-based deserts.

To the north of the subtropical gyre is a subpolar gyre with upwelling in its center due to an Ekman layer divergence there. Hence, the thermocline, the halocline and the nutricline (that generally coincides with the pycnocline, the region of rapid density increase with depth) are all elevated, putting increased levels of nutrients into the euphotic zone in support of photosynthesis. This results in a spring to summer bloom to the north of the oceanic desert region once both light intensity and duration increase after the winter solstice gives way to the spring equinox and summer solstice. Similar, may be gleaned to the south of the subtropical gyre where upwelling also occurs between the North Equatorial Current and the North Equatorial Countercurrent owing to an Ekman layer divergence there. The phytoplankton distributions in the southern hemisphere follow the same explanations, only with the sign of the Ekman transport revered (i.e., where the Ekman transport is to the left of the wind direction, versus to the right).

Why the Ekman transport is to the right of the wind in the northern hemisphere and to the left of the wind in the southern hemisphere follows from the same explanation given in the previous chapter. By standing upside down in the southern hemisphere, what you perceive there may look different relative to your viewpoint, while being quite the same relative to the Earth's rotation vector, which points in the same direction in both hemispheres.

The foregoing discussions leave the coastal regions and the equator requiring further explanation. Ekman divergence also applies to the coastal regions. If surface water is driven away for the coast via Ekman transport, then these waters must be replaced by new water that upwells from greater depth. Northerly winds along eastern boundaries and southerly winds along western boundaries cause such upwelling. It is for this reason that northwest Africa and the west coast of the United States have relatively cold water, with ample phytoplankton, and hence zooplankton that feeds upon these and therefore vibrant fisheries.

The equator is unique in that the Coriolis force acts to the right of the fluid motion just to the north of it, whereas it acts to the left of the fluid motion just to the south of it. Thus, a westward directed current along the equator (the South Equatorial Current), as driven by easterly winds, is divergent, thereby causing the thermocline, halocline, hence the pycnocline and the nutricline to rise up toward the surface at the equator. The result is a thin green line along the equator, with the intensity of color bleeding off to the north and south, just as if an artist had painted it.

We actually have observations of the divergence and estimates of the upwelling, both in the Atlantic and the Pacific Oceans using instrumentation deployed from stationary moorings. These observations show that the equator acts as a line of symmetry, owing to the Earth's rotation. Because of this, the equator also serves as a waveguide, supporting the propagation of large-scale, planetary waves that adjust the temperature distribution (and other water property distributions) to changes in winds. It is for this reason that we experience the El Nino—Southern Oscillation (ENSO) and its worldwide climate ramifications. As an example of how such coupled ocean–atmosphere oscillation can affect primary productivity consider a comparison between Chlorophyll-A, as seen from space in June 2009 (during an El Nino year) and in June 2010 (during a La Nina year), as given in Fig. 3.7. Equatorial upwelling is diminished during an El Nino, when the easterly winds over the eastern equatorial Pacific Ocean tend to weaken, whereas equatorial upwelling is intensified during a La Nina, when the easterly winds over the eastern equatorial Pacific Ocean tend to intensify. From these findings, it is easy to understand that while the local climate may be forced by essentially the same variations in sunlight each year, the coupling between the ocean and the atmosphere can result in much different seasonal variations in any given year. Such climate variations are referred to as interannual variability due to the coupling between the ocean and the atmosphere, and ENSO is merely one such coupled mode of oscillation. There are other coupled ocean–atmosphere modes of oscillation that we know about and still more to be discovered, and these occur on time scales ranging from interannual to decadal and longer, some longer than human lifespans. Climate is complex; much more so that you may think by only listening to popular media outlets and politicians.

From the foregoing chapters, it is now evident that the ocean not only initiates climate by setting in motion the currents and winds that are required to balance the Earth's heat budget, but also that the ocean's circulation determines where primary productivity can flourish over nearly three quarters of the Earth's surface. Moreover,

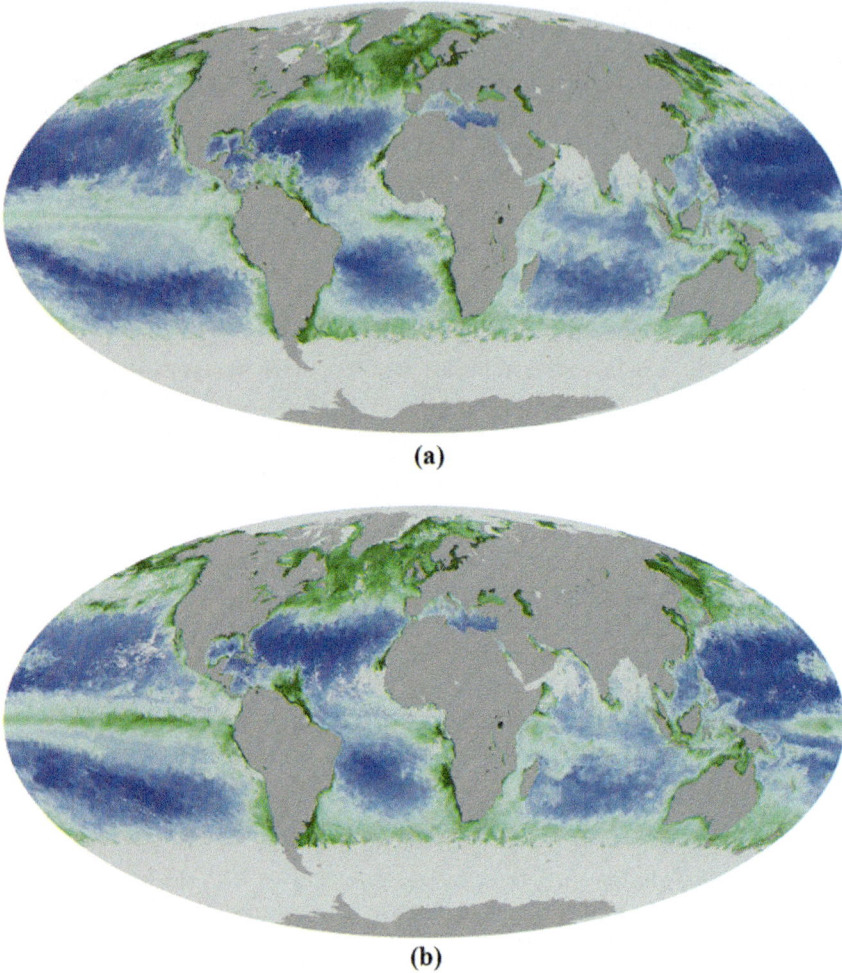

Fig. 3.7 Upper panel (**a**) is the global distribution of Chlorophyll-A as seen from space and averaged over the month of June 2009. The lower panel (**b**) is the global distribution of Chlorophyll-A as seen from space and averaged over the month of June 2010 (courtesy of NASA Earth Observatory, https://earthobservatory.nasa.gov/global-maps)

through its coupling with the atmosphere, the ocean's circulation is what determines where primary productivity can flourish on the remaining quarter of the Earth's surface that is land. From a global perspective, we may now appreciate that the ocean's circulation is existential, which provides ample justification to study and understand it.

Addendum: Baked Cod

We spend part of our summers in Maine escaping the Florida heat. A Boothbay Harbor staple is cod, baked with a crumb topping. Again, a ½ lb per person seems to be about right, although I usually make this with a whole (skinned) filet, whatever that tends to weigh. Rinse, dry and add salt and black pepper (sometimes I also use a yellow or red spice mix consisting of lots of different stuff that you may find online or in a kitchen store). Place the filet in a Pyrex baking dish or something similar that has been lightly greased with olive oil or butter. Put a whole sleeve of Ritz crackers in a small plastic zip-lock bag, seal and then gently pound it on the countertop with the soft side of your fist. Turn and repeat until the crackers are broken into something just short of a powder. Sprinkle these pulverized crackers onto the fish until the entire filet is covered with no flesh showing. Cut up a few thin pats of butter and place these at one-inch separations over the crumbed fish. Bake at 375° for about 25 min. You will know when the fish is done when you begin to see a whitish liquid appear through the crumbs. Cod does need to be cooked throughout until flakey, but be careful not to overcook (that goes for any fish).

The crumbs provide the carbohydrate so any medley of vegetables provides a great accompaniment. You might try roasting cauliflower or zucchini with some salt, pepper and an olive oil drizzle. Alternatively, boiling cauliflower and then pureeing in a blender to make a cauliflower mash is also terrific. When I do this, I usually add some ricotta cheese to give it a little body.

A sauvignon blanc or a vermentino appeals to me.

Chapter 4
The Coastal Ocean: How It Is Driven

The coastal ocean is where society meets the sea. As such, coastal ocean regions tend to have multiple, competing utilizations. As a consequence of both land runoff and deeper ocean interactions, most coastal ocean regions tend to be nutrient rich and hence ecologically productive, with abundant commercial and recreational fisheries. Coastal ocean beaches provide the basis for tourism, which supports the associated hospitality and restaurant industries, and its deep-water ports facilitate active marine commerce hubs, each with vibrant population centers. Given such confluence between natural resource and human endeavor, it is important to understand just how the coastal ocean system works so that its various utilizations may remain compatible with one another.

What distinguishes the coastal ocean from the larger ocean basins is the water depth. Coastal ocean regions tend to be relatively shallow, 10–100 s of meters deep, versus 1000 s of meters deep for the deeper ocean basins. As an example, consider the west coast of Florida with its broad, gently sloping West Florida Continental Shelf (WFS) that extends offshore by a distance comparable to that of the subaerial, land-based State of Florida (Fig. 4.1). The bottom slope for much of the WFS tends to be nearly constant out to distances of about 120–200 km, where the slope then suddenly changes at what is termed the shelf break. Seaward of the shelf break, the water depth then increases more rapidly into the abyss. For much of the central WFS, the shelf break is located at about the 75 m isobath, beyond which the bottom slope continually steepens down to the abyssal Gulf of Mexico at a depth of about 3000 m. The region between the shelf break and the abyss is called the continental shelf slope.

Looking northward from around the Tampa Bay vicinity, the shelf width first increases into the Florida Big Bend region before decreasing to a minimum width near Pensacola, FL. The depth of the shelf break also decreases in this direction becoming as shallow as 30 m near Pensacola, FL. These geometrical differences impact both the shelf circulation behavior and how the shelf and the deeper ocean

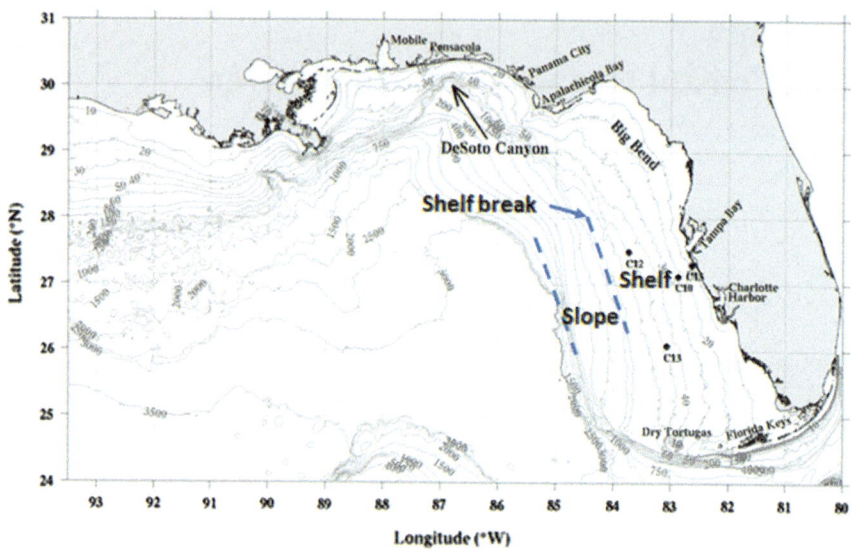

Fig. 4.1 Geometry of the West Florida continental shelf. The relatively shallow continental shelf extends from the coast to the shelf break, and this is followed by the steeper shelf slope that extends from the shelf break to the abyssal ocean. Also shown are the locations of presently deployed (2024) moored instrumentation

may interact with one another. We will first begin our discussion of the continental shelf circulation with how it is forced by local winds before concerning ourselves with how it is impacted by interactions with the adjacent deeper ocean.

4.1 The Wind-Driven Coastal Ocean Circulation

The three primary forces that determine the coastal ocean response to external factors are the pressure gradient, the Coriolis and the frictional forces. As discussed previously, the pressure gradient derives from variations in sea level and water density, and the Coriolis force is due to the Earth's rotation. Friction is somewhat more involved. Friction is generally owing to the variations of the horizontal velocity vector with depth, causing fluid parcels to rub against one another throughout the entire water column and also to rub against the bottom.

For the case of the wind-driven coastal ocean circulation, water is set in motion by the friction of the wind acting upon the surface (the wind stress), i.e., the fluid atmosphere rubbing against the fluid ocean. In accordance with Ekman theory, as discussed in previous chapters, coastal ocean water may either converge upon, or diverge from the shoreline (Fig. 4.2), causing the sea surface to either slope up or

4.1 The Wind-Driven Coastal Ocean Circulation

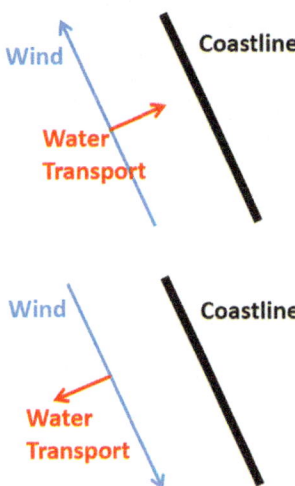

Fig. 4.2 A schematic showing how alongshore directed winds result in a net water transport at a right angle to the shoreline

down against the shoreline, thereby inducing a pressure gradient force. The water beneath the surface is then set in motion by this pressure gradient force, and as the subsurface-induced flow feels the bottom, its motion is opposed by the frictional force of the water rubbing against the bottom. The entire process is referred to as the Ekman-geostrophic spin-up process, where spin-up is in recognition of the role played by the rotation (or spinning) of the Earth.

Understanding the Ekman-geostrophic spin-up process is the key ingredient to understanding how material properties (such as the freshwater derived by river inflows, nutrients, phytoplankton, zooplankton, fish larvae, plus other materials that may be contained in the coastal ocean water) are transported across the continental shelf. The role of the Coriolis force is pivotal.

Let's start again from the beginning by considering a wind blowing parallel to the shore with the land to the left, i.e., with the wind blowing toward the SSE with respect to the WFS (see Fig. 4.1). The surface water that is set in motion by the frictional stress of the wind will be accelerated to the right by the Coriolis force, thereby moving away from the coast and causing sea level to decrease there. Farther seaward, as the water depth increases, a depth is reached where the total transport of water, as summed from the surface down to the base of the Ekman layer (i.e., the depth where direct wind forcing is no longer felt by the ocean, as in Fig. 2.3), is at a right angle to the wind direction (see the lower panel in Fig. 4.2).

While the total transport of water being at a 90° angle to the right of the wind velocity vector may seem non-intuitive, this consequence of the Earth's rotation has profound influence on the ocean circulation and hence on both the Earth's climate and ecology, as discussed in previous chapters. With the wind-driven water moving away from the shore and sea level dropping there, an onshore-directed pressure gradient force develops that will tend to push water back toward the coast over the entire water column. However, in the same way that the wind-driven surface water is accelerated to the right by the Coriolis force, the pressure gradient-driven water is also accelerated to the right by the Coriolis force. Thus, in the middle of the water column, and away from the direct frictional effect of the wind stress at the surface, or the bottom stress, the Coriolis force will tend to balance the pressure gradient force, and the water will flow parallel to the coast in the direction of the wind, which in this case is toward the SSE. Upon approaching the bottom, friction will cause this interior flow to slow down, which decreases the Coriolis force so that it can no longer balance the pressure gradient force. Once this imbalance occurs, the overpowering pressure gradient force will turn the velocity vector to the left, and toward the shore.

The regions of the water column where the Coriolis and frictional forces are comparable are referred to as the Ekman layers, which are found both at the surface and near the bottom, where the frictional forces tend to be the largest. Given a concomitant slope to the sea surface that occurs when water either converges upon or diverges from the coast, there is also a pressure gradient force that adds to the Coriolis and frictional forces. If the water is deep enough so that the surface and bottom Ekman layers become separated, then the force balance in between these two Ekman layers will be between the Coriolis and the pressure gradient forces alone. We refer to this interior region as the geostrophic interior where the force balance is a purely geostrophic balance.

These combined effects of the pressure gradient, the Coriolis and the frictional forces give rise to the Ekman-geostrophic spin-up of the coastal ocean to wind forcing as previously stated. With the Coriolis parameter (with units of $\sec^{-1}$) being the only determinant of the time scale for which this Ekman-geostrophic spin-up process occurs, the response time for the coastal ocean, when acted upon by wind, is a pendulum day (24 h divided by the sine of the latitude), and all of these three forces (friction, Coriolis and pressure gradient) build up nearly in unison over the course of a pendulum day.

To recap this sequence of events: a shore parallel wind sets the surface water in motion, upon which the Coriolis force causes a turning to the right and hence a resultant pressure gradient force as water either diverges from, or converges upon, the coast; the across-shelf pressure gradient then sets the entire water column in motion, which is acted upon by the Coriolis force, causing a shore parallel flow, which is slowed near the bottom by friction such that the pressure gradient deflects the near-bottom flow to the left. It is in this manner that the full water column is set in motion by the wind acting on the surface, and this entire interconnected process evolves over the course of a pendulum day.

Direct observations of wind, sea level and coastal ocean currents for the WFS, provide a demonstration of this Ekman-geostrophic spin-up process, as shown in Fig. 4.3. Nature provided, in this instance, what applied mathematicians refer to as an initial value problem. Following several days of very light winds, the winds intensified rapidly from the NNE (or toward the SSW). Each one of the lines in Fig. 4.3 represents a wind or a water velocity vector, wherein the length of the line provides the speed and the orientation provides the direction in which the wind (or the coastal ocean current) is moving. We see that in response to these winds, the near-surface currents turn to the right of the wind velocity vector, causing an

4.1 The Wind-Driven Coastal Ocean Circulation 35

offshore-directed transport over the surface Ekman layer and an associated decrease, or set down, of sea level along the coast. Occurring almost in unison is a systematic left-hand turning of the water velocity vectors with depth owing to the effects of the geostrophic interior flow response to the pressure gradient force, as the coastal sea level decreases. The combined surface Ekman layer and geostrophic interior flows, by rubbing against the bottom, cause a further systematic left-hand turning of velocity vector with depth upon approaching the bottom. With these surface Ekman layer, geostrophic interior and bottom Ekman layer flow all overlapping across the coastal ocean water column, and all occurring over the course of a pendulum day, it is difficult to distinguish these responses separately. Nonetheless, the combined result is an initial right-hand turning of the water velocity vectors at the surface from the direction of the wind velocity vectors, followed by a systematic left-hand turning of the water velocity vectors from the surface down to the bottom. Later, over the course of the Ekman-geostrophic spin-up, we see that coastal sea level stops dropping once the landward flow in the bottom Ekman layer is sufficient to balance the seaward directed flow in the surface Ekman layer. Then, upon the abatement of the wind, the pressure gradient force relaxes as sea level returns to its original, undisturbed state.

This process of coastal ocean response to wind forcing occurs repeatedly as the winds vary with each successive weather front passage. For the WFS, a SSE directed, along-shelf wind results in an upwelling circulation, whereas a NNW directed along-shelf wind results in a downwelling circulation. The terms upwelling and downwelling refer to the direction of flow in the vertical. If coastal sea level is set down due to an offshore transport of water within the surface Ekman layer, then there must eventually (within a pendulum day) be a compensating onshore flow within the bottom Ekman layer and hence an upwelling of water from the bottom to the surface, or an upwelling response, and conversely for a downwelling response.

Along with the upwelling and downwelling responses to along-shelf winds, the shallow, gently sloping WFS geometry also affords upwelling and downwelling responses to (ENE and WSW) across-shelf winds because as the depth shoals to less than the Ekman depth, the direction of the wind-driven flow becomes ever more downwind. These Ekman-geostrophic, coastal ocean wind-driven responses are what account for the higher than normal high tides and the lower than normal low tides with the passage of cold fronts that generally occur from fall through spring months. Downwelling occurs in advance of the front, when the winds have a southerly component, and upwelling occurs in the wake of the front when the winds have a northerly component. Downwelling causes sea level to rise, and upwelling causes sea level to fall.

Such responses to successive frontal passages are evident in both actual sea level and velocity observations and in numerical circulation model simulations. As an example, sea level and wind observations from St. Petersburg, FL for the month of April 1998 are shown in Fig. 4.4 (lower two panels), along with coastal ocean velocity observations obtained across the water column from the near surface to the near bottom at a station located on the 20 m isobath offshore of Sarasota, FL. Three different frontal passages occurred during April 1998. As shown by the wind velocity record (lower panel), southerly winds at the beginning of the month changed

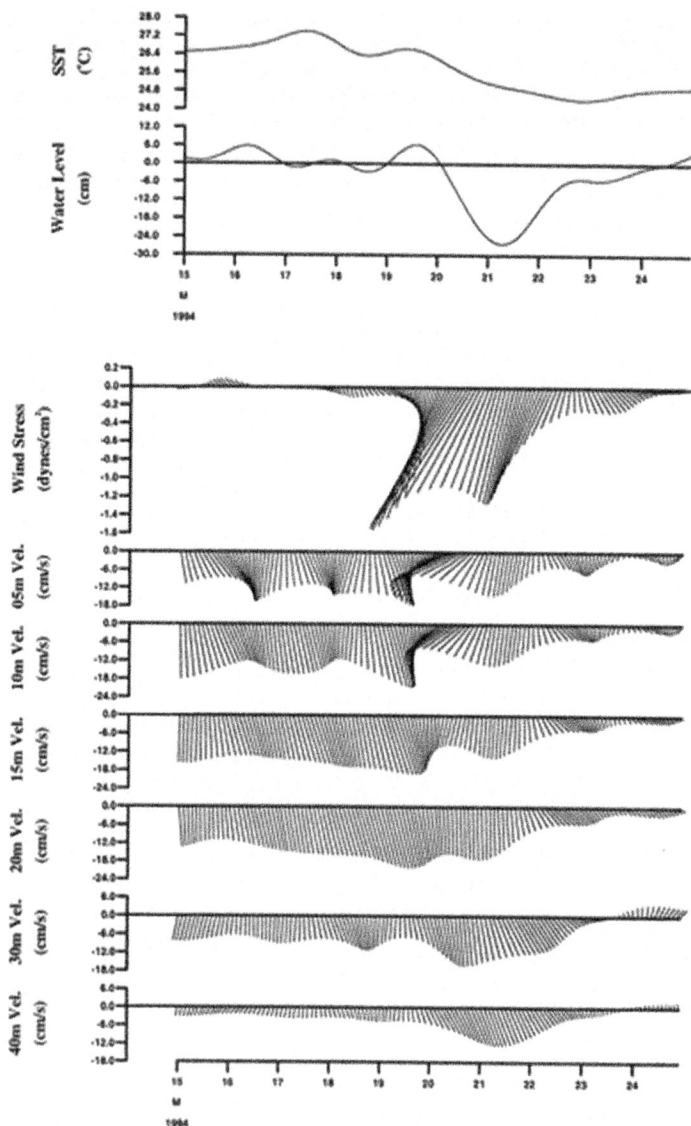

Fig. 4.3 Wind velocity and ocean current velocity vectors, plus sea surface temperature observed on the 50 m isobath in the Florida Big Bend region from May 15–25, 1994, along with sea level at St. Petersburg, FL. The time series are all low-pass filtered to remove variations at time scales shorter than 36 h. The orientation of the velocity vectors is such that north is up and east is to the right, and the magnitude of the vectors is given on the ordinate (adapted from Weisberg, R. H., B. Black, Z. Li (2000). An upwelling case study on Florida's west coast, *J. Geophys. Res.*, 105, 11,459–11,469)

to northerly and so on, and with each successive change, sea level either trended up or down. The upper two panels show the across-shelf and the along-shelf components of water velocity, respectively. As an example of the downwelling and upwelling responses, first consider the time interval from April 7th to April 13th. With the onset of southerly winds on April 7th, sea level rises, as the across-shelf component of ocean velocity near the surface is directed shoreward. Sea level peaks around April 9th, when the across-shelf component of ocean velocity near the bottom is directed seaward and of magnitude sufficient to offset the landward flow, thereby stopping the convergence of fluid along the coast. In synchrony with the across-shelf component of velocity, we see the development of a NNW directed along-shelf current across the entire water column, just as explained in the previous discussion on the Ekman-geostrophic spin-up process.

Then, as the winds decrease and reverse to become northerly, sea level drops as the across-shelf and the along-shelf components of velocity adjust in the opposite sense. Offshore flow at the surface accompanied by onshore flow at depth demonstrates upwelling along with its SSE directed along-shelf flow, just as the opposite occurrence demonstrated downwelling. This month-long record shows three successive realizations of such wind-driven, coastal ocean responses.

Given these actual observations to gauge the veracity of a model simulation, we can also use model simulations to investigate the force balances, as previously explained. The simplest example to consider is that for which the water density is held constant, as occurs in wintertime, when the entire water column is well-mixed from the surface down to the bottom. Systematically summing across the water column from top to bottom each of the individual terms that comprise the force balances in either the along-shelf or the across-shelf directions after the flow field reaches a relatively steady state results in the force balance illustration of Fig. 4.5. In the along-shelf direction, we see that beyond some distance from the shoreline, here at about 80 km where the 50 m isobath is located, there is an approximate balance between the wind stress and the vertically averaged Coriolis force, just as Ekman theory suggests. In other words, with the total transport beyond that location being directed offshore in the surface Ekman layer, the Coriolis force associated with that Ekman transport must be directed along-shelf in the direction opposite from the wind stress, in this case toward the NNW because the wind is directed toward the SSE.

Going shoreward from that 80 km (50 m isobath) location, we see that the Coriolis force term begins to diminish toward zero at the shoreline. As the role of the Coriolis force term in the alongshore momentum balance diminishes, it is replaced by an increasing bottom friction force term, plus a smaller contribution from an alongshore pressure gradient force term because the shoreline is not exactly straight (and there is also a change in the magnitude of the flow in the along-shelf direction).

These force balance transitions allow us to define a portion of the continental shelf as being the inner-shelf. In its simplest sense, the inner-shelf is the region where the pure Ekman balance (that may exist when the water is deep enough) transitions to one in which bottom friction plays an ever-increasing role as the water depth shoals

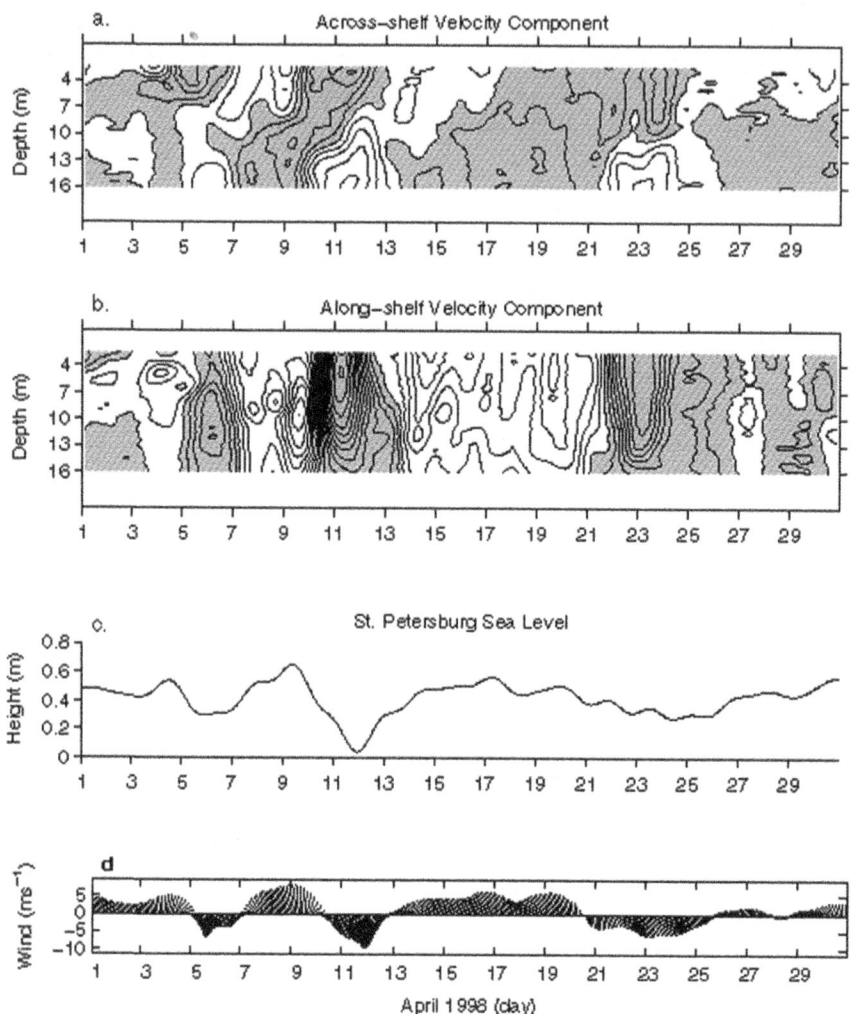

Fig. 4.4 From top to bottom are the across-shelf and the along-shelf components of velocity observed on the 20 m isobath offshore from Sarasota, FL for the month of April 1998, plus sea level and wind velocity from St. Petersburg, FL. All of the time series are low-pass filtered to exclude variations on time scales shorter than 36 h (adapted from Weisberg, R. H., Z. Li, and F. E. Muller-Karger (2001), West Florida shelf response to local wind forcing: April 1998. *J. Geophys. Res.*, 106, 31,239–31,262)

toward the shoreline. Referring to Fig. 4.5, this region extends shoreward from about the 50 m isobath. Concurrently, the inner-shelf is also the region where an across-shelf surface slope sets up due to surface Ekman layer divergence (convergence) that builds up until this becomes balanced by a bottom Ekman layer convergence (divergence).

4.1 The Wind-Driven Coastal Ocean Circulation

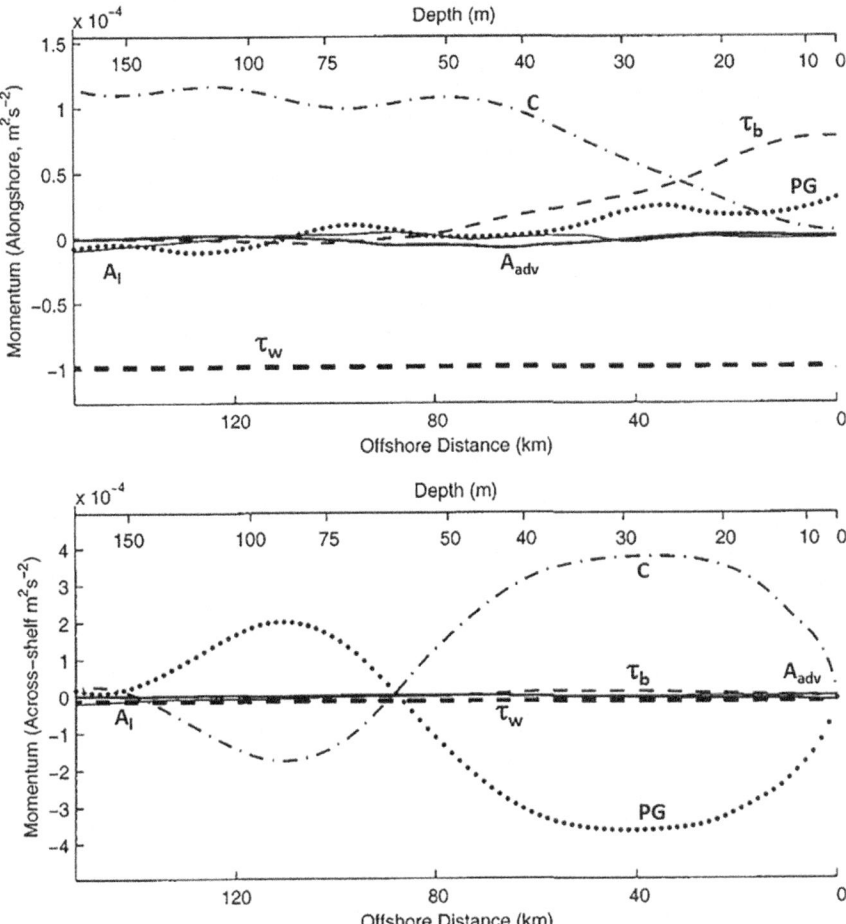

Fig. 4.5 Terms in the vertically integrated momentum balance at steady state for an initial value problem in which the WFS is forced from rest by an upwelling favorable wind stress. The top panel is for the along-shelf direction, and the bottom panel is for the across-shelf direction. The dotted line is the pressure gradient term (PG), the dot-dashed line is the Coriolis term (C), the plain-dashed line is the bottom stress (τ_b), and the bold-dashed line is the surface stress (τ_w). The bold-solid and the plain solid lines are the advective (A_{adv}) and local (A_l) acceleration terms, respectively (adapted from Li, Z. and R. H. Weisberg (1999). West Florida continental shelf response to upwelling favorable wind forcing, 2: Dynamics, *J. Geophys. Res.*, 104, 23,427–23,442)

As contrasted with the along-shelf direction, in the across-shelf direction, the vertically integrated (summed) force balance is between the pressure gradient and the Coriolis forces. In other words, the force balance is geostrophic because the frictional forces near the surface and the bottom, by acting opposite to one another tend to cancel out upon vertical integration (summation). There is also an across-shelf structure to these force balances. With the Coriolis force in the across-shelf direction being proportional to the speed of the current in the along-shelf direction, we see that the inner-shelf current is strongest roughly between the 20 and 40 m isobaths (the second peak farther offshore is a consequence of the model geometry and can be ignored for the purpose of this discussion). We refer to this as the coastal jet, which peaks where the pressure gradient force is large and the bottom friction, while a factor, is much less than what it is closer to the shoreline.

Building upon the observations of Fig. 4.4, a better sense of the spatial structure of these wind-driven responses, both vertically and horizontally, is achieved by utilizing more of the available data. By using a pattern recognition technique known as the Self Organizing Map, Fig. 4.6 shows the largely wind-driven, coastal ocean current response patterns that ensued over the three-year interval spanning October 1998 to September 2001. During this time interval, a total of 11 different locations were occupied by moorings situated between the 10 m to the 50 m isobaths, each with instruments measuring velocity over nearly the entire water column. After filtering out the tides to focus on the wind-driven circulation, the most frequent of the upwelling and downwelling patterns are given in panels 1 and 11, respectively. For the most recurring of the upwelling patterns, we see the near-surface (green) arrows (velocity vectors) oriented a little offshore, the near-bottom (blue) vectors oriented a little onshore and the middle of the water column (green) vectors oriented along isobath to the SSE, as previously explained. For the most recurring downwelling patterns we see just the opposite, with near-surface vectors turned shoreward and near-bottom vectors turned seaward relative to the middle of the water column vectors oriented alongshore toward the NNW, as previously explained. All of the other patterns shown in Fig. 4.6 behave similarly, but of lesser magnitudes owing to lesser winds, and the ones with the least magnitude tend to be those occurring in summer months. These observed patterns and their temporal evolution (not shown) demonstrate that the inner-shelf response to wind forcing is in accordance with the Ekman-geostrophic response mechanism as conceptually explained and as borne out by physics-based numerical ocean circulation models.

4.2 Deeper Ocean Forcing

Section 4.1 introduces the concept of the inner-shelf, and how, in response to wind forcing, the Ekman-geostrophic spin-up process may set the inner-shelf into either an upwelling or a downwelling motion. It was further shown that the seaward extent of the inner-shelf is determined by the water depth at which direct wind forcing no longer extends all the way down to the bottom. Thus, if the shelf break occurs at a deeper depth, thereby extending farther seaward than the extent of the inner-shelf, then there must be an outer-shelf whose circulation is determined by other means. So in addition to the direct effects of wind forcing that may result in across-shelf transport of water and other materials via the surface and bottom Ekman layers, we must now also consider the role of the deeper ocean in driving a continental shelf circulation.

4.2 Deeper Ocean Forcing

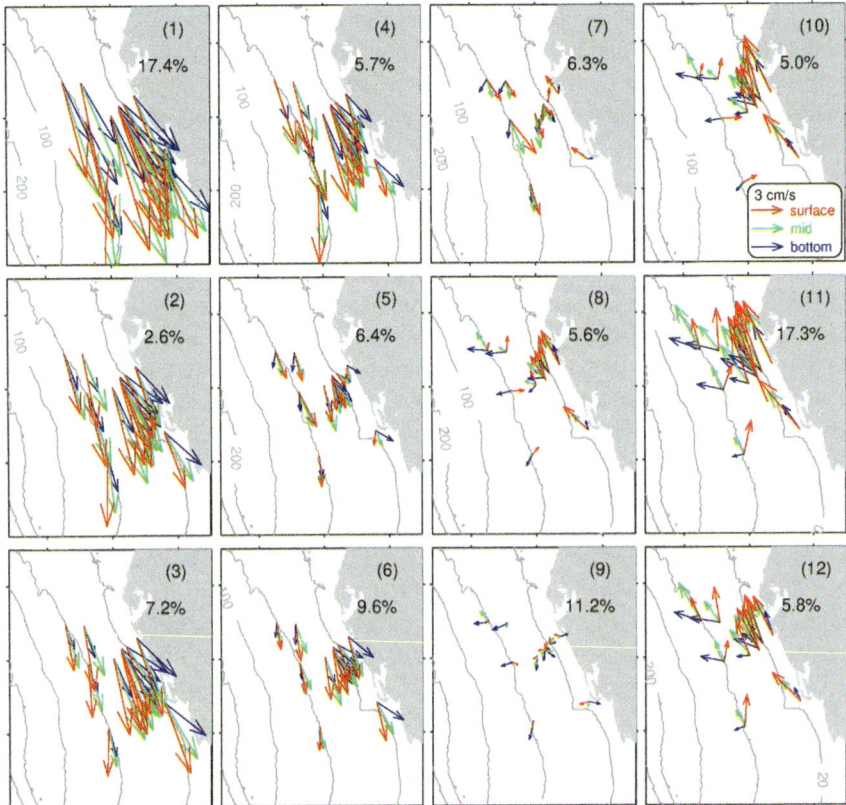

Fig. 4.6 How the velocity vectors organized between the near surface and near bottom over the succession of upwelling and downwelling events observed using moored instrumentation deployed between the 10 m to 50 m isobaths from October 1998 to September 2001. The arrows show the velocity vector speed and direction, red being near the surface, blue being near the bottom and green being at the middle of the water column. The frequency of occurrence for each of these patterns is provided in the upper right of each panel (adapted from Liu, Y. and R. H. Weisberg (2005), Patterns of ocean current variability on the West Florida Shelf using the self-organizing map. *J. Geophys. Res.*, 110, C6, C06003)

Interactions of deeper ocean currents with the continental shelf slope will typically only affect a short distance between the shelf break and the shoreline because it is difficult for deeper ocean water to flow into a shallower region. The reasons are twofold. The first of these involves mass conservation. For a given volume of water to be conserved upon going from a deeper to a shallower depth, the horizontal area would have to increase as the vertical dimension (the water depth) decreases. So conceptually, if this were to occur, then a large spatial area of fluid would have to be displaced from the continental shelf to accommodate the water of deeper origin, which conceptually seems impractical.

The second of these involves angular momentum conservation. By virtue of the Earth's rotation, the relative angular momentum of the fluid would have to change dramatically if deeper ocean water were to flow into a shallower region. This is because all particles that are rotating with the Earth possess a large amount of planetary angular momentum (planetary vorticity). Increasing or decreasing the horizontal area of the fluid is akin to changing the radius of rotation and hence the spin of the fluid particles contained within that enclosed area. In analogy to figure skaters who may spin either faster or slower by moving their arms and/or legs either in, or out, the fluid particles contained within a column of water must also spin slower or faster (in the clockwise direction) upon moving into regions of shallower or deeper depths, respectively. This second, angular momentum constraint is even more restrictive than the first, mass conservation constraint. When combined, these two constraints result in what is referred to as the conservation of potential vorticity, and a related concept (known as the Taylor–Proudman theorem) states that fluid tends to flow along isobath (along lines of constant depth), versus across isobath.

The rotation of the Earth also affects the distance over which the pressure gradient force will adjust to keep the flow in an approximate geostrophic balance, i.e., a balance between the Coriolis and the pressure gradient forces. The distance over which this extends is referred to as the Rossby radius of deformation, which for a temperature and salinity (hence density) stratified fluid on a continental shelf tends to be only a few tens of kilometers.

These GFD (rotating and gravitating fluid) concepts are particularly important for the WFS because of the existence of a powerful western boundary current, known as the Gulf of Mexico Loop Current, which flows adjacent to the WFS slope. The Gulf of Mexico Loop Current, which forms in the Caribbean Sea west of the Nicaraguan Rise, enters and exits the Gulf of Mexico through the Yucatan Strait and the Straits of Florida, respectively, before continuing up the east coast of the United States as the Gulf Stream. While within the Gulf of Mexico, the Loop Current may penetrate northward by some distance before looping around and heading back south to the Straits of Florida, hence the Loop Current moniker. In so doing, the Loop Current occasionally comes in contact with the WFS slope, and being that the Loop Current is a very powerful current that is capable of transporting material properties of every form, how it interacts with the WFS is quite important.

An example of how the aforementioned mass and angular momentum constraints work to keep the Loop Current from overriding the WFS is illustrated by Figs. 4.7 and 4.8. The orientation of these Figures is such that Sarasota, FL is at the coast on the far left, the deeper Gulf of Mexico is on the far right, and the shelf break is located at about the 75 m isobath. Derived from shipboard, across-shelf transects conducted on June 6, 2000, and again on June 28, 2000, Fig. 4.7 shows distributions of temperature, salinity and sigma theta (a measure of density calculated from the observed temperature and salinity) on these two dates. In both instances, the water column is stratified with lighter water near the surface and denser water at depth. Note that the coldest water is straddling the shelf break and that the temperature

4.2 Deeper Ocean Forcing

isolines are all sloping up along the shelf slope. The salinity and density isolines also show upward slopes along the shelf slope. In other words, these water properties appear to be upwelling from deeper depths.

Through the application of the geostrophic approximation, Fig. 4.8 shows the along-shelf currents for these two transects estimated from the temperature and salinity measurements by equating the pressure gradient force (determined from

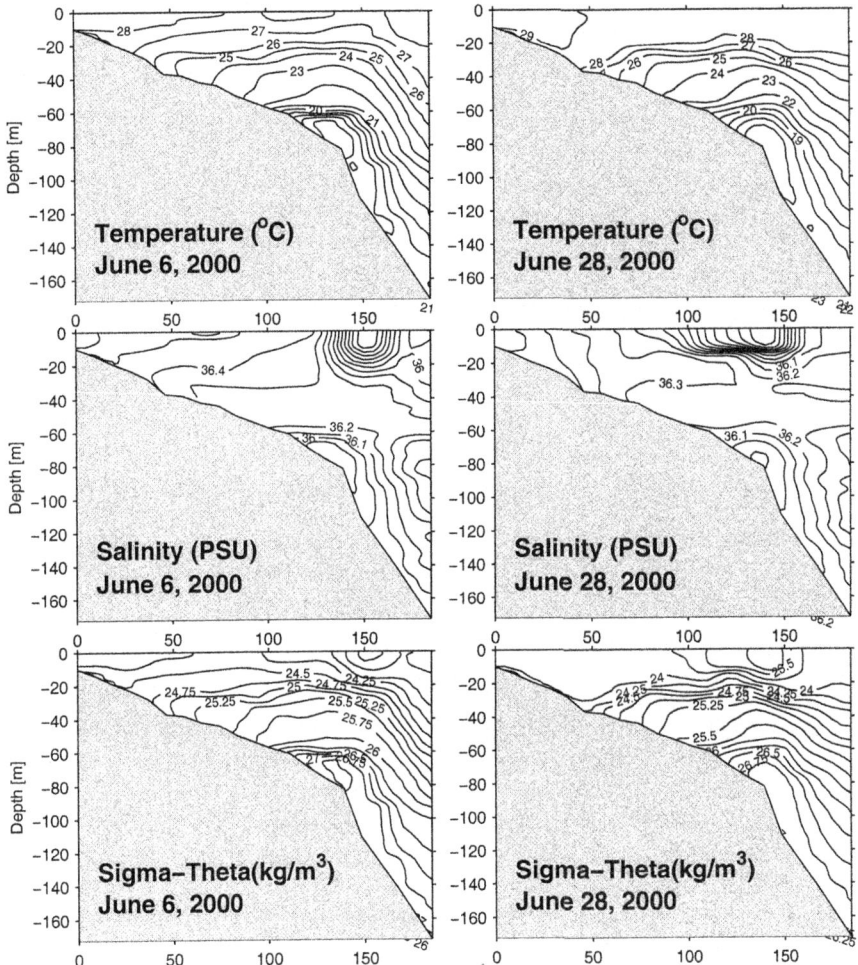

Fig. 4.7 Temperature, salinity and sigma theta (a measure of density) measured from shipboard transects on June 6th and June 28th [adapted from He, R. and R.H. Weisberg (2003), A Loop Current intrusion case study on the West Florida Shelf. *J. Phys. Oceanogr.*, 33, 465–477, © American Meteorological Society. Used with permission]

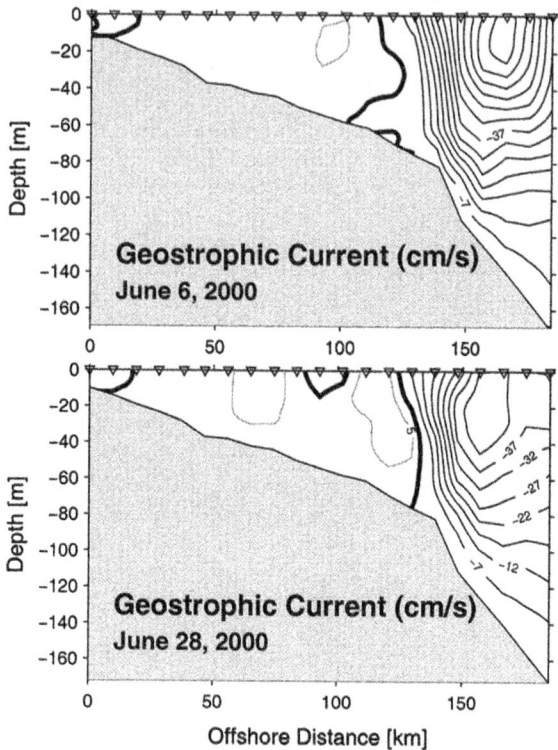

Fig. 4.8 Geostrophic current transects on June 6th and June 28th determined from shipboard sampled temperature and salinity at each of the stations denoted by triangles at the top of each panel. Isotach (speed) contours are in cm/s, and the bold lines denote the zero-speed contour [adapted from He, R. and R. H. Weisberg (2003), A Loop Current intrusion case study on the West Florida Shelf. *J. Phys. Oceanogr.*, 33, 465–477, © American Meteorological Society. Used with permission]

the observed temperature and salinity distributions) with the Coriolis force to arrive at the component of velocity that is normal to the pressure gradient force. Note that the Loop Current, while thrust up along the shelf slope, does not override the WFS itself. Owing to the constraints previously mentioned, the Loop Current can only extend a short distance (a Rossby radius of deformation) onto the shelf, as seen by the zero-speed (bold) contour line.

How the Loop Current-driven outer-shelf may combine with the wind-driven inner-shelf is further illustrated in Fig. 4.9. Rather than observations as in Fig. 4.7 and an analytical (geostrophic) approximation in Fig. 4.8, Fig. 4.9 is from a numerical ocean circulation model simulation. The top panel shows how the Loop Current, by itself, may interact with the WFS. As with the observations, this simulation also shows the Loop Current being constrained to flow along the shelf slope. Adding winds in either upwelling (middle panel) or downwelling (bottom panel) configurations further demonstrates how the inner-shelf (primarily wind-driven) and the outer-shelf (primarily deeper ocean-driven) currents may co-exist distinctly separate from one another.

4.2 Deeper Ocean Forcing

Fig. 4.9 A model simulation of the Gulf of Mexico Loop Current constrained to flow along the WFS slope by itself (top panel A), or together with wind-driven upwelling (middle panel B) or downwelling (bottom panel C) circulations [adapted from He, R. and R. H. Weisberg (2003), A Loop Current intrusion case study on the West Florida Shelf. *J. Phys. Oceanogr.*, 33, 465–477, © American Meteorological Society. Used with permission]

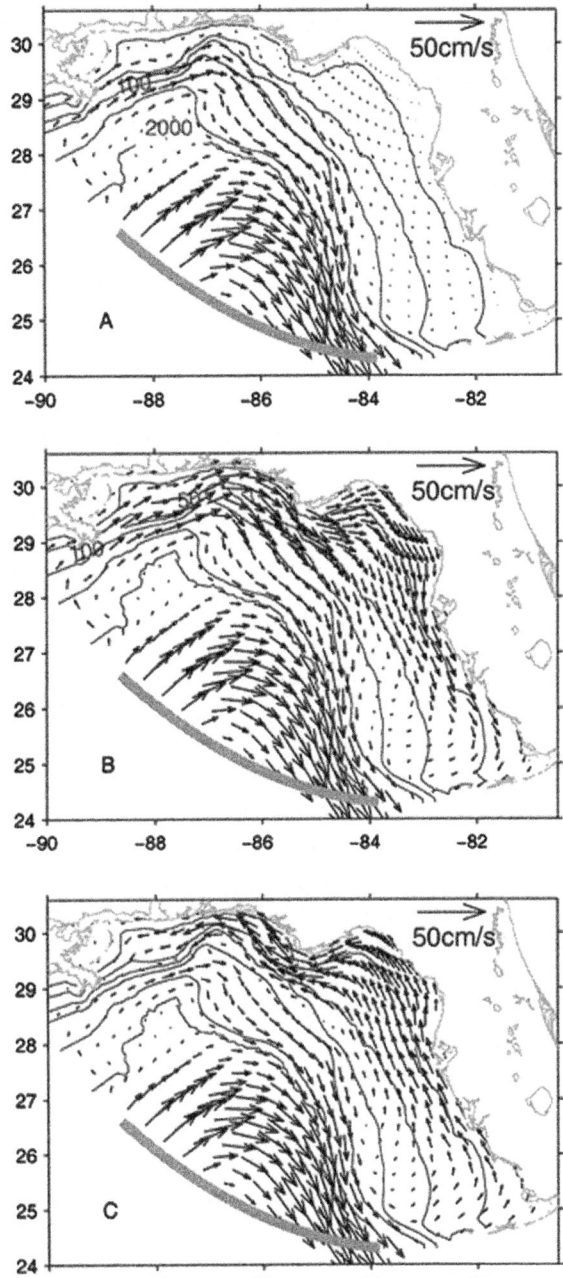

Each of these circulations (outer-shelf Loop Current-driven, and inner-shelf wind-driven) can affect across-shelf transports, but only over distinct regions of the continental shelf slope and the continental shelf. If the shelf is wide enough, as much of the WFS is, then the inner-shelf may be isolated to some extent from the deeper ocean, and as will be discussed in later chapters, this may have profound ecological consequences.

There is an exception to the above finding on the separate nature between Loop Current and wind forcing for the WFS. Appreciating this requires one additional GFD concept, the direction that large-scale waves may propagate on a rotating Earth. For reasons similar to western intensification, as discussed in Chap. 2, large-scale waves, or ones for which either the spherical shape of the Earth or variations in bottom depth may affect the fluid's relative angular momentum, there occurs a directionality to their propagation. Such waves are referred to as planetary waves, Rossby waves and topographic Rossby waves, and unlike the smaller-scale waves that we see on the sea surface while boating or going to the beach, these large-scale waves in both the ocean and the atmosphere are constrained by angular momentum considerations. Rossby waves propagate westward when topography is not a consideration, and topographic Rossby waves propagate with shallow water to the right in the northern hemisphere (shallow water to the left in the southern hemisphere).

A wave may be thought of as nature's way of adjusting a particular medium to a perturbation (or displacement) to that medium. For example, sound waves (pressure perturbations in the air) will propagate away from a generating source. Similarly, if a stone is tossed into a lake, surface gravity waves will propagate way from the splash (a sea surface displacement perturbation). Larger surface gravity waves like tsunamis will propagate away from an underwater Earthquake (a sea surface displacement perturbation due to a vertical shift in the ocean bottom). Thus, if a large-scale, sea surface displacement is imposed upon on the continental shelf slope, then it may propagate away as a topographic Rossby wave. Such effect may be seen in Fig. 4.9a. Whereas the simulated Loop Current only comes in direct contact with the shelf slope at latitudes to the south of about 27° N, we see that an along-shelf current exists farther north along the shelf slope, i.e., the disturbance imposed on the shelf slope to the south of 27° N is seen to have propagated farther to the north with shallow water on the right. The disturbance is actually a pressure gradient (a sea surface slope) owing to the Loop Current whose southward flow is in geostrophic balance with higher pressure to the right and lower pressure to the left. Being that this pressure gradient is constrained to within a Rossby radius of deformation from the shelf break (e.g., Fig. 4.7), such a Loop Current-induced perturbation, by propagating to the north, can only affect the shelf slope and shelf break region over the extent that it is in contact at the perturbation origin, i.e., by only a Rossby radius of deformation shoreward from the shelf break.

Given the directionality associated with large-scale wave propagation, the exception to the separable responses between Loop Current and wind forcing is if the pressure gradient perturbation occurs where the shelf width is narrower than a Rossby radius of deformation. For the WFS, this occurs at the southwest corner of the domain in the vicinity of the Dry Tortugas. With the Dry Tortugas being the westernmost

4.2 Deeper Ocean Forcing

set of islets of the Florida Keys chain, Fig. 4.1 shows that all isobaths as shallow as about 25 m must wrap around the Dry Tortugas. Therefore, a Loop Current contact near the Dry Tortugas will necessarily be in contact with these shallow isobaths, and hence when the associated pressure perturbation propagates away, it will impose an across-shelf pressure gradient force essentially across the entire WFS. So unlike contacts at other locations to the north of the Dry Tortugas (as in Figs. 4.7 and 4.8) resulting in a current extending shoreward to only within a few 10 s of km from the shelf break, a Loop Current contact near the Dry Tortugas, if prolonged enough, will affect the entire WFS.

The summer of 2010 provides an excellent example of such deeper ocean forcing on the WFS. The Loop Current initiated its contact with the shelf slope near the Dry Tortugas (a region that we refer to as the pressure point) in mid-May 2010, and the Loop Current remained affixed to the pressure point through December 2010. As a result, a protracted upwelling circulation ensued, which had a marked impact on WFS water properties, as demonstrated by the observations of water column temperature, salinity, chlorophyll and colored dissolved organic matter (CDOM), all shown in Fig. 4.10. These variables were all sampled using a robotic glider, a device that repeatedly dives and ascends via buoyancy control and is propelled forward by the lift and drag forces exerted on its wings.

In a typical year, the water column over the inner-shelf tends to be uniformly warm between the surface and the bottom by July, but here we see cold water extending across the entire inner-shelf, along with relatively high values of salinity. Additionally, chlorophyll and CDOM exhibit near-bottom maxima at mid-shelf, a region that is generally depleted of these substances.

Such anomalous conditions are attributed to a protracted, Loop Current-induced upwelling circulation that was both observed using moored instrumentation and simulated using a numerical ocean circulation model. Figure 4.11 shows the simulated near-bottom currents superimposed on temperature when averaged over the entire month of June 2010. The Loop Current at his time was rubbing up against the southwest corner of the shelf slope, and this pressure point contact set up a protracted upwelling circulation. The result was that deeper ocean upwelled across the shelf break and was subsequently transported landward across the WFS within the bottom Ekman layer, consistent with the observations made both by the robotic glider transects and moored instrumentation.

The significance of these findings is clear from Fig. 4.10. Whereas the middle of the WFS tends to be nutrient deplete (oligotrophic) in a typical year, if deeper ocean water, which is colder and richer in inorganic nutrients, broaches the shelf break and is subsequently transported across the entire shelf, instead of just being constrained to within a Rossby radius of deformation of the shelf break, then the entire shelf may be ventilated by new, deeper ocean water that is replete with nutrients.

The issue of ventilation by new water raises additional questions on both the origins and the pathways for this ventilation process, and the protracted nature of the deeper ocean-induced upwelling during 2010 provided impetus to answer this question, again using a numerical ocean circulation model simulation. One experimental approach is to seed the model with particles at given depths and then track where

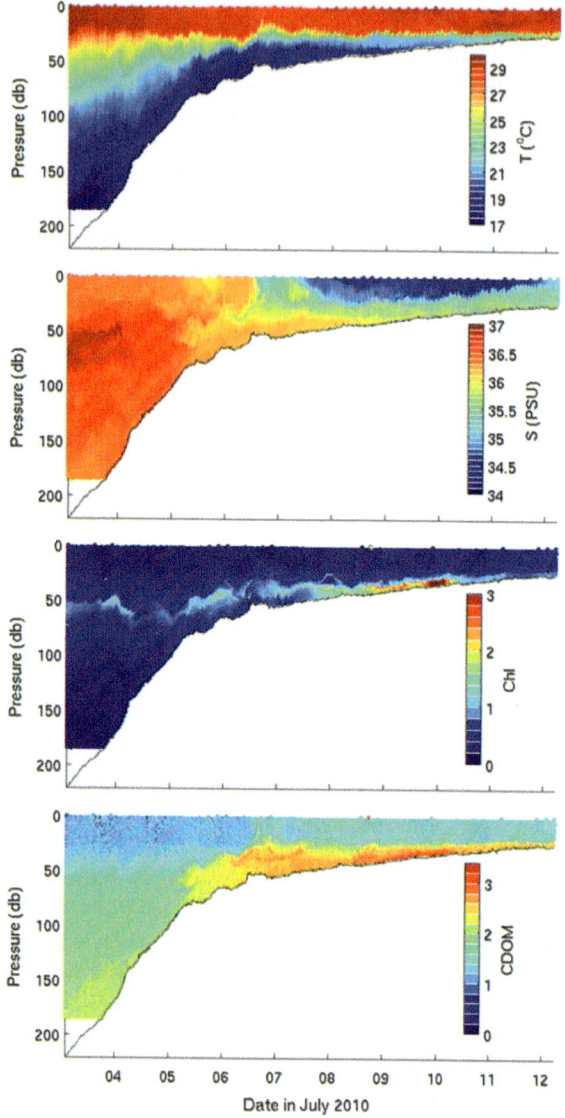

Fig. 4.10 An across-shelf transect of temperature, salinity, chlorophyll and Colored Dissolved Organic Matter (CDOM) sampled from July 3 through July 12, 2010, using a robotic glider (adapted from Weisberg, R. H., L. Zheng, Y. Liu, C. Lembke, J. M. Lenes and J. J. Walsh (2014), Why a red tide was not observed on the West Florida Continental Shelf in 2010. *Harmful Algae*, 38, 119–126. https://doi.org/10.1016/j.hal.2014.04.010)

these particles go. Choosing start times and durations are important for such experimentation. It is known from satellite imagery just when the Loop Current started contacting the pressure point, and it is also known from observed speeds on the WFS that 45 days is a reasonable time for fluid to be transported across the WFS. Using this information, Fig. 4.12 shows the results of such experimentation. Particles were distributed near the bottom along the 100 m isobath, with start times of April 1, 2010, May 1, 2010, and June 1, 2010, with the first of these corresponding to a time period (45 days) when there was no Loop Current contact with the pressure point, the

4.2 Deeper Ocean Forcing

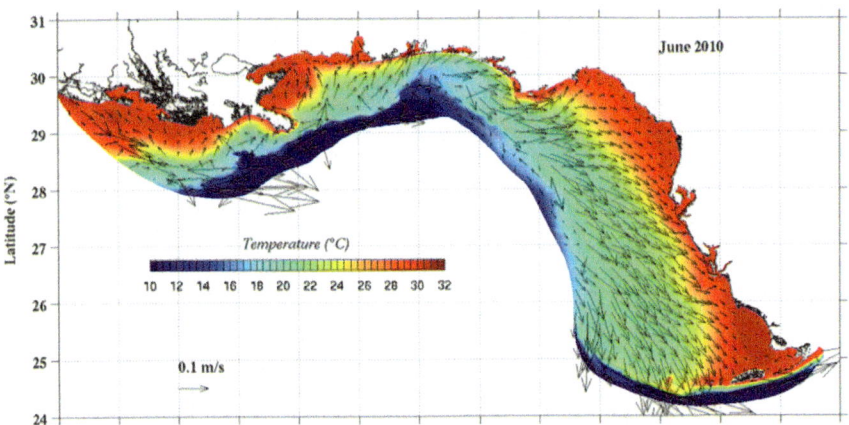

Fig. 4.11 Numerical circulation model simulated near-bottom currents superimposed upon temperature averaged over the month of June 2010 (adapted from Weisberg, R.H., L. Zheng, Y. Liu, C. Lembke, J. M. Lenes and J. J. Walsh (2014), Why a red tide was not observed on the West Florida Continental Shelf in 2010. *Harmful Algae*, 38, 119–126. https://doi.org/10.1016/j.hal.2014.04.010)

second one having pressure point contact for a portion of the interval, and the third one having pressure point contact throughout the entire interval. Without pressure point contact there is no systematic movement across the shelf, with some pressure point contact we see such movement initiate, and with full duration pressure point contact, the particles are capable of transiting across the entire WFS, and in particular making landfall between the Tampa Bay and the Charlotte Harbor estuaries, a point that we will return to in later chapters.

Similar experiments with starting depths along the 75, 150, 200, 250 and 300 m isobaths show that the upwelling of new, deeper ocean water across the shelf break and then across the entire WFS to the near shore have origins over the upper portion of the shelf slope, particularly between the shelf break and about the 200 m isobath. This finding matches expectation based upon the coldest water that may be found on the WFS along the bottom. The upper shelf slope origin is also deep enough to be below the euphotic zone so that the upwelled water indeed has elevated nutrient concentrations. With the origin of the upwelled water so determined, the pathway toward the shore is within the bottom Ekman layer that is activated by the geostrophic interior flow set up to balance the Loop Current-induced across-shelf pressure gradient force once the Loop Current comes into protracted contract with pressure point. A subsequent chapter will show how this is related to the occurrence of red tide, and the final chapter will show how this even leads to a good grouper (fish) sandwich.

I will close this chapter with a discussion of use by those who like to go shelling at the beach. Recall from the earlier discussions on wind-driven upwelling and downwelling responses that northerly winds result upwelling, whereas southerly winds result in downwelling. Upwelling responses have an onshore flow within the bottom Ekman layer such that an upwelling flow may carry materials shoreward along the

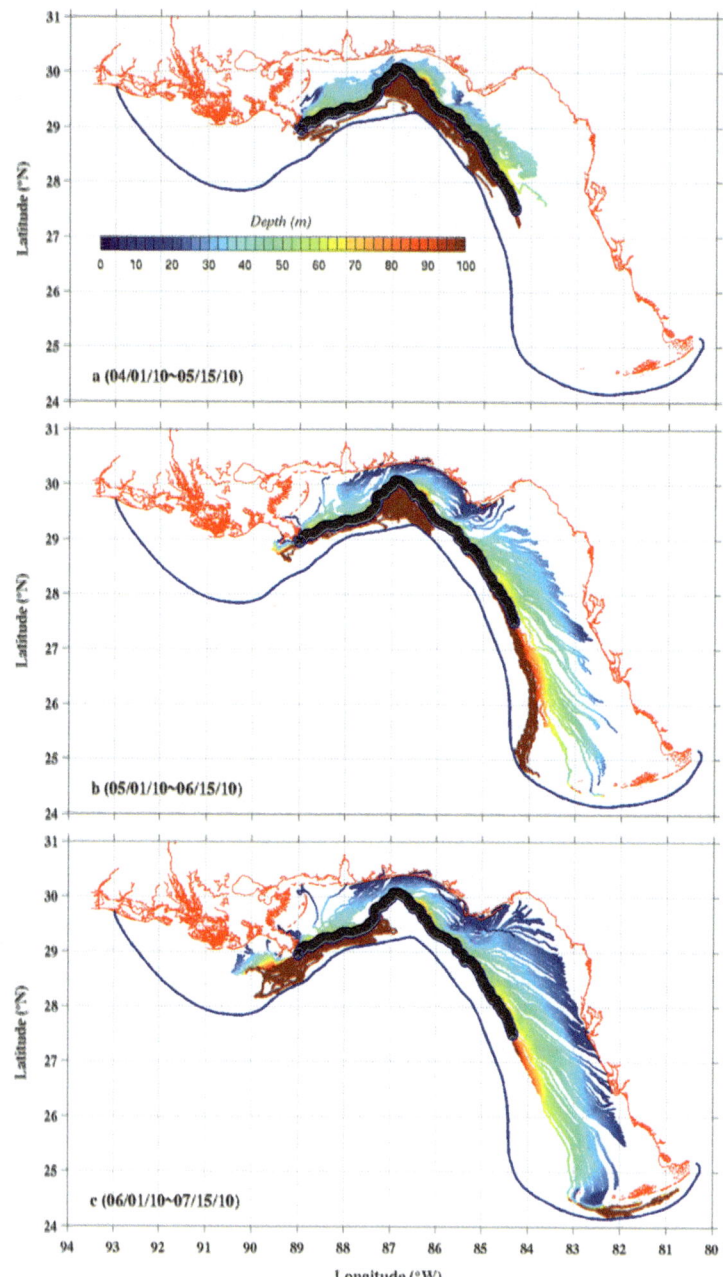

Fig. 4.12 Simulated particle trajectories over a 45 day duration from the 100 m isobath with start times of 4/1/10, 5/1/10 and 6/1/10. The color coding denotes the particle depth (adapted from Weisberg, R. H., L. Zheng, and Y. Liu (2016), West Florida Shelf upwelling: Origins and pathways, *J. Geophys. Res.*, 121, 5672–5681. https://doi.org/10.1002/2015JC011384)

bottom. Northerly (and oftentimes strong northwesterly) winds after fall through winter cold front passages also tend to produce larger than normal waves. The waves, once they interact with the bottom, can dislodge whatever may be along the bottom and repeated wave passage can systematically move such dislodged materials toward the beach. Thus, the combination of the shoreward directed currents and the shoreward directed wave induced particle motions results in all sorts of materials ending up on the beach after the passage of a cold front. It is common to see shells, sponges and even old tires (from the days when someone thought it to be useful to make artificial reefs out of discarded tires). To prove that these materials originated along the bottom simply pick up a sponge and toss it back in the water. It will sink because when wet, it is heavier than the water. Thus the only way that it could have arrived on the beach is along the bottom driven by a combination of bottom Ekman layer currents and wave effects. I once provided this explanation to the Marco Island Shell Club while giving and invited seminar there. The members were delighted to learn just why their best shelling days were after a cold front passage.

Addendum: Pesce Delicioso

This is a dish inspired by a dinner that I had in Fortaleza, Brazil. It is basically a banana fish. Snapper, of any variety, seems to be about right, although I've also used a small grouper or any firm white fish whose filet is not too thick. Lightly sear the fish filet (with salt and pepper) on both sides in a saute pan with butter and set aside (it will be further cooked in the oven). Peel and slice in half lengthwise as many bananas as necessary to cover the fish filet(s), i.e., one banana for two people with one fish filet, etc. Add more butter to the saute pan and cook the banana halves until soft with some brown sugar sprinkled over them. The amount of sugar is a matter of taste.

Place the fish in a buttered baking dish, add the bananas atop the fish and spoon on the butter/brown sugar mixture, and then top all of this with a béchamel sauce (a roux and milk mixture with just a little brown sugar sprinkled in to complement the sauteed bananas). Note that you can also find a béchamel sauce mix on the condiment aisle of the grocery store if you do not want to make this yourself. If you do make it from scratch, then be sure to stir the roux (butter and flour) for a few minutes so that it does not taste like pasty flour.

Bake the entire concoction in the oven at about 400° for maybe 20 min until the béchamel starts to slightly brown, the fish cooks through and the flavors combine, which means that you may have to place a fork into the fish to test if it is done.

Rice or couscous would go well with the sauce of this dish, and to accompany the sweetness of the bananas, a vouvray or any other not too dry white or rose wine will work. A prosecco or a sparkling rose might even be better.

Chapter 5
Estuarine Circulation

Estuaries are the regions of intersections between rivers and oceans. Before the introduction of roads and rail as the lines of communication, rivers provided the basis for commerce conveyance, explaining why almost all major coastal cities are collocated with estuaries. Given their competing maritime commerce, ecological and recreational utilizations, plus the sanitation requirements for large population centers, the understanding of how estuaries function is an important oceanographic and societal concern.

Just like all other ocean regions, the Geophysical Fluid Dynamics influences of tidal and wind forcing, as modified by the Coriolis force and bottom friction are all important determinants of the circulation within an estuary. In addition to these factors, the density contrast between fresh, river water and salty, ocean water results in a very distinctive mode of circulation that we refer to as estuarine circulation, which, on average, plays a controlling role in the distribution of an estuary's water properties and consequently its water quality and ecology.

Morphologically, estuaries take on many shapes and origins. Most, especially those with deep-water ports were formed from drowned river valleys as sea level rose after the last ice-age (e.g., New York's Hudson River estuary). Some estuaries were also formed tectonically by shifts in the Earth's crust (e.g., California's San Fransisco Bay estuary) and others were formed through sedimentation resulting in barrier islands (e.g., North Carolina's Pamlico Sound estuary). Despite disparate origins, they all share the attribute of an estuary circulation, whereby the river water makes its way from the entry point at the head of the estuary to the exit point at the mouth of the estuary, where the river water enters the adjacent ocean.

Thus, we may think of the overall circulation of an estuary as generally having three contributions, those driven by: (1) tides, (2) winds and (3) buoyancy (density) gradients. The tidal circulation is generally the largest of the three because large volumes of water must flow in and out of the estuary as sea level goes up and down roughly twice per day. The wind-driven circulation may be equally large with the passage of frontal systems that occur roughly every two to ten days. The buoyancy-driven (or estuary) circulation, while perhaps the smallest of these, is actually the most important because it persists throughout the year, varying somewhat slowly with river inflow rates, but, nonetheless, being an ever-present factor.

The tidal circulation is the easiest to understand. With the adjacent ocean rhythmically rising and falling twice per day in a deterministic manner, a tidally varying pressure gradient force drives water into and out of the estuary. Because tidal waves have very long wavelengths (distances between peaks or troughs), the pressure gradient forces for tidal waves tend to be uniformly distributed across the entire water column, and we refer to these long wavelength wave pressure gradient forces as barotropic pressure gradient forces. Vertically uniform forcing results in vertically uniform flows of water into the estuary with every flood tide and out of the estuary with every ebb tide, with these floods and ebbs modified by bottom friction. A "back of the envelope" estimation of the tidal current speed may be made through consideration of an estuary's geometry. As an example, consider an estuary that is 10 km wide, 50 km long, and 10 m deep. If the tidal range is 1 m, then the volume of water exchanged with each tidal cycle is roughly 500 million cubic meters. If this amount of water must flow into or out of the mouth of the estuary in about six hours, then the speed of flow must be roughly 1 ms^{-1} or about 2 kts, which is not an uncommon estuary tidal flow magnitude.

Wind-induced speeds are generally of much lesser magnitude. As a "rule of thumb," with water speed being about 2% of the wind speed, a 10 kt wind may produce a 0.2 kt current, or about 0.1 m/s. Yet, despite a much lesser speed, the fact that a wind-driven flow is not uniformly distributed over the water column may have profound impacts upon water property distributions, even more so than the much faster tidally driven currents.

The buoyancy-driven estuarine circulation is even slower, but generally more impactful than either the tidal or the wind-driven flows. It arises as follows. At a given temperature, freshwater is less dense (lighter) than salt water so equal masses of freshwater have larger volumes than salt water. For this reason, if we average out the tidal and wind-driven sea level variations, we find that sea level at the head of the estuary, where the water is fresher, tends to be higher than that at the mouth, where the water is saltier. This sea level gradient (or slope) results in a pressure gradient force that tends to drive fluid seaward at the estuary surface. But, if we sum up the weight of the water from the surface to the bottom at the mouth of the estuary, we find that the bottom pressure is higher at the mouth than it is at the head because saltier water is heavier than fresher water. So while the pressure gradient force is directed seaward at the surface, it is directed landward at the bottom. We refer to this density-induced variation in the pressure gradient force as baroclinic, and it tends to drive fluid into the estuary along the bottom. Thus, it is this vertical change in the pressure gradient force, owing to the density differences between fresh and salt water, that drives an inflow at the bottom and an outflow at the surface, or a two-layered circulation that we refer to as the net estuarine circulation. While the tidal and wind-driven flows reverse with time tending to cancel out their net effects, the estuarine circulation is omnipresent and only slowly varying so, on average, it is what largely determines the distribution of water properties within an estuary.

5.1 A Narragansett Bay, Graduate School Digression

After graduating from Cornell University with a BS in Engineering in May 1969, I arrived at the Graduate School of Oceanography, University of Rhode Island in August 1969 to begin my studies in physical oceanography, rather clueless of what that really meant. I was met by my assigned major Professor, Dr. Wilton (Tony) Sturges III, who at the time, despite about a ten-year age difference, did not look to be too much older than me. Our formal relationship only lasted about three years before he left for the Florida State University, where he remained for the duration of his career, but our friendship and my deep respect lasted a lifetime. Tony had a profound influence on my career, how I interacted with my own students and colleagues, and I am forever indebted to him for his kindness, patience and encouragement.

Tony at that time was a blue water oceanographer with emphasis on the North Atlantic, but having come from the Johns Hopkins University, where the Chesapeake Bay Institute played a leading role in early estuary studies, he had an interest in estuary circulation and in particular initiating a program of study within Narragansett Bay. Not knowing what I wanted to do, I was tagged for this project. Without specifically directed research funding, Tony agreed to facilitate it as best he could. Serial measurements over long durations were just beginning, as new recording devices were being implemented. Estuarine observations, up until that point, were generally made from small boats, which meant that they were generally made during periods of calm winds. Averaging over several such sets of observations indeed led to the realization of a two-layered estuary circulation, but how this circulation may have varied and for how long one would have to average to actually observe it were unknown. We set out to remedy this, which is what my MS and Ph.D. research focused upon.

Without directed support, this had to be done on the cheap. Tony borrowed some now antiquated current meters from the Wood Hole Oceanographic Institute that were being replaced by newer instrumentation. A current meter is a device that measures the speed and direction of a flow at a given point and stores the information internally over some designated time interval. How to deploy these was another matter. With no moorings or tower devices available, we had to improvise. A telephone pole, a 55-gallon drum, some steel rods and rope were my design elements. We secured a pole (donated for free) from the local telephone company, a drum from a scrap yard and rods and rope (the most expensive items) from a hardware store. The concept was to sharpen one end of the pole so that it would sink into the muddy bottom of Narragansett Bay with a drum full of scrap steel and concrete as ballast and with rods positioned at regular intervals to affix the current meters to, supported by ropes. So I sharpened a point at one end of the pole with an axe, cut holes in the drum through which to thread the pole leaving room to fill it with scrap steel and concrete. The pole was then predrilled for the support rods and off we went on a chartered fishing boat (the Billie II) to deploy it. After hoisting the pole vertically and letting it go, it was indeed stuck in the mud as planned. We then affixed a four-point anchor system to stabilize it and began the process of building out the supports and securing the current meters using SCUBA. It was my first introduction to underwater work, and despite being a robust 22-year-old, by the end of the afternoon, I was exhausted. I was at an orchestral performance of Beethoven's 5th Symphony that evening and slept through most of it. It was then that I decided that an academic career, as difficult as that may be, was better than one of hard labor. With the current meters deployed, we then hammered some prepainted red and white wooden slats onto the pole, and put a light on top. With these finishing touches completed, we succeeded in having the first "barber's pole" deployment of current meters in an estuary (and probably the last).

The deployment lasted for 39 days, yielding very insightful time series observations of circulation over the full water column. Winds were indeed found to be important, and while a two-layered, mean estuarine circulation occurred, the variations about the mean were too large to fully define the net estuarine circulation. We also found that multiple connections with the adjacent ocean could lead to differential flows between connected passages and that the tides themselves could even express a two-layered flow owing to bottom friction, just as predicted analytically. Not only was this enjoyable and scientifically illuminating, in contrast with the remainder of my career, which entailed large amounts of research funding, what we learned was in essence done for next to nothing, maybe a few hundred dollars, versus millions. Illustrations of my first two moorings are given in Fig. 5.1.

When Tony departed after my MS degree work was completed, I was picked up by Professor Kern Kenyon, another great mentor who imparted what became my adopted mode of teaching. Like Tony, Kern had an unassuming, modest and kind nature, and his classroom skills were inspiring. He had both a knack for and patience to explain complex topics in ways that were understandable, and this made a great impression on me—I endeavored to do similarly when I later became a professor.

Kern saw me through my Ph.D. comprehensive examination and my initial foray in the Ph.D. fieldwork that was a follow on to what I did for my MS degree. Instead of the West Passage of Narragansett Bay, which had an East Passage counterpart that made the overall circulation more complex, I focused on the Providence River section of Narragansett Bay to explore the estuarine circulation and how it may be modified by wind forcing. The Providence River was an active shipping lane so I was only able to deploy a near-bottom current meter, versus another "barber pole." That deployment lasted for 51 days, and the results were striking. Whereas an obvious net inflow occurred, the wind-induced fluctuations about the net inflow were large, and from this, we learned that a certain record length was necessary to begin discussing the non-tidal estuarine circulation and that the wind-induced estuary circulation fluctuations were describable using standard time series analysis techniques. What I learned is that a well-thought-out experiment was more important than something expensive and elaborate, and this experience remained a guiding principle for every future scientific campaign that I led.

As soon as the Ph.D. comprehensive examinations were over and the fieldwork was complete, it was time for my sojourn to Fort Huachuca Arizona for a three-month, US Army Reserve active duty for training stint. It was during this time that Kern Kenyon left to return to the Scripps Institute of Oceanography, leaving me without a major professor. Upon my return, Dr. John Knauss, the Dean of the Graduate School

5.1 A Narragansett Bay, Graduate School Digression

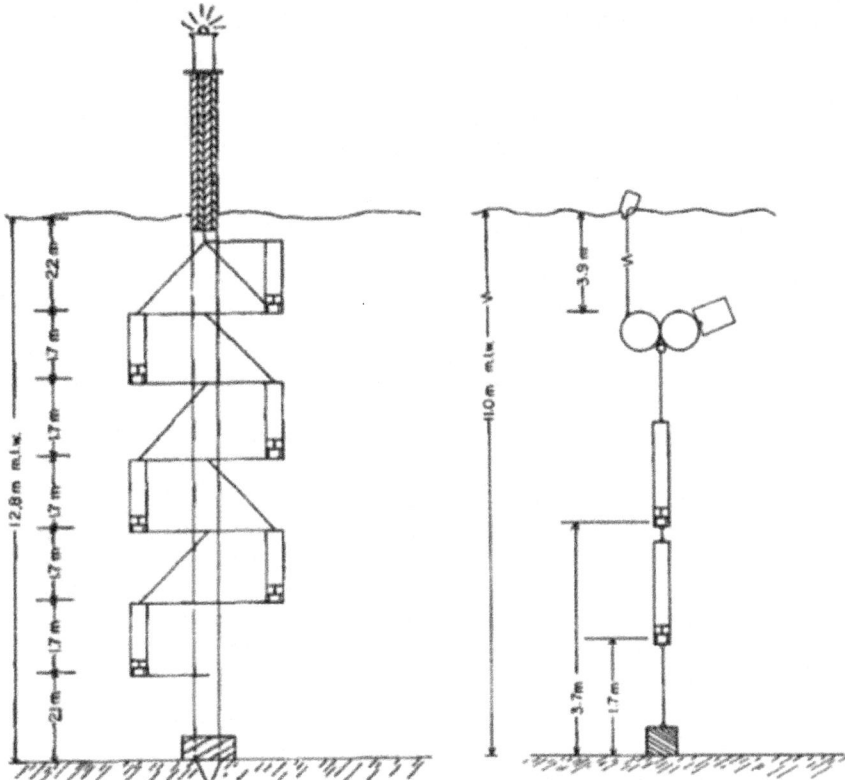

Fig. 5.1 My first field campaign observing currents in Narragansett Bay using moored instrumentation [adapted from Weisberg, R. H. and W. Sturges (1976), Velocity observations in the West Passage of Narragansett Bay: A partially mixed estuary, *Jour. Phys. Oceanogr.*, 6, 345–354, Published 1976 by the American Meteorological Society]

of Oceanography, agreed to assume that role, being familiar with me because I served as teaching assistant in his Introduction to Physical Oceanography course. I warned John that he was putting himself at risk being that my two prior major Professors ended up leaving, but he was undaunted. John, like Tony and Kern, imparted traits that I took very seriously throughout my career. I benefitted greatly by having all three incredibly good mentors during my graduate student days.

John Knauss honed his research acumen studying the currents of the tropical Pacific Ocean, leading expeditions to describe the Equatorial Undercurrent and the North Equatorial Countercurrent, the first being a subsurface current directly on the equator, the second being a surface current to the north of the equator and both flowing in directions opposite to the prevailing winds. Upon return from Fort Huachuca and with the Ph.D. work having an end in sight, there was at that time (1973) a large, international effort being planned for the Tropical Atlantic Ocean called the Global Atmospheric Research Program (GARP), Atlantic Tropical Experiment, or GATE. As an equatorial oceanographer, John was keenly aware of this, and he encouraged his two graduate students, Laury Miller and me, to get involved. Along with Laury and John, I wrote my first National Science Foundation (NSF) proposal, we were funded and so while still a Ph.D. student completing the dissertation, I was also thrust into a new research direction that further defined my career.

Instead of research on the cheap, I was now doing it the more traditional way, with NSF funding and realizing that whenever I deployed an instrument over the side of a research vessel, it was as if I was tossing something valued at as much as my house into the ocean. Doing this also required the audacity to think that months or a year later, when I came back to retrieve it, that it might actually be there and, if so, it might actually have recorded the data as it was designed to do. Oceanography is not for the faint of heart, nor is it for anyone who does not take the work seriously enough because you do not get many (if any) second chances should you fail—it is too expensive and the competition is too keen.

GATE was an amazing experience for a young scientist, and the fact that John had the confidence in me to design and carry out such an expedition made a lasting impression. Doing for my mentees what he did for me became an ongoing professional goal. The GATE experience launched me into the second phase of my career, that of an equatorial oceanographer, but that is a topic for another chapter. I'll close this digression by saying just how important mentorship is in any profession. I was fortunate to have three of the best Tony, Kern and John, each of whom, in their own way, guided my approach to personal interactions, teaching and doing science.

5.2 Back to Estuaries

I departed estuaries after graduate school to study the equatorial ocean, but the interest did not wane. After spending nine years at the North Carolina State University, I was recruited by the University of South Florida, Department of Marine Science (now College of Marine Science) in Saint Petersburg, Florida situated on Tampa Bay, the largest of the Florida estuaries and the one with the only deep-water port on the west coast of Florida. Naturally, it was time to think about building a program of estuary studies while still engaging in the tropics. I quickly found, as did Tony Sturges for Narragansett Bay, that there was little local interest in funding estuarine physical oceanography.

With the ocean sciences nominally consisting of the applications of biology, geology, chemistry and physics to the workings of the ocean, the breakdown of disciplinary interest seems to occur in that order. Additionally, whereas blue water (the deep ocean) oceanography traditionally has strong national defense support by virtue of naval operations, the support for green water (the continental shelves and estuaries) oceanography is traditionally more locally driven. Consequently, blue water oceanography tends to have a balanced, multidisciplinary funding base, whereas green water oceanography tends to lean more heavily toward biology, and with so many more biological oceanographers seeking employment, plus a new cadre of environmental scientists being added to the market in recent years, the state and local agencies that manage estuarine resources are heavily populated by biologists and environmentalists, many of whom lack a full appreciation for the importance of the ocean physics, which is one reason why writing this book was undertaken. Another factor affecting estuarine studies is one of purview. Each agency operating within an estuary, and there are many, has a certain mandate. They also tend to be very protective of their perceived purviews. Together, disciplinary bias and purview protection make it quite difficult to find local support for estuary circulation studies.

My breakthrough came when NOAA initiated a Tampa Oceanography Project, and I received (in partnership with a NOAA colleague whom I met during the GATE Program) a very small award from NOAA SEAGRANT to analyze velocity profile data obtained in the main shipping channel under the Sunshine Skyway Bridge. Until that time, Tampa Bay was considered to be well-mixed and the net buoyancy-driven estuarine circulation was ignored.

The initial velocity profile observations demonstrated two things. The first is that a well-defined, buoyancy-driven estuarine circulation indeed existed, with a mean landward-directed inflow observed over the entire portion of the water column for which data were acquired (between 3.6 m down from the surface and 2 m up from the bottom). The second is that a very well-defined wind-driven circulation also occurred, with winds directed toward the head of the estuary driving a current directed toward the mouth of the estuary within the main shipping channel. In other words, along with tides, the Tampa Bay circulation was found to be fully three-dimensional and both buoyancy (density)-driven and wind-driven. Along with two other colleagues, we also provided the first three-dimensional and density-dependent numerical ocean circulation model simulations for Tampa Bay, bearing out these observed findings. However, these accomplishments held no sway for additional funding. NOAA took purview of the circulation, and the local agencies maintained their biological and water quality concentrations, as if advancing on these topics could be done independent from the circulation. I then gave up on Tampa Bay for the next 14 years and instead concentrated elsewhere. As the expression goes: "You can't fight City Hall."

As discussed in Chap. 4, coastal ocean circulation studies are best performed when combining observations with numerical circulation model simulations. There are never enough observations to fully describe a complex system, and simulations without observations for veracity testing may be quite misleading. The two approaches are best performed in concert with one another. I was fortunate to build a research team with excellent sea-going technical support, a gifted numerical ocean circulation modeler and some dedicated graduate students. Being that the continental shelf connects with both the deeper ocean and the estuaries, it is necessary to include the estuaries in such modeling. Not being dependent upon local agencies, my research group's national and state support was stable, enabling us to build a substantive program of study, including (albeit indirectly) both the Tampa Bay and the Charlotte Harbor estuaries. Scientific curiosity should always trump scientific funding (as I learned from Tony Sturges) so with an excellent collaborator (Dr. Lianyuan Zheng at that time), we set out to do modeling studies not only for the WFS that we were funded for, but also for the Charlotte Harbor estuary, the Naples Bay/Rookery Bay estuary system and eventually for Tampa Bay.

If the time averaged, estuary circulation is buoyancy-driven by the horizontal salinity gradient (the salinity differences between the estuary's head and mouth) and that gradient increases with increasing vertical mixing (as salinity isolines rotate from being horizontal with no vertical mixing to vertical with increasing vertical mixing), why should increasing vertical mixing first increase the estuary circulation before eventually decreasing it? This conundrum was unexplained by the existing literature.

Conceptually, without vertical mixing, fresh river water could simply flow atop heavier salty water while en route to the sea. With vertical mixing, the river water becomes mixed together with the underlying saltier water and a combination of river and some salt water now proceeds out to sea. To compensate for this, more ocean water must flow into the estuary along the bottom. Increased vertical mixing further increases these outgoing and incoming flow regimes, but only up to a certain extent; hence the conundrum. This led to my first foray back into estuary circulation with a publication entitled: "How estuaries work." Together with Dr. Lianyuan Zheng, we explored this enigma by considering the sources and sinks of mechanical energy derived from river, tide and wind forcing. How the energy partitions between work against buoyancy (that sets the driving force) and the production of turbulence (that acts as a brake) provided the answer. Once the salinity isolines rotate to become nearly vertical, there is little opposition to turbulence production and hence the breaking action by turbulent mixing increases and estuarine circulation slows down.

Funding aside, it is discovery that always motivated my science so stimulated by the above finding, and with an excellent collaborative partner, it was time, after a 14-year hiatus, to get back into Tampa Bay. Previous Tampa Bay circulation models ended at the bay mouth so these were incapable of considering the exchange of materials between the adjacent continental shelf and the estuary. They also lacked adequate resolution to include Sarasota Bay, the Intra-Coastal Waterway and all of the inlets connecting these with each other and with the adjacent Gulf of Mexico. Like anything else, numerical ocean circulation models evolve, and, over the past few decades, those skilled in the field of computational fluid mechanics began instituting what are called community models, wherein user groups developed and shared their expertise, leading to model improvements. Thus, short of being an expert in model development, one could apply existing community model codes to a specific region with technical support available from the community of users. This offered major improvement to ocean modeling at all scales, estuarine to global, and it is how my group was able to include circulation modeling in our various research studies.

We published a Tampa Bay modeling study in 2006 that considered the influences of tides, winds and rivers on the Tampa Bay circulation and its interactions with the adjacent WFS over the three-month interval spanning September through November 2001 over which we had observations for comparison with the model simulation. Finding excellent agreement, we then described in detail how Tampa Bay responds to tide, wind and river forcing. Oddly, this was the first paper on the circulation of Tampa Bay to be published in a refereed professional journal. Prior works were in conference proceedings, technical reports or other gray literature, and yet, for reasons of purview and disciplinary bias, we remained without any local agency

5.2 Back to Estuaries

support until around 2016 when Pinellas County, in receipt of Federal RESTORE Act money, awarded us a competitive grant.

I found it fascinating, some 47 years after my first introduction to estuaries, when Tony Sturges and I had to engage in an unfunded manner, that local funding for estuary circulation physics remained so difficult to obtain. Science requires perseverance and with that comes risk, which, in part, led to Lianyuan Zheng moving on to NOAA where he no longer had to worry about his salary. As an aside, the public is generally unaware that research scientists, to a large extent, rely on soft money; in other words, if they do not have adequate grant support, they can actually go without salary. So not all research scientists, even tenured professors, have stable positions; many must raise their own salaries, and failures at thus occur. Having a tenured professorship is not as comfortable as some might think.

Lianyuan Zheng and I remain in close contact and collaborations persist. He also trained another associate, Yonggang Liu, now a tenure track professor who took over my former group once I became professor emeritus. With the arrival of a new graduate student, we decided to increase the resolution of our Tampa Bay model so that we could more accurately resolve the main shipping channel and all of the inlets connecting Tampa Bay, the Intra-Coastal Waterway and Sarasota Bay with the adjacent Gulf of Mexico. A few years later, with another new graduate student, we coupled this higher resolution Tampa Bay model with another one for the adjacent Gulf of Mexico and automated it to provide daily nowcasts and forecasts available to the general public via the internet. Figure 5.2 provides an orientation to the Tampa Bay estuary region, and Fig. 5.3 provides an overview of the Tampa Bay Coastal Ocean Model (TBCOM) domain, its numerical grid geometry and its horizontal resolution.

Models, of course, are only useful if they accurately portray what actually occurs in nature. As demonstrations of TBCOM veracity, Figs. 5.4, 5.5, 5.6 and 5.7 provide direct comparisons between sea levels and currents as actually observed and TBCOM simulated at locations for which such observations are available. The time interval considered is September through November 2001. The first of these (Fig. 5.4) pertains to sea level. The observations are from tide gauges and the TBCOM simulation is sampled at these same tide gauge locations: Mckay Bay (at the head of Tampa Bay), Clearwater (on the adjacent Gulf of Mexico), Port Manatee (in the middle of Tampa Bay on the east side) and St. Petersburg (in the middle of Tampa Bay on the west side). The top panel provides the wind vectors observed at St. Petersburg. With the simulations (red lines) nearly overlaying the observations (black lines), the model veracity for sea level is quite good, with the correlation coefficients being above 0.9 and the root mean square deviations being around 10 cm.

The origin of the root mean squared deviations is not from the tides, which are much more accurately portrayed, but instead from the wind-induced fluctuations. For the reasons explained in Chap. 4, each reversal in wind direction is accompanied by either a set up, or a set down of sea level due to both the WFS inner-shelf response (the Ekman-geostrophic spin-up) and the local response that is internal to Tampa Bay. These wind-induced sea level variations are evident in Fig. 5.5 after removing the tidal variations. Whereas the red (observations) and black (simulations) lines

Fig. 5.2 Orientation to Tampa Bay, its subregions, river inflows and observing stations (adapted from Chen, J., Weisberg, R. H., Liu, Y., Zheng, L., Law, J., Gilbert, S., Murawski, S. A. (2023), Tampa Bay coastal ocean model (TBCOM) nowcast/ forecast system, *Deep-Sea Research Part II*, 211, 105,322. https://doi.org/10.1016/j.dsr2.2023.105322)

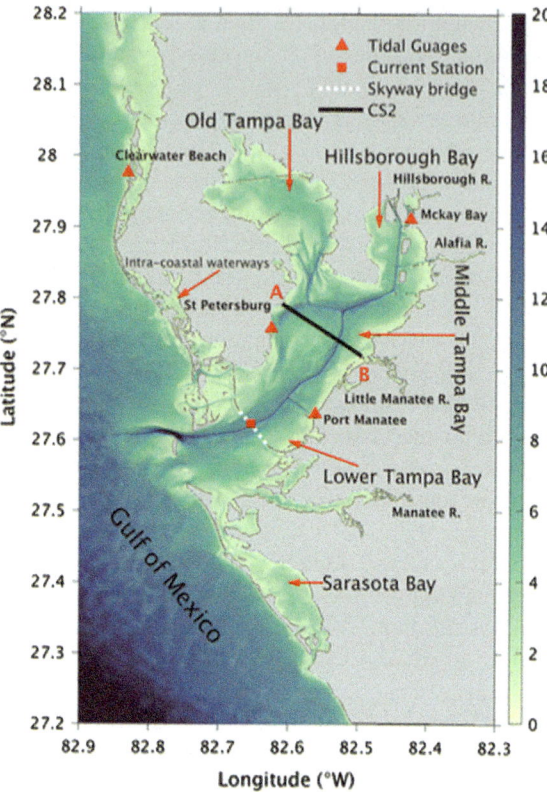

track each other reasonably well, there still exists a root means squared deviation of about 6 to 9 cm, which accounts for most of the deviations that were found for Fig. 5.4.

The currents are also reasonable well-simulated by TBCOM, as shown in Fig. 5.6, but there is only one station available for comparison, that one being in the main shipping channel beneath the Sunshine Skyway Bridge. As with sea level, the red and black lines nearly overlay one another resulting in correlation coefficients of 0.95 and root means squared deviations of 10 to 16 cm/s, which is not bad considering that the currents may be as large as 100 cm/s (or about 2 kts). Also note that the coaxial currents (the component that is perpendicular to the shipping channel orientation) are miniscule compared with the axial currents (the component that is parallel with the shipping channel orientation), i.e., the water is constrained to flow along the channel axis, versus across it.

As was the case for sea level, the root mean squared deviations for the currents are also primarily associated with the wind-induced flow. While not shown, after removing the tidal variations, the wind-induced fluctuations are smaller at about 10–15 cms^{-1}, and the differences between the observations and simulations are mainly due to the winds themselves being less than what actually occurred. This is because

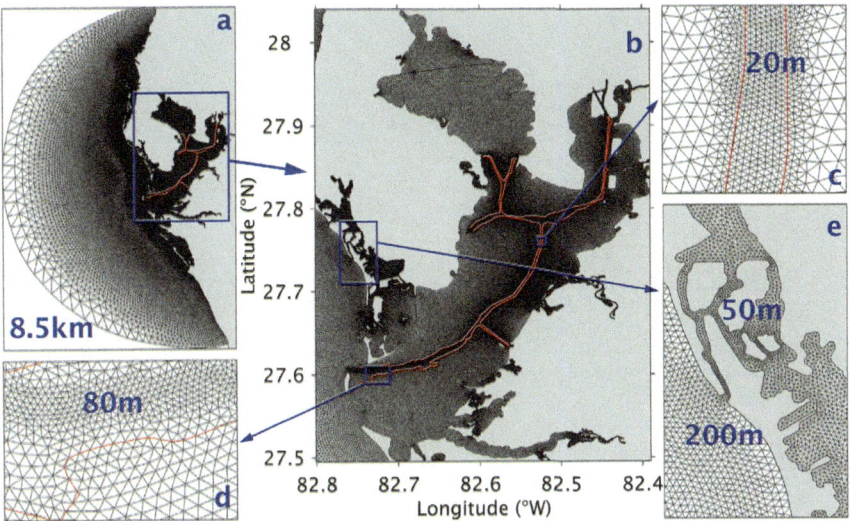

Fig. 5.3 Tampa Bay Coastal Ocean Model grid and zoomed views of selected regions. The red lines indicate the shipping channel where the horizontal resolution is as fine as 20 m. Resolutions for other regions are as indicated (adapted from Zhu, J., R. H. Weisberg, L. Zheng, and S. Han (2015), On the salt balance of Tampa Bay. *Cont. Shelf Res*, 107, 115–131. https://doi.org/10.1016/j.csr2015.07.001)

the wind observations used to drive the model simulation were made over land, whereas the winds over water which actually drive the bay circulation tend to be larger than those over land because there is less friction to impede the winds over water. The wind-induced fluctuations observed in the main shipping channel (again not shown) also agree with the observations from a decade earlier, wherein the current direction in the shipping channel from several meters below the surface down to the bottom is opposite from the axial direction in which the wind is blowing. With most of the estuary being shallower than the main shipping channel, downwind flows over the shallow flanks must be compensated for by upwind flows in the deeper channel.

Whereas marine commerce interests must contend with the effects of tides and winds for safe navigation, anything of an ecological nature must contend with the water properties, which are largely determined by the time averaged, or long-term, mean estuarine circulation, i.e., the net circulation that remains after both the tidal and wind-forced contributions are removed. How well TBCOM addresses this part of the circulation is shown in Fig. 5.7. The results from two different simulations are provided, one with winds so that a direct comparison could be made with observations and the other without winds, to assess what the buoyancy-driven circulation might be by itself. The case with winds shows both observations and simulations nearly overlaying each other. With the observations only extending up to 4 m from the surface, the remainder of the water column flow field had to be estimated by extrapolation. In either of the cases, with and without wind, we see a classical two-layered estuarine circulation with a net outflow at the surface and a net inflow at

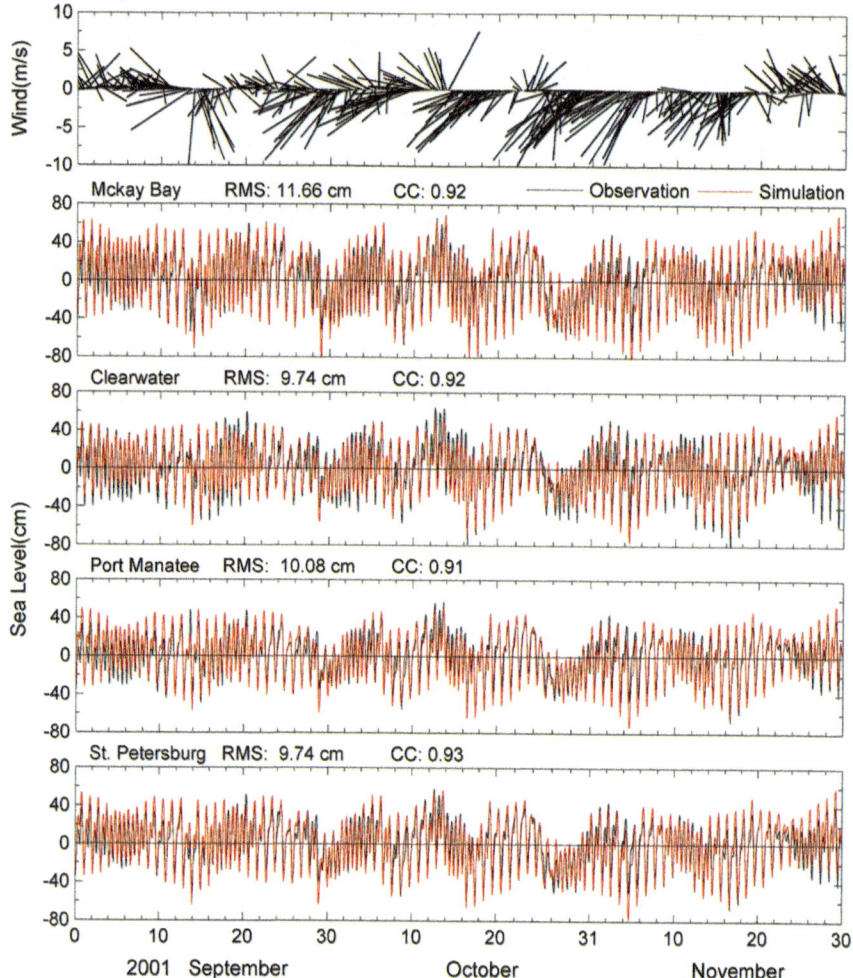

Fig. 5.4 Comparisons between hourly modeled (red) and observed (black) sea levels at the McKay Bay, Clearwater, Port Manatee, and St. Petersburg tide gauges along with the wind vectors subsampled every 6 h. Also provided are root mean square (rms) deviations and correlation coefficients (cc) (adapted from Weisberg, R. H. and L. Zheng (2006), Circulation of Tampa Bay driven by buoyancy, tides, and winds, as simulated using a finite volume coastal ocean model. *J. Geophys. Res.*, 111, C01005. https://doi.org/10.1029/2005JC003067)

the bottom. With the shipping channel being much deeper than the remainder of the estuary, the inflow within the shipping channel is larger than the outflow there. The rest of the outflow necessary to conserve mass occurs over the shallower flanks beyond the channel.

The spatial distributions for both the mean sea level and near-surface and near-bottom net estuarine circulations are shown in Fig. 5.8, which in this case is from a

5.2 Back to Estuaries

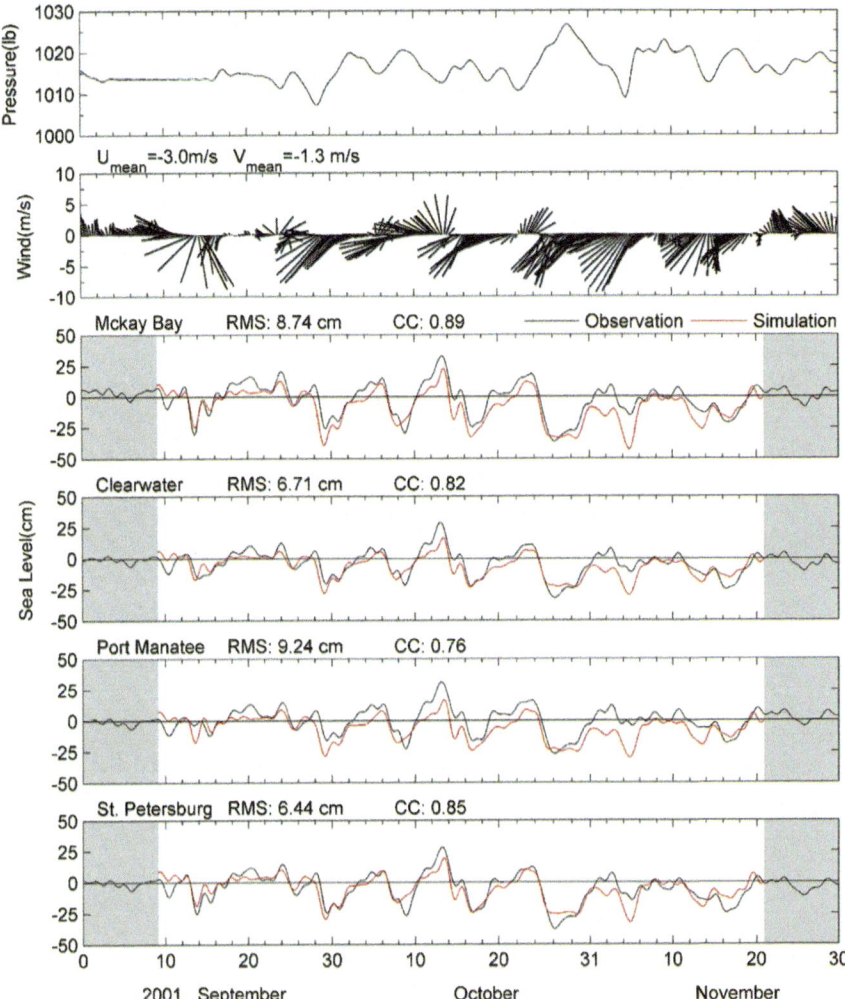

Fig. 5.5 Comparisons between hourly modeled (red) and observed (black) sea levels filtered by using a 36-h low-pass filter at the McKay Bay, Clearwater, Port Manatee, and St. Petersburg tide gauges along with the low-pass filtered wind vectors and air pressure. Shaded regions indicate the end effects of low-pass filtering. Also provided are root mean square (rms) deviations and correlation coefficients (cc) (adapted from Weisberg, R. H. and L. Zheng (2006), Circulation of Tampa Bay driven by buoyancy, tides, and winds, as simulated using a finite volume coastal ocean model. *J. Geophys. Res.*, 111, C01005. https://doi.org/10.1029/2005JC003067)

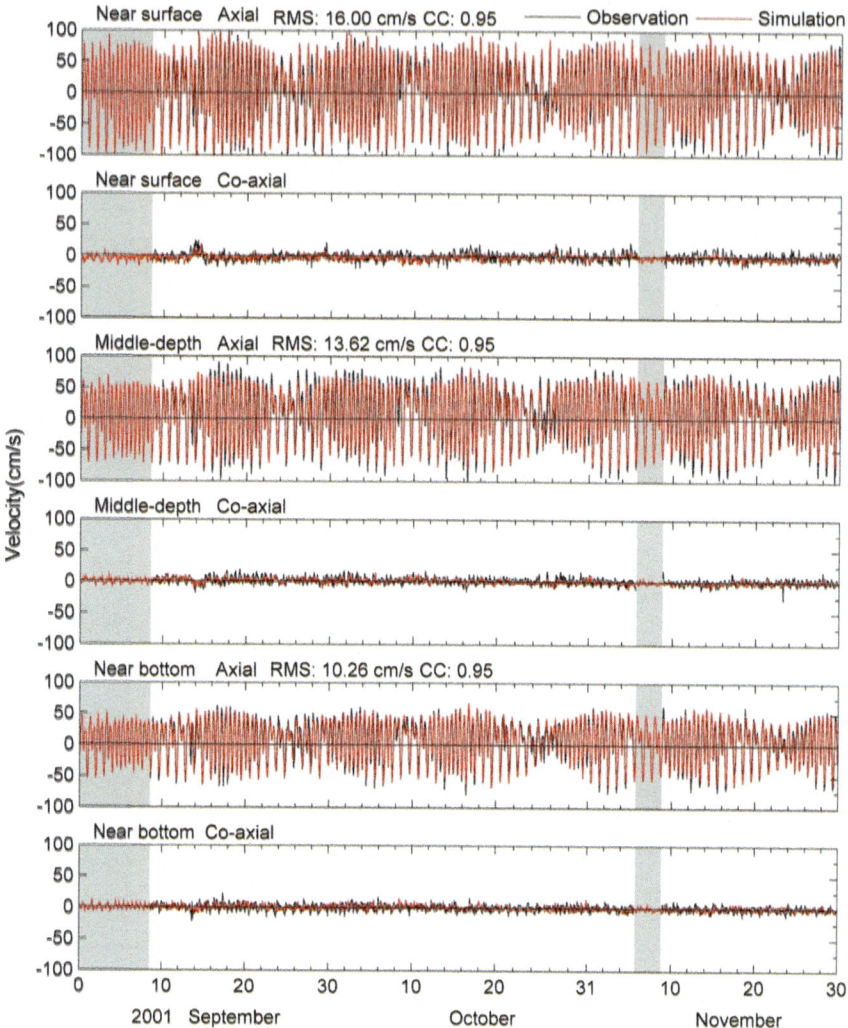

Fig. 5.6 Comparisons between hourly modeled (red) and observed (black) axial (parallel with the channel) and coaxial (perpendicular to the channel) velocity components at the near surface, middle depth, and near bottom beneath the Sunshine Skyway Bridge. Shaded regions indicate observational data gaps. Also provided are root mean square (rms) deviations and correlation coefficients (cc) (adapted from Weisberg, R. H. and L. Zheng (2006), Circulation of Tampa Bay driven by buoyancy, tides, and winds, as simulated using a finite volume coastal ocean model. *J. Geophys. Res.*, 111, C01005. https://doi.org/10.1029/2005JC003067)

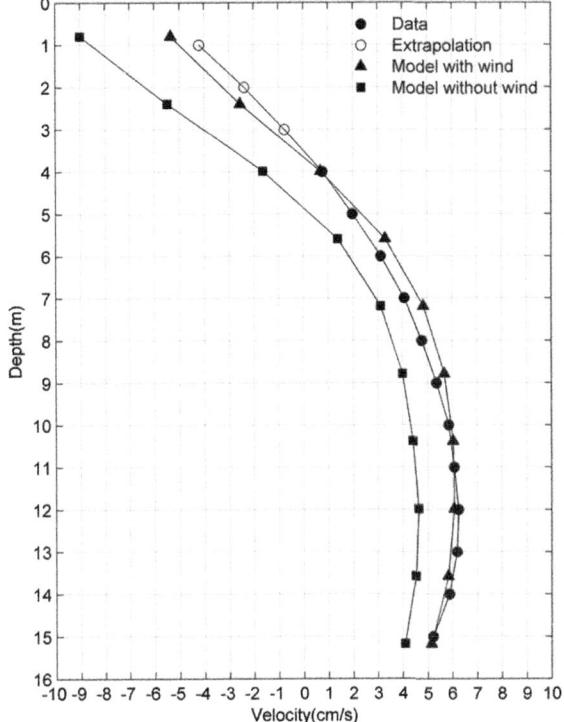

Fig. 5.7 Comparisons between the observed and TBCOM simulated net estuarine circulation within the main shipping channel beneath the Sunshine Skyway Bridge, as averaged for the three-month interval September through November 2001. The observations (made with a bottom-mounted acoustic Doppler current profiler) sampled at one-meter intervals from within 4 m of the surface and 1 m from the bottom. Simulations were performed with wind to compare directly with observations and without wind to estimate what the buoyancy-driven flow would be by itself without wind (adapted from Zhu, J., R. H. Weisberg, L. Zheng, and S. Han (2014). Influences of channel deepening and widening on the tidal and non-tidal circulation of Tampa Bay. *Estuaries and Coasts*, 38, 132–150. https://doi.org/10.1007/s12237-014-9815-4)

simulation excluding the winds. As discussed in the beginning of this chapter, note how sea level (the left-hand panel a) stands higher at the head of Tampa Bay than at the mouth, with the difference being about 4 cm. This is enough to produce a pressure gradient force to drive water seaward on average, thereby aiding the transport of the freshwater being input to Tampa Bay by the rivers. The middle panel b shows the spatial distribution of the near-surface outflow superimposed on the surface salinity. The largest speeds are in the deep channel into which the surface flow also converges. Similarly, the largest near bottom inflows (the right-hand panel c) are also in the channel because this is where the landward-directed pressure gradient force is the largest. It is also where the highest salinity water, originating in the adjacent Gulf of Mexico, is transported landward. As a corollary, this is where any of the water properties of WFS origin are transported landward. So unlike either the tidally driven

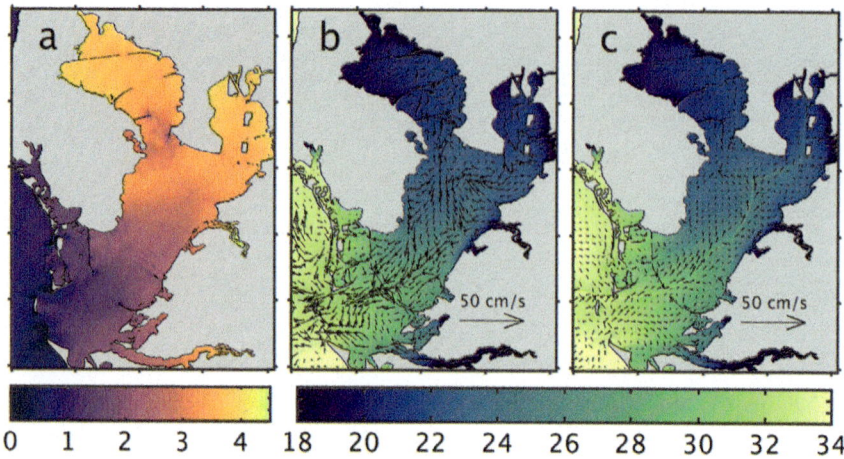

Fig. 5.8 a Non-tidal water level distributions of non-wind experiment in color coding (units in cm). **b** Distributions of non-tidal current (arrows) and salinity units in practical salinity unit (color coding) throughout Tampa Bay at the near surface. **c** Same as (b) but for the near bottom

or the wind-driven flows that move water to and fro, these net buoyancy-driven estuary circulation flows are much more persistent, and hence, it is this buoyancy-driven or net estuarine circulation that is responsible for the ultimate distributions of water properties and hence the ecology throughout the estuary.

Having now looked at a single vertical profile within the main shipping channel and at planar views at the surface and bottom, Fig. 5.9 shows how salinity transports (and hence the transport of any other material property) are distributed across the estuary at a cross section sampled just to the south of St. Petersburg in the middle of Tampa Bay. The flow field is found to be fully three-dimensional, with the main shipping channel favoring inflows of salinity across most of the water column with outflowing salinity occurring within the shallower sides in compensation for the inflows. The divergences and convergences of the horizontal flow fields are connected by the vertical circulation, thus conserving the masses of both water and of salt. This numerical model exercise demonstrates that it is the physics of the estuary circulation that determines the water property distributions within the estuary and hence the organization of the estuary ecology, here in Tampa Bay and similarly elsewhere.

Along with Tampa Bay, we pursued similar applications for the Charlotte Harbor/ Sanibel estuary system and the Naples Bay/Rookery Bay estuary system. At lesser resolution, all of the West Florida estuaries are included in our West Florida Coastal Ocean Model (WFCOM), whose publicly (internet) available, nowcast/forecast model simulations have various applications, as will become evident in subsequent chapters.

I will close this discussion of estuaries with a specific application that demonstrates why an understanding of estuarine circulation and the ability to model it accurately are important. Among its many utilizations, Tampa Bay is a major origin for the export of phosphate materials used in fertilizer production. Waste products from the production of ammonium phosphate are stored in phosphogypsum stacks, one of which was

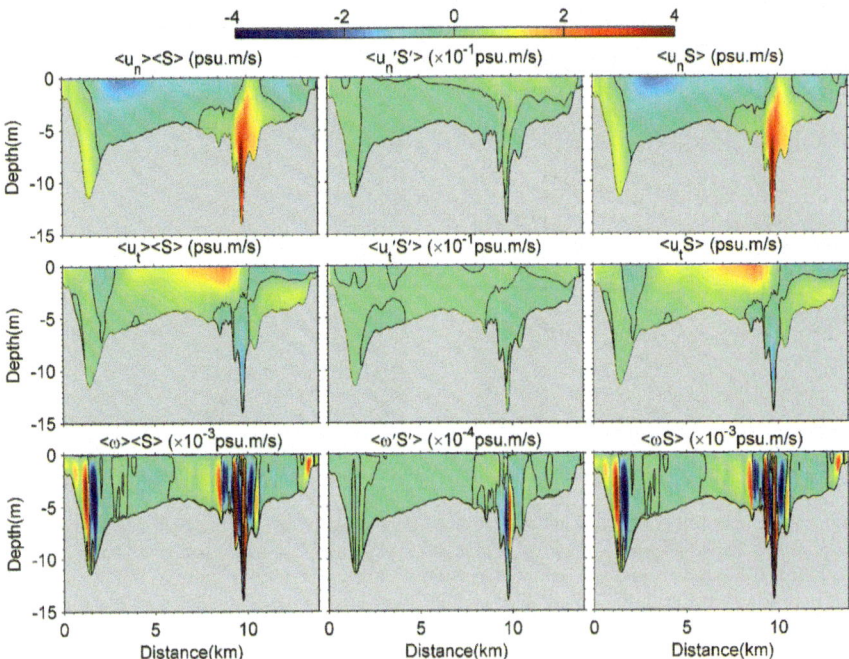

Fig. 5.9 TBCOM sampled, across estuary section just to the south of St. Petersburg, FL. The top set of panels are the fluxes of salinity (psu, or practical salinity unit times meters per second) by the mean horizontal flow directed normal to the cross section (left-hand panel), the perturbation correlations (middle panel) and their sum or the total salinity flux (right-hand panel) directed normal to this mid-bay cross section. The middle set of panels does similarly for the transverse direction, and the bottom set of panels is for the vertical direction (adapted from Zhu, J., R. H. Weisberg, L. Zheng, and S. Han (2015). On the salt balance of Tampa Bay. *Cont. Shelf Res*, 107, 115–131. https://doi.org/10.1016/j.csr2015.07.001)

abandoned at Tampa Bay's Port Manatee in 2001 when the Mulberry Corporation declared bankruptcy. When dangerously overfilled with water in spring 2021, and with the fear of levee collapse, the wastewater from this Piney Point stack was released into Tampa Bay from March 30 through April 8 via a pipeline that terminated at Port Manatee. This emergency point source release resulted in an estimated 205 tons of nitrogen compounds being dumped into the bay at Port Manatee, as compared with an estimated 175 tons of nitrogen compounds that enter the entirety of Lower Tampa Bay over a full year (most of which comes via atmospheric deposition). The ecological consequences of this Piney Point spill were immediate and long lasting.

To orient the reader, Fig. 5.10 shows the Port Manatee/Piney Point location and how this portion of Tampa Bay is resolved by the TBCOM nowcast/forecast modeling system. The highest TBCOM spatial resolution (as fine as 20 m) is within the port and its shipping channel.

The wastewater release occurred over a nine-day period at varying release rates. Figure 5.11 shows these rates, the cumulative amount of effluent volume and just how

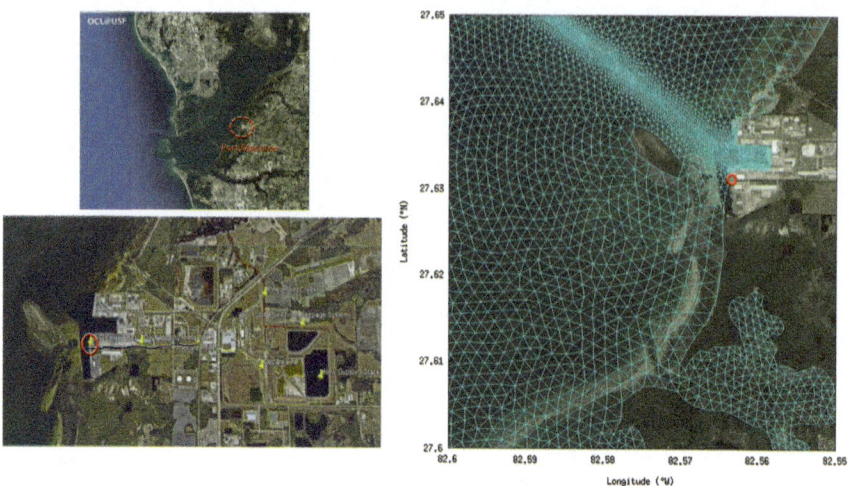

Fig. 5.10 Upper left panel shows the location of Port Manatee within Tampa Bay; the lower left panel shows the Port Manatee vicinity, where the Piney Point stack is located (the large pond on right side of the panel) and the point source pipe location (red circle); and the right-hand panel shows the TBCOM grid that nicely resolves this Tampa Bay vicinity (along with the entirety of Tampa Bay and its adjacent waters)

well TBCOM performed in modeling the simulated amount of effluent discharge [that was tagged with a tracer of initial normalized concentration (tracer per unit volume) equal to 1.0] that was input to the model at the point source (the red dot in Fig. 5.10 right-hand panel).

Given the demonstrated fidelity between the model tracer and the actual amount of effluent input to the bay (about 800,000 m^3), it is fair to ask just where the effluent went as driven by tides winds and the net buoyancy-driven estuarine circulation and how quickly it was diluted by mixing with the ambient Tampa Bay waters. Initially, both the actual wastewater and the model tracer remained in the vicinity of Port Manatee, and the phytoplankton response to the wastewater nutrients was immediate, as demonstrated by Fig. 5.12, spanning the period from April 6 to April 10. The top panels provide daily snapshots of chlorophyll remotely sensed by satellite. The highest values (encircled in red) are in the immediate vicinity of Port Manatee, and the bottom panels show the simulated tracer concentrations sampled at 1600 UTC on April 8, 9 and 10. Comparing the observations with the simulations again demonstrates the model fidelity. The initial and immediate phytoplankton bloom was also quite evident by eye. To see for myself, I took one of my graduate students and an associate there aboard my sailboat. The color contrast across the bloom was striking, with green water outside of it and brown water in the midst of it. Such an intensive bloom would have secondary consequences, as we found out some six weeks later.

Using our daily, automated TBCOM, we produced 3.5-day forecasts of where the tracer (the diluted effluent) would go each day, and we provided these (via the internet) to the public in the form of movie loops, which were of value to those who

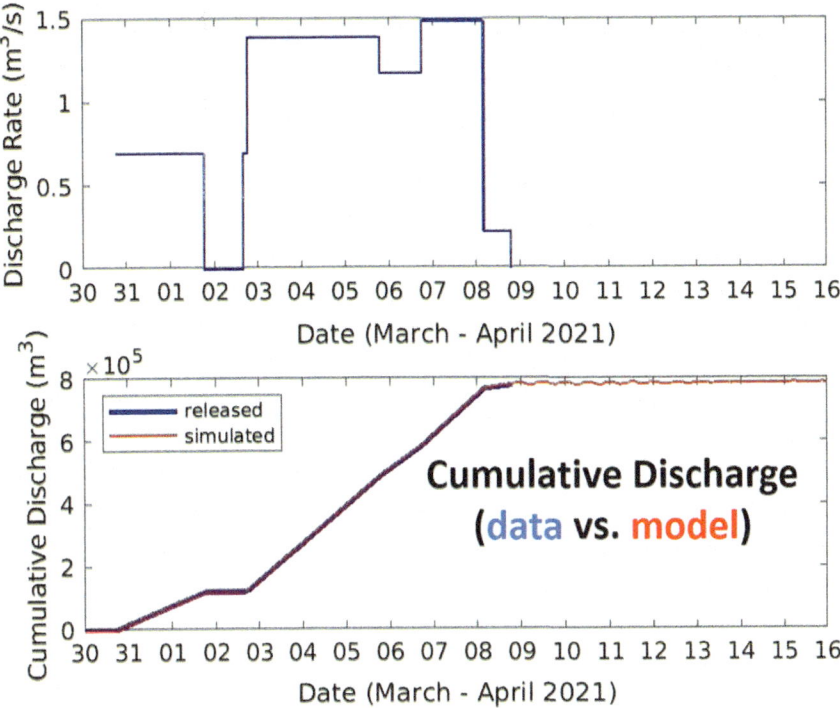

Fig. 5.11 Piney Point wastewater discharge rates, the cumulative discharge and how well the TBCOM tracer simulation accounted for the cumulative discharge (adapted from Liu, Y., Weisberg, R. H., Zheng, L., Sun, Y., Chen, J., Law, J. A., Hu, C., Cannizzaro, J. P., Frazer, T. K. (2024), A tracer model nowcast/forecast study of the Tampa Bay, Piney Point effluent plume: Rapid response to an environmental hazard, *Marine Pollution Bulletin*, 198, 115,840. https://doi.org/10.1016/j.marpolbul.2023.115840)

were planning shipboard sampling of the effluent in the field. With time, the effluent eventually spread throughout the entirety of Tampa Bay and along the adjacent Gulf of Mexico coastline once the effluent slowly flushed out of the bay. Figure 5.13 demonstrates this spreading and dilution by showing tracer concentration snapshots at 0000 UTC on April 9, April 30, May 30 and June 30. After the initial spread of effluent along the eastern side of the bay, the combination of winds, tides and buoyancy gradients facilitated the crossing of the effluent to the western side of the bay and eventually throughout the bay. Flushing, primarily by the wind and the buoyancy-driven circulation, transported the effluent into the Gulf of Mexico and along its coastline.

The consequences of such effluent transport were visually evident in shallow water regions. I could even see discolored water behind my house located on Tierra Verde, an island community in the southernmost part of Pinellas County located to the north of the bay mouth and Fort DeSoto Park. By continuing to keep a quantitative tally of the effluent concentration throughout the bay, we were able to estimate the flushing

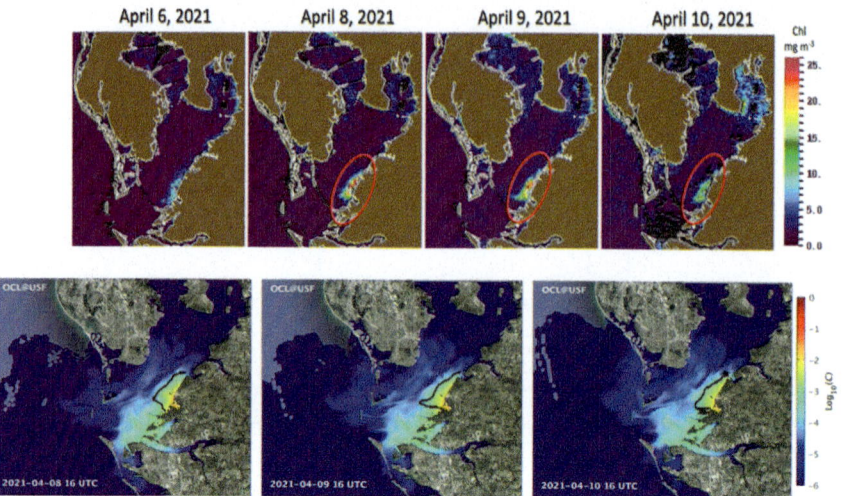

Fig. 5.12 Top panels are remotely sensed satellite images showing elevated values of chlorophyll (i.e., a phytoplankton bloom) on April 8, 9 and 10. The bottom panels are model simulated tracer concentrations, where the color coding provides the dilution, using a $\log_{10}$ scale, i.e., 0 means an initial normalized concentration of 10^0 or 1.0, -1 means a 10^{-1} reduction of 0.1, -2 means a 10^{-2} reduction or 0.01 and so on (adapted from Liu, Y., Weisberg, R. H., Zheng, L., Sun, Y., Chen, J., Law, J. A., Hu, C., Cannizzaro, J. P., Frazer, T. K. (2024), A tracer model nowcast/forecast study of the Tampa Bay, Piney Point effluent plume: Rapid response to an environmental hazard, *Marine Pollution Bulletin*, 198, 115,840. https://doi.org/10.1016/j.marpolbul.2023.115840)

time for the bay to be about 127 days. Flushing time is conventionally defined as the time that it takes for the total amount of the input tracer to decrease to an amount equal to 0.37 of its initial value, and this flushing time of 127 days agreed reasonably well with prior published estimates. The June 30th panel of Fig. 5.13 shows that effluent was transported to the northern-most part of Pinellas County where its effects were also evidenced by Pinellas County environmental personnel.

Coincident with the Piney Point wastewater discharge, an unrelated harmful algae bloom, or red tide, was located to the south of Tampa Bay in the vicinity of the Charlotte Harbor/Sanibel estuary. Persistent southerly winds eventually transported some of these red tide cells along the coast to the mouth of Tampa Bay, where tides and winds then quickly transported these red tide cells into the bay. By this time, the initial phytoplankton bloom at Port Manatee (Fig. 5.12) had abated, with these cells sinking to, and decomposing along, the bottom. Given the recycling of nutrients, now in an organic (versus inorganic) form, the newly transported red tide cells had an abundant nutrient source upon which to increase their population. Thus, by the end of May, 2021, and long after the Piney Point effluent release had ended, Tampa Bay was beset by a major *Karenia brevis* red tide outbreak (see Chap. 11) causing a massive fish kill, which, through the decay of dead fish, resulted in a new nutrient load to Tampa Bay that even exceeded what Piney Point initially injected. It was estimated that some 55 tons of nitrogen compounds were removed from the bay by

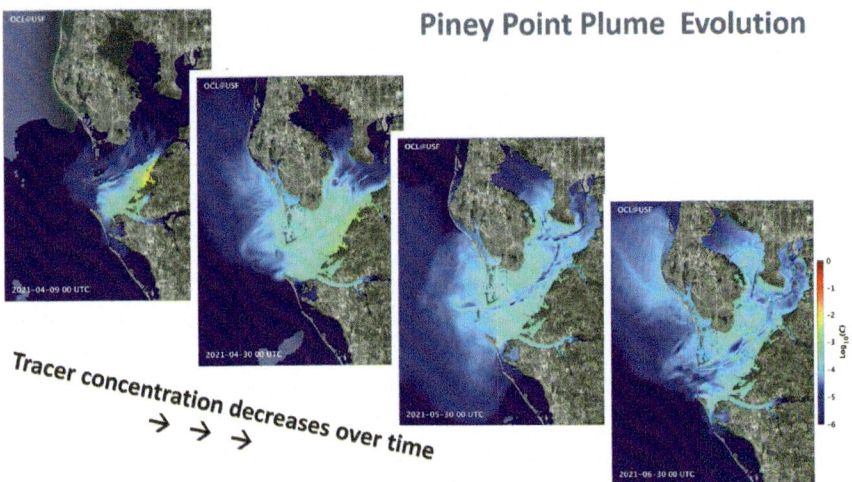

Fig. 5.13 Successive snapshots of simulated tracer concentrations on April 9, April 30, May 30 and June 30 showing the Piney Point wastewater plume evolution, all using a $\log_{10}$ concentration scale (see Fig. 5.13) (adapted from Liu, Y., Weisberg, R. H., Zheng, L., Sun, Y., Chen, J., Law, J. A., Hu, C., Cannizzaro, J. P., Frazer, T. K. (2024), A tracer model nowcast/forecast study of the Tampa Bay, Piney Point effluent plume: Rapid response to an environmental hazard, *Marine Pollution Bulletin*, 198, 115,840. https://doi.org/10.1016/j.marpolbul.2023.115840)

cleaning up dead fish, but this was only a fraction of the fish that were killed. The Piney Point effluent discharge thus became a gift that kept giving.

An additional consequence was a widely distributed bloom of bottom-dwelling, mat-forming algae that became entwined with the ambient seagrass. As these encroaching algae died and became buoyant (via decomposition gas formation), the entwined seagrass was yanked from the bottom, resulting in a massive seagrass die-off and a consequent loss of habitat for species that rely on seagrass. It is fair to say that Tampa Bay continued to suffer the consequences of the Piney Point effluent discharge even several years after the occurrence.

Where the effluent went, how quickly it dispersed and became diluted were all determined by the circulation, plus the biochemical utilization and recycling. Additionally, why *K. brevis* cells were coincidentally at the mouth of Tampa Bay at an inopportune time was also owing to the circulation. From this Piney Point experience, it is clear that anything of an ecological nature within an estuary is critically dependent upon the circulation. The environmental stewardship of our estuaries must therefore include adequate provisions for observing and modeling the movement of water within the estuary, i.e., the estuarine circulation as driven by tides, winds and river inflows (buoyancy).

Addendum: Pasta with Scallops and Calamari

For a little change from fish, let's try a pasta dish. Linguini is fine, but if you want to be somewhat adventuresome given the title, try some squid ink linguini. The smaller bay scallops (that cost about a third of the larger sea scallops) are the way to go. Half and half between these and calamari is the suggestion. The ingredients are shallots, garlic, mushrooms, ricotta cheese, salt, pepper and some red chili flakes. A half-pound (half-box) of linguini for two people and a full-pound for four should suffice. For the sauce, I use about a ½ lb each of scallops and calamari, plus a package (½ lb) of baby bella mushrooms, a large shallot or two small ones and about four cloves of garlic.

The pasta is easy—just put in a pot of salted boiling water and cook al dente as directed on the package (dry takes about 9 min, fresh only a few minutes) so make the sauce first or at least start it before starting the pasta.

I usually get cleaned calamari tubes at the fish store, but if you are starting from whole ones, then be sure to clean them thoroughly, discard the outer coloring and the stiff, pencil-like structure; then cut into ¼ inch rounds. The tentacles are also good; keep whole if small; cut in half if they look too large. Rinse all of the seafood prior to cooking.

The first step is to dice the shallots and garlic and saute lightly until just soft (don't brown) in a high-quality olive oil. Clean and slice the mushrooms and add to the shallot/garlic and cook until soft, adding salt and pepper to taste. The salt will help to release some water from the mushrooms, adding to the sauce. Once the mushrooms are soft, add the seafood, which will cook rather quickly. You can tell when these are done by the calamari beginning to curl and turn opaque. The release of their juices will also add to the sauce. A splash of white wine (~1/2 cup) will further extend the sauce. Cook for a short time to let the alcohol evaporate. Use a vermentino (Tuscan white), or a roero anais (Piedmont white).

The final step is to use some ricotta cheese to thicken the sauce. Spoon in a little at a time and stir to combine (maybe three tablespoons in total). It should take on the consistency and color of a cream sauce (you could also just use cream, but I prefer the ricotta). Then taste, and add more salt and pepper if needed.

To serve, place the pasta on a plate, spoon and toss in the sauce mixture, add some grated parmigiano reggiano, garnish with some fresh chopped parsley, capers and some red chili flakes (if you like a little heat). All that this needs at the side is a salad and maybe some garlic bread.

The same wine used for cooking is the accompaniment, a vermentino (from Tuscany) or a roero anais (from Piedmont).

Chapter 6
Sea Level: Why It Goes Up, Down and Where It Might Be Heading

Have you ever wondered why there are two high tides and two low tides each day and why each is so different, or why the overall tidal ranges within each month tend to vary between being relatively large (spring tides) and relatively small (neap tides)? If so, you are not alone. In every generation, humans have pondered these questions. Also of wonder are why sea level tends to be higher in summer and lower in winter, and why with the passage of weather fronts the shore parallel winds set sea level up or down; whereas, with the passage of a hurricane, it is the shore normal winds that do this. Finally, in view of mass media discussions on global warming and sea level rise, even verging upon hyperbole, should you sell your near-shore property now and head for the hills, or might you and your descendants be safe for another generation?

Several topics require discussion in order to answer these questions. These topics include: tides, the effects of winds, atmospheric pressure and water density variations through seasonal heating and cooling, plus the ocean mass exchanges that may occur by ice accretion or loss from land and other net freshwater accumulations or losses from land. In addition, the slow uplifts or downward movements of land relative to mean sea level through glacial rebound, sediment consolidation, or ground water pumping, must also be considered, and, at even longer, geological time scales, variations in the geometry of the ocean basins are a factor. These effects are all in contrast with the fastest variations in sea level that result from rapid vertical shifts in the Earth's crust due to earthquakes, volcanos and landslides, all of which can cause tsunamis, plus storm surges caused by hurricanes, both of which will be discussed in subsequent chapters.

Let's begin with tides. For rotating and gravitating Earth-moon and Earth-Sun systems, there are ever-changing balances between the attractive forces by gravity and the centrifugal forces by rotation. For the case of the Earth's rotation about its own axis, the centrifugal force acting on a parcel of water combines with the Earth's self-gravitation to yield a slightly modified gravitational potential and hence a local vertical direction that is slightly askew from what it would be without the Earth's rotation. This is not the centrifugal force that we concern ourselves with as regards

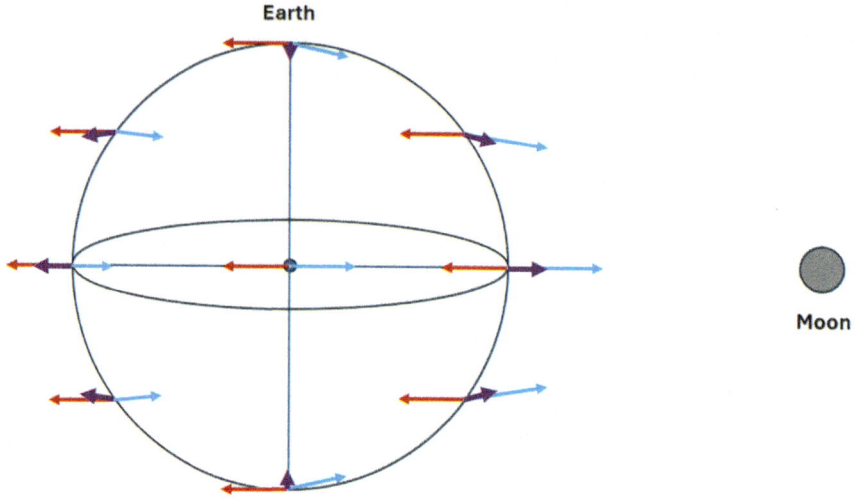

Fig. 6.1 Uniformly distributed centrifugal force (red lines) due to the coupled Earth-moon revolution, the gravitational attraction by the moon at any point of the Earth's surface (light blue lines) and their difference (purple lines), which is what we call the tide producing force

tides. Instead, tides result from the respective centrifugal forces due to the revolving Earth-moon and Earth-sun systems.

For the Earth-moon system, this coupled pair of celestial bodies revolves around a common axis every 27.32 days, and with every point on the Earth transiting the same circular distance, the centrifugal force associated with this coupled system revolution is uniform over the entire Earth. Moreover, for the Earth-moon system to remain stable, this centrifugal force must equal the gravitational attraction by the Moon for a point at the Earth's center. A tide-generating force arises because the gravitational attraction of the moon on any other point on the Earth is slightly different from that at the Earth's center. For instance, the distance between the moon and the surface of the Earth on either side in line with the moon is twice the radius of the Earth. Hence, whereas the centrifugal force due to the coupled Earth-moon revolution is uniform over the entire Earth, the gravitational pull of the moon at any location on the Earth will vary by the distance between that location and the moon, as depicted in Fig. 6.1.

Such a static analysis for an Earth without any distributed land masses is referred to as an equilibrium theory of tides, and it results in a bulging out along the equatorial plane and a flattening at the poles. The Earth, in reality, has continents, it rotates daily and it is also acted upon by the sun (in a fashion similar to that of the moon). These additional complications require what is referred to as a dynamical theory of tides for which a complete tidal potential is derived and for which we must seek an empirical determination of the various tidal constituents (or harmonics) at each point on the Earth to accurately account for the tides as observed around the Earth.

Regardless of these complications and the mathematics necessary to describe them, the tidal generating force is proportional to the celestial body mass divided by the cube of its distance from the Earth. For this reason, the tidal generating force by the moon is about twice that by the sun (the sun is much more massive, but it is also much farther away), and all other celestial bodies, being even farther away, can be ignored relative to the tidal forcing by the moon and the sun.

From Fig. 6.1, we also see that it is the component of the tidal generating force (the thick arrows) that is parallel to the Earth's surface that causes ocean water to move (it also causes the atmosphere's air to move and even the solid Earth to move at a much lesser extent). By being set in motion by these tidal generating forces, ocean water locally is either set up or down, resulting in pressure gradient forces that, along with the acceleration of the water (including the Coriolis acceleration), tend to balance the tidal generating forces, and this process goes on in perpetuity. In other words, the oceans, the atmosphere and even the land and the liquid magma beneath are never fully at rest; instead, they have tides that vary continually. Over time, the friction associated with these perpetual tidal motions acts to slow the rotation of the Earth, thereby lengthening the duration of a day, but only by a few milliseconds per century. The length of the day also changes due to any variation in its mass distribution; but, again with durations of only milliseconds. These are topics of Earth Geophysics, versus oceanography, but they do point out the interconnectivity of the various scientific disciplines.

An additional complication is that the orbital geometries of the Earth-moon and Earth-sun systems are themselves varying, the result being three different species of tides, which are referred to as long-period (longer than a day by virtue of orbital changes), diurnal (daily) and semi-diurnal (twice daily), and while we can define very specific periodicities for these species, because their individual periodicities are not integer multiples of one another, the overall resulting tidal variations are aperiodic. They are close to being periodic but not exactly so. Consequently, whereas tidal prediction is quite accurate, no two tides are ever exactly the same; they are nearly so, but not exactly so.

To begin answering the questions posed at the beginning of this chapter, it is instructive to consider the principal tidal constituents as determined at a given location and see how they sum up to account for the observed sea level variations. For this exercise, we will use observations and tidal constituent results calculated for a tide gauge station at St. Petersburg, Florida. There, in descending order, the four largest tidal constituents are the principal lunar semi-diurnal (M2), the luni-solar diurnal (K1), the principal lunar diurnal (O1) and the principal solar semi-diurnal (S2), where the periodicities, or their repeat intervals are 12.42 h, 23.93 h, 25.82 h and 12.00 h, respectively. Figure 6.2 compares each of these individual constituents with the actual observed sea level variations for the month of January 2017. Note that individually, none of these constituents resembles the sea level observations.

What happens when we start adding these together? Two different and distinctive behaviors emerge when adding either two species of similar periodicities together, or adding two species with distinctly different periodicities together, as shown in Fig. 6.3. Consider the different periodicity cases first. The top panel of Fig. 6.3

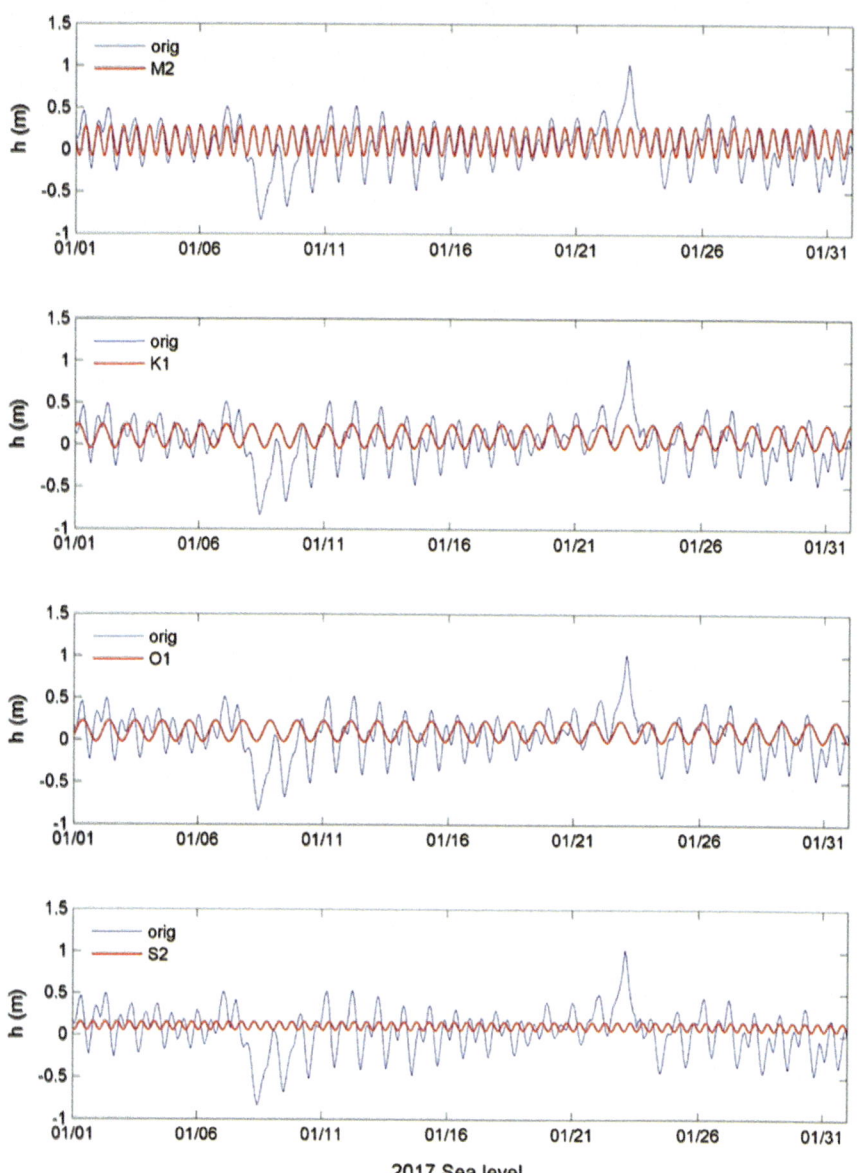

Fig. 6.2 From top to bottom are the four principal tidal constituents, M2, K1, O1 and S2 (in red) superimposed on the observed, hourly averaged sea level variations at St. Petersburg Florida for the month of January 2017

compares the sum of M2 and K1 with the observed sea level record at St. Petersburg, Florida. We can now begin to understand why the two high and low tides of a given day have different amplitudes. During one of the high tides, the two constituents add constructively with one another, whereas for the second high tide, they add destructively, and similarly for the low tides. This is because the diurnal constituent has a period that is nearly twice as long as the semi-diurnal constituent so nearly once per day they will be in phase (with peaks occurring nearly at the same time, i.e., adding constructively) or out of phase (with a peak and a trough occurring nearly at the same time, i.e., adding destructively to somewhat cancel each other out). Thus, it is not surprising that by adding M2 with O1 (as shown in the second panel from the top), a similar behavior emerges. This explains the origin of what is referred to as the diurnal inequality, the fact that the two highs and lows in any given day are not the same.

Now consider adding species of similar periodicity. The third panel from the top shows the results for M2 + S2, and the last panel for the top shows the result for K1 + O1. For each of these, we see an overall tidal amplitude that varies over the month such that there are two intervals during the month for which the tides are either of larger, or smaller amplitude. The intervals of larger amplitude tides are referred to as spring tides and the intervals of smaller amplitude tides are referred to as neap tides. The explanation is similar to that for the diurnal inequality. Spring and neap tides are also the result of constructive and destructive interference between tidal constituents, but since the periodicities for similar tidal species are nearly the same, it takes a longer time for their relative phases to change such that they add either constructively or destructively.

Armed with an understanding of both the diurnal inequality and the spring and neap tide cycles, let's see what happens when we systematically add all four of these constituents together, as shown in Fig. 6.4. From top to bottom are (M2), (M2 + S2), (M2 + S2 + K1) and (M2 + S2 + K1 + O1). By continually adding constituents, the tidally simulated sea level (the red lines) comes in ever-closer agreement with the observations. With the inclusion of all four major constituents, the tidally simulated portion of sea level is in very close agreement with the observations. However, once we subtract the sum of the four constituents from the actual observed sea level, there still exists some significant deviations as evident in Fig. 6.5. What might be the cause of these residual sea level variations?

A portion of the residual, the small fluctuations that appear to be varying at tidal periodicities, is also due to tides. Whereas we only diagnosed the four major constituents for this location, analyses show that there are a total of eight tidal constituents required to more fully account for the St. Petersburg, Florida tides. Nonetheless, the larger and also longer time scale fluctuations, particularly the large swings between January 6–11 and January 21–26, are not tide related. For further explanation, we must leave the realm of tides and look to another factor that causes sea level to vary, the winds.

Recall from Chap. 4, particularly the discussion regarding Fig. 4.2, that the Coriolis force accelerates ocean currents to the right in the northern hemisphere. In deep enough water, this results in a net transport of water at a right angle to the wind. Being

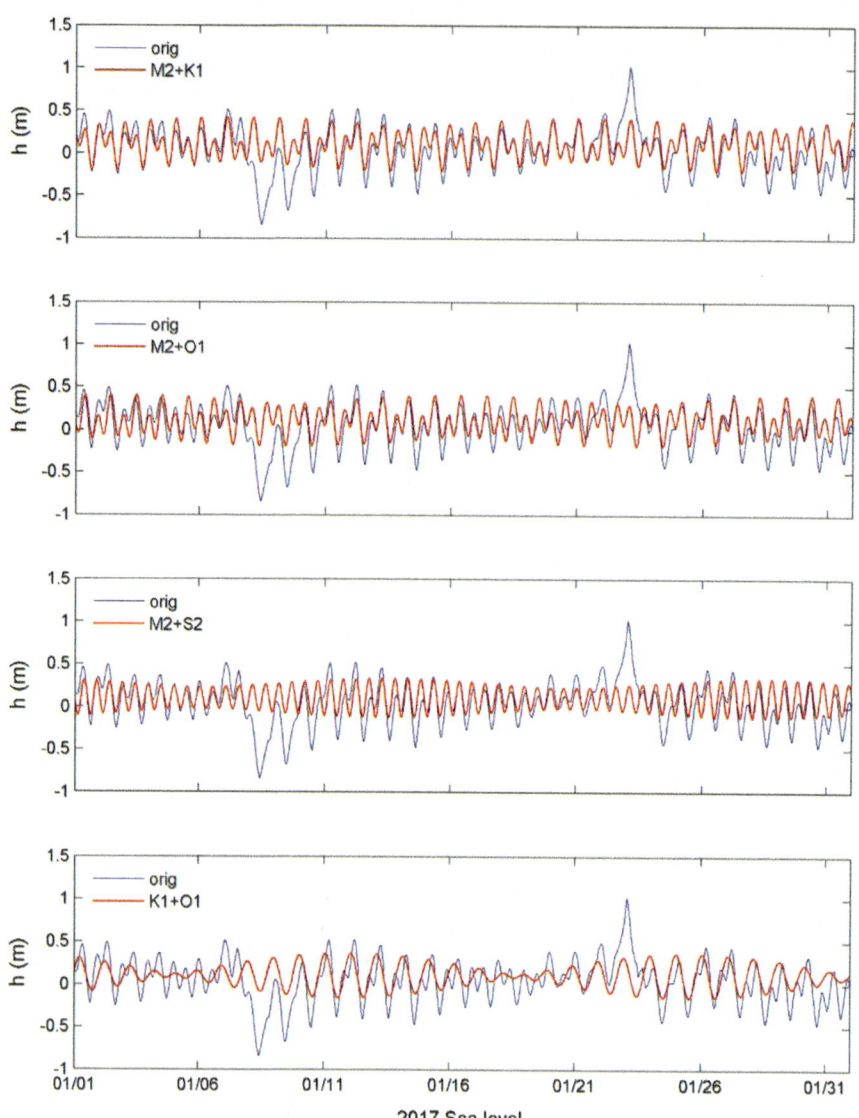

Fig. 6.3 From top to bottom are the sums of M2 + K1, M2 + O1, M2 + S2 and K1 + O1 (in red) superimposed on the observed, hourly averaged sea level variations at St. Petersburg Florida for the month of January 2017

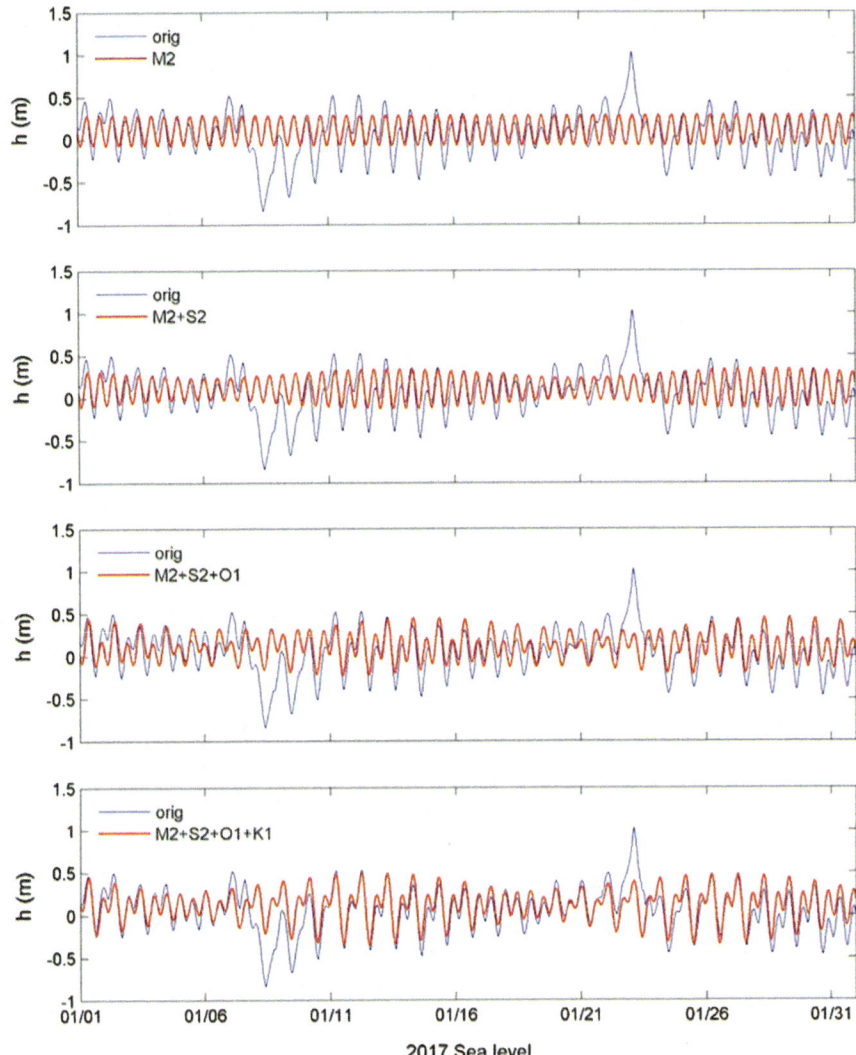

Fig. 6.4 From top to bottom are M2, M2 + S2, + M2 + S2 + O1 and M2 + S2 + O1 + K1 (in red) superimposed on the observed, hourly averaged sea level variations at St. Petersburg Florida for the month of January 2017

that the west coast of Florida is oriented along a SSE to NNW line, winds blowing parallel to the coastline from the SSE cause water to pile up along the coastline, whereas winds blowing parallel to the coastline from the NNW cause water to move away from the coastline.

As the water depth shoals, thereby allowing bottom frictional effects to compete ever more with the Coriolis force and eventually even overtake the Coriolis force in

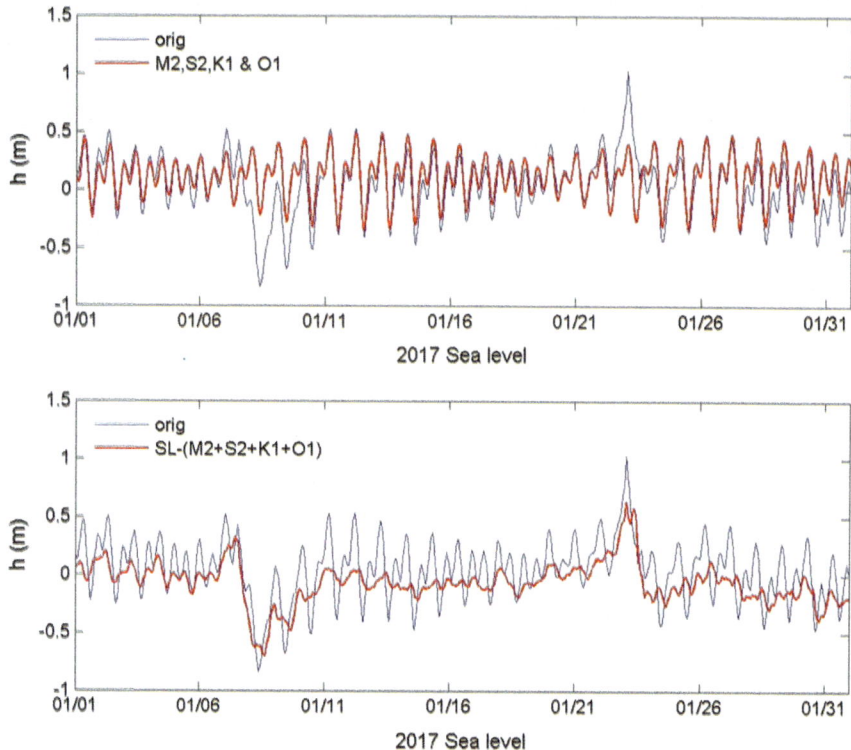

Fig. 6.5 Top panel is the sum of M2 + S2 + O1 + K1 (in red) superimposed on the observed, hourly averaged sea level variations at St. Petersburg Florida for the month of January 2017, and the bottom panel is the residual sea level (in red) when subtracting the sum of the four major tidal constituents from the observed sea level (in black)

shallow enough water, the transport of water, instead of being at a right angle to the wind, turns ever more downwind. This transition, from waters deep enough so that the wind-driven transport (for a shore parallel wind) changes from being shore normal to shore parallel, occurs over the inner-shelf, which, for the WFS, exists roughly between the 50 m isobath and the shoreline (see Fig. 4.5). A similar argument exists for winds that are directed normal to the shoreline. Far enough from the coast, these winds transport water parallel to the coast, whereas at shallow enough depth, these winds transport water downwind, or shore normal. Thus, both components of wind must be taken into consideration when trying to interpret sea level variations observed at the St. Petersburg tide gauge. As determined empirically, the wind direction that evinces the best correlation with the St. Petersburg tide gauge is at an orientation along the line directed from 170 to 10°, i.e., sea level sets up most efficaciously if the winds are blowing from just west of southerly, and sea level sets down most efficaciously if the winds are blowing from just east of northerly. It is in that manner

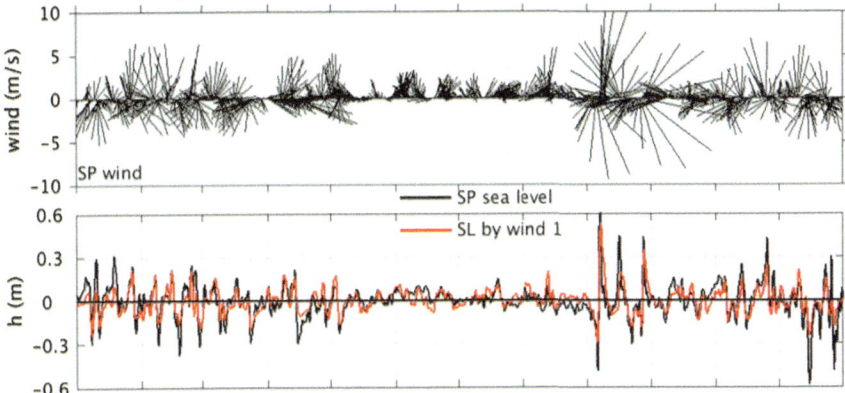

Fig. 6.6 Wind velocity as observed at St. Petersburg Florida (top panel) and the non-tidal (residual) sea level after removing the eight principal tidal constituents from the observed sea level at St. Petersburg Florida for the entire calendar year of 2004. The red line is the portion of the residual sea level that is accounted for by the component of wind in the 170° direction, which is the direction of maximum sea level/wind correlation for this location, and the black line is the residual sea level

that both the shore parallel contribution from deep enough water and the shore normal contribution from shallow enough water add constructively with one another.

Given the above explanation on the coastal ocean physics, a statistical analysis (referred to as a linear systems analysis) reveals just how much of the residual sea level can be accounted for by wind forcing, and the results for an entire year (this example being for calendar year 2004) are shown in Fig. 6.6. By comparing the overlain red and black lines, we see how the wind-driven (red line) portion nearly mimics the observed (black line) residual, non-tidal sea level.

Whereas most of the local sea level variations may be accounted for by a combination of tides and winds, there are still other factors that must also be considered. One of these is atmospheric pressure, with its various confusing units of measure. The average sea level pressure, or the weight per unit area of the overlying atmosphere, is about 1013 mbars (or 14.7 lbsin^{-2}, or 29.92 in of mercury, depending historically on how and by whom it was measured, i.e., metric units, English units and instrument type), where a mbar is 1000 dynescm^{-2}. Given the density differences between air and water, a 1 mbar change in atmospheric pressure is equivalent to the change in pressure attributed to 1 cm of water level. Each time a weather front passes over a region, it is common for the atmospheric pressure to vary by about 5 to 10 mbar, which translates to about 5 to 10 cm, or 2 to 4 inches. We call this effect the inverted barometer effect because as the mercury in the barometer falls, sea level rises, and conversely. Water moves in response to pressure differences so if the pressure is larger on one side of the front and smaller on the other side then water will flow from the larger to the smaller pressure region, causing sea level to rise where the atmospheric pressure is smaller; hence, again the inverted barometer effect.

Density provides another influencing factor. Summertime temperatures tend to be warmer than wintertime temperatures, and by changing temperature, we change

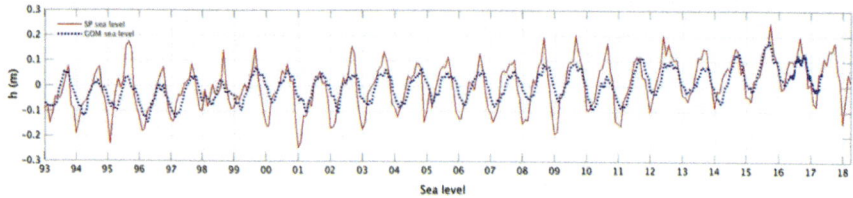

Fig. 6.7 Monthly averaged sea level time series from the St. Petersburg, Florida tide gauge (red line) and as spatially averaged over the Gulf of Mexico using satellite altimetry observations (blue dots) for the 25-year period, 1993 to 2018

density and hence the specific volume of seawater. Lower density means larger volume, and conversely, so for a given mass of water, sea level actually stands higher in summer than in winter, and this is observable both in tide gauges and in satellite altimetry measurements, as shown in Fig. 6.7 for the time interval from 1993 to 2018.

Note that for this 25-year record, there exists a trend in sea level amounting to about 3 mm per year, plus a regularly occurring annual cycle that ranges between about 20 and 40 cm. The trend is consistent with the global rise in sea level associated with a warming Earth, and the seasonal cycle is largely attributed to seasonal heating and cooling. Other factors that remain to be more thoroughly studied are gyre scale mass adjustments to winds over the Atlantic Ocean, plus various modes of coupling between the ocean and the atmosphere that ocean and atmosphere scientists are just beginning to describe and understand.

Knowing how sea level responds to tides, winds, atmospheric pressure and seasonal heating and cooling, what might the future bring? The answer to this question is uncertain, although history provides some insight. The first historical realization is that the configuration of the Earth's land masses and ocean basins has changed considerably over Earth's history. Only in the past few million years of its order 4.6-billion-year existence, has the Earth been in its present configuration following a slow evolution from the Pangaea supercontinent to what exists today. Thus, it is difficult to even define what sea level is relative to land and ocean geometries that continually change. Even today there are regions where land is rising by virtue of glacial rebound (coastal Alaska, for instance) and sinking by the consolidation of thick sediments (coastal Louisiana, for instance). Whereas actual distributed sea level gauge observations exist from the late nineteenth century onwards, recently augmented by accurate global estimates using satellite altimeter records that began in 1993, sea level estimates from prior years and eras are based on proxy records such as geological evidence, layered sediments and their skeletal remains and geochemical records that can be related to temperature and ice volume. Thus, over the past 500 M years, sea levels have been estimated to have been as much as 250–400 m higher than the present level, although much of this is conjecture with different estimates deviating from one another. Nonetheless, a rough estimate of several hundred meters in response to changes in continental land masses, ocean basin geometries and the Earth's temperature appears to be robust.

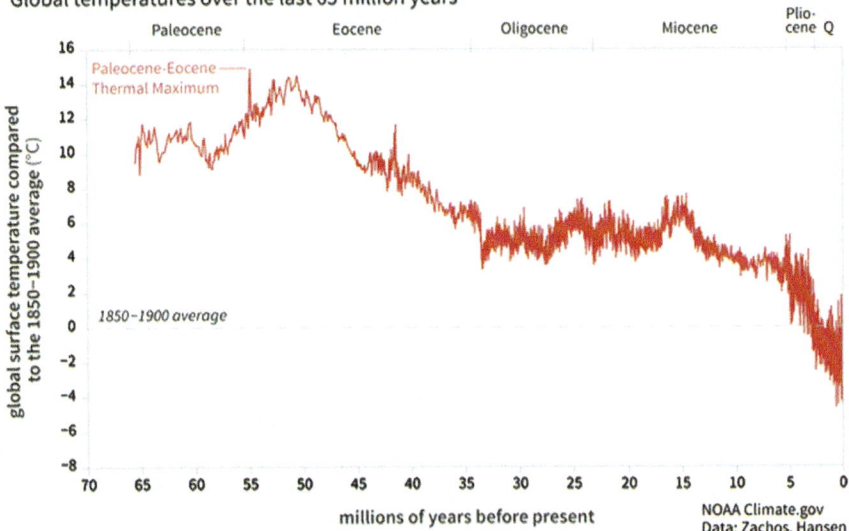

Fig. 6.8 Global average surface temperatures estimated from geochemical proxy records compared to the average estimated temperature from 1850 to 1900 (courtesy of NOAA Climate.gov, https://www.climate.gov/media/15006)

The uncertainties in historical sea level estimation diminish as we approach the present time. The present Earth configuration roughly dates back some 25 M years when the Southern Ocean became a contiguous water body effectively isolating the Antarctic continent, thereby initiating the beginnings of the present ocean circulation, which we found in the first few chapters to be largely controlling of the Earth's climate and ecology. Given this onset, it still took many millions of years for the Earth to settle into its present climate rhythm. Geochemical proxy records going back some 65 M years suggest that the Earth at that time was much warmer than at the present time (Fig. 6.8), with the corresponding sea level being about 250 m above the present level.

As the global temperature slowly decreased a combination of ice formation, land-based water and ice formation gradually diminished sea level until a more rapid decrease set in around 5 M years ago when it may be theorized that the deeper ocean waters became sufficiently cool, raising the thermocline so that seasonal and episodic upwelling could facilitate cooling the sea surface. It then took a few more million years before the Earth settled into its present relatively cool period consisting of rhythmic ice-age to interglacial cycles starting about 1 M years ago. Whereas the ice-age to interglacial cycles remains to be fully understood, there are strong correlations between these climate variations and certain astronomical factors, specifically the eccentricity of the Earth's orbit around the sun, the tilt of the Earth's rotation axis and the precession of the rotation axis around the orbital plane with the sun. Together, these three factors comprise what is referred to as Milankovitch cycles (Fig. 6.9),

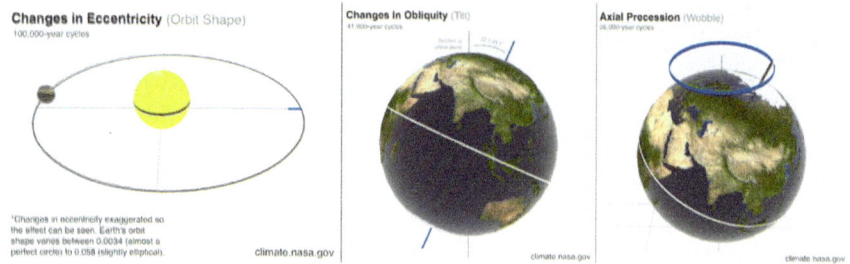

Fig. 6.9 Milankovitch cycle contributors owing to the eccentricity of the Earth's elliptical orbit around the sun, which varies over a roughly 100,000-year period, the tilt of the Earth's axis relative to the orbital plane with the sun, which varies on a roughly 41,000 year cycle and the precession of the rotation axis about a line normal to the orbital plane, which varies on a roughly 26,000 year cycle. The first of these factors determines the insolation received from the sun, the second determines how the insolation is distributed around the Earth and hence the seasons and the third determines the relative distance of the winter and summer hemispheres with respect to the sun (adapted from and courtesy of Science.NASA.gov, https://science.nasa.gov/science-research/earth-science/milankovitch-orbital-cycles-and-their-role-in-earths-climate/)

which affect the incoming shortwave radiation that the Earth receives from the sun. For instance, if the northern hemisphere tilts toward the sun at apogee, its summer will be cooler than if it titled toward the sun a perigee, and conversely, and the larger the tilt (varying between 22.1 and 24.5°), the larger the seasonal variation. Eccentricity provides the approximate 100,000-year pacing of the ice-age to interglacial cycle, which also contains the other two periodicities. By adding either constructively, or destructively (like tides), the overall ice-age to interglacial variations are both modulated and aperiodic. Other factors are also in play such as the weight of the ice, once massive glaciers form and how the coupled ocean and atmosphere along with greenhouse gas concentrations respond to the changes in global temperature. Climate science, despite its advances, remains wide open to the next generation of scientists.

The state of climate today is that presently we are in an interglacial period, as evident from Fig. 6.10 showing the past four such periods, the one on the right being the present one. Evident from this figure is that the interglacial periods appear to coincide with peaks in insolation, although the climate variations are much more complicated, as seen by the characteristic differences between insolation and the saw-tooth nature of the ice-age to interglacial cycles. Not shown is the estimated sea level, although this can also be gleaned from proxy oxygen isotope records indicative of ice volume. In essence, sea level at the end of the last glacial maximum, which occurred about 20,000 years ago, was about 120 m lower than presently observed. Sea level begins to rapidly rise with the abrupt glacial maximum termination. This rise was very steep until around 8000 years ago when the Earth's climate settled into a more slowly varying regime that persists through the present time. Interestingly, the Hebrew calendar assigns 5785 as the year of this book being written. Considering the rapid variations in climate that would accompany the collapse of an ice-age, it may

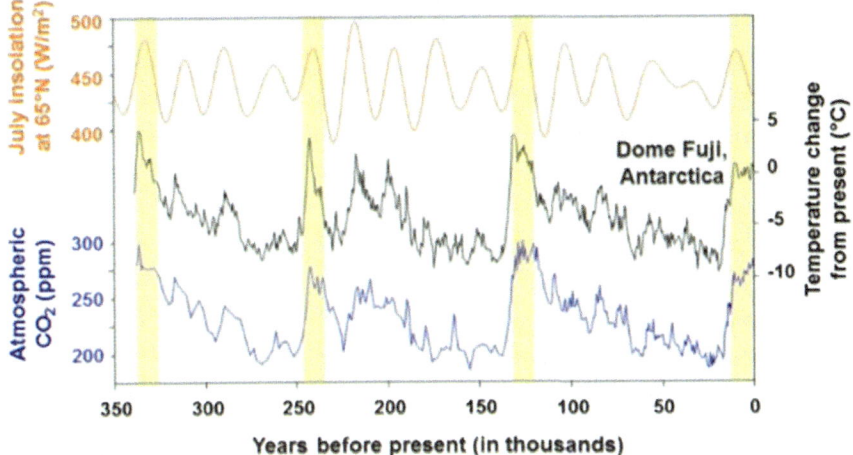

Fig. 6.10 Past three interglacial to ice-age cycles ending with the present interglacial period to the far right. The top panel shows the average July insolation at $65°N$ as determined by Milankovitch astronomical forcing, the middle panel is temperature at an Antarctic station and the bottom panel is the atmospheric CO_2 determined from ice core records (courtesy of NOAA/NCEI, https://www.ncei.noaa.gov/sites/default/files/2021-11/1%20Glacial-Interglacial%20Cycles-Final-OCT%202021.pdf)

not be surprising that organized societal advancements might have to await climate settling into a more stable regime. While the beginning of the present interglacial period saw the development of various civilizations, just what transpired remains a topic for historians and scientists alike. As such it is difficult to draw firm conclusions on just where present climate is heading. For example, history speaks of a Roman optimal period followed by the dark ages when climate may have been less conducive to thriving populations. There is agricultural evidence for wineries in England that eventually failed due to the decreasing temperatures of the Little Ice Age, which even impacted George Washington's crossing of the Delaware River during the American Revolution, and there is also evidence of barley production in Iceland under warmer conditions that cannot be replicated today, plus the known rise and fall of indigenous populations throughout the Americas. Historical accounts do not always agree with some of the climate proxy records.

Nonetheless, the proxy records, when looked upon in composite, do provide some range of sea level variations that are germane as shown in Fig. 6.11. The details for each of the panels shown are available in the Intergovernmental Panel on Climate Change (IPCC) source listed. Of the five illustrations provided in Fig. 6.11c, d are noncontroversial because they are from actual tide gauge records, i.e., they stem from actual observations. Panels b and e covering 1700 to 2010 are derived from proxy records, and thus, their bounds of uncertainty are relatively large. These suggest a period of nearly constant sea level from 1700 to about 1850 followed by a continual rise of about 0.3 m in about 150 years, which seems consistent with the actual sea level records from 1880 onward. So far so good; the proxies make sense. It is the period dating back 3000 years that offers considerable confusion. Based on the composite of all estimates, one could conclude that there was no appreciable sea level rise during this interval. However, if one wants to give credence to any of the groupings, one could surmise that sea level may have varied either up or down during several intervals by amounts comparable to what is seen in the other panels. If history speaks of periods of warming and cooling over the course of civilized human endeavor, then for the same reasons that sea level may be varying today (thermal expansion, ice formation, land storage), it may have also varied back then.

More recent records based on actual tide gauges and other means for recording sea level are more robust, and the rates of global sea level rise gleaned from these are in agreement with estimates of ocean mass and volume. Figure 6.12 shows these as global averages, the left-hand panel as determined from globally distributed tide gauges dating back to 1900 and the right-hand panel from satellite-borne altimeters dating back to 1993. Thus, over the modern era of actual recorded observations, sea level appears to be increasing at a rate of about 2 mm/year, with the more recent satellite altimeters suggesting an acceleration to about 3 mm/year. This larger value is equivalent to 0.3 m, or 1 ft, in 100 years.

Along with satellite altimeters that accurately track the sea surface height to within a few cm, there are also satellite sensors that can determine area-averaged mass by how this affects the local gravitational potential. It is in this way that the masses of the Antarctic and Greenland ice sheets have been monitored since 2002 (Fig. 6.13). When combined, the loss of ice mass for Greenland and the Antarctica is 411 GT per year. Noting that 1 GT = 1012 kgm, and using a water density of 103 kgmm^{-3} and an ocean surface area of 361 × 106 km^2, this ice loss amounts to a globally averaged sea level rise of 1.14 mm per year. With the year 2100 being 76 years off from now, this rate of rise yields 87 mm, or 3.4 inches by 2100. While by itself, this is not an alarming figure, it is much less than some recent reports might suggest. It is possible that the losses from Greenland and the Antarctic will accelerate, but that remains to be seen, and the total potential for sea level rise from Greenland and the Antarctic (if all of the ice were to eventually melt) is ~7 m and ~58 m, respectively. With that in mind, we know that sea level has exceeded its present level in the past. For instance, during the last interglacial period, some 130 thousand years ago, sea level was about 6 m higher than at present without any help from humans burning fossil fuels. Granted, Milankovitch cycle insolation was estimated as being higher back then (Fig. 6.10), but other factors may also have come into play.

Consideration of other glacial meltwater in addition to the Greenland and the Antarctic ice sheets, plus the addition to sea level rise by the thermal expansion of seawater based on observed ocean temperature changes, when combined can account very well for the observed sea level rise as evident from Fig. 6.14. The scientific community does indeed have the ability to apportion rates of rise of sea level to its contributing factors. What remains less well defined is how to apportion the changes in each of these factors between natural and anthropogenic (human induced) causes.

Efforts are underway to estimate the future of sea level. Most notably, an Intergovernmental Panel on Climate Change (IPCC) was established in 1988 by the

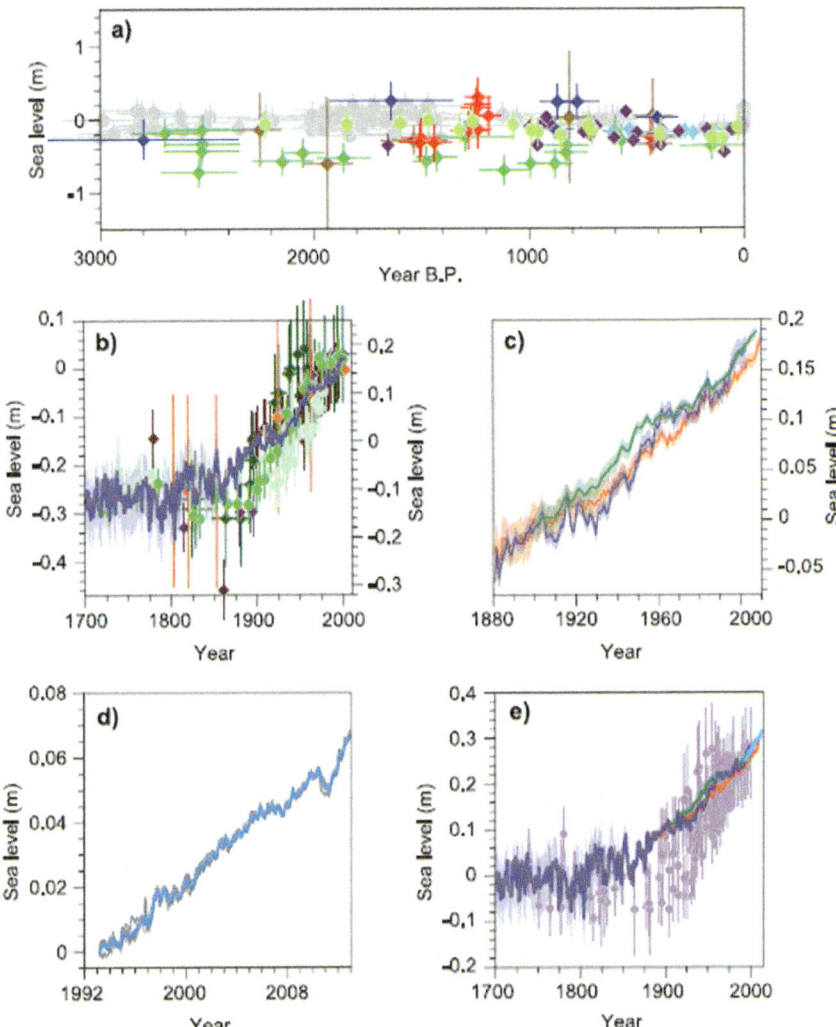

Fig. 6.11 Compilation of various proxy sea level records from 3000 years ago through 2010 excerpted from sea level change (Church, J. A., P. U. Clark, A. Cazenave, J. M. Gregory, S. Jevrejeva, A. Levermann, M. A. Merrifield, G. A. Milne, R. S. Nerem, P. D. Nunn, A. J. Payne, W. T. Pfeffer, D. Stammer and A. S. Unnikrishnan, 2013: Sea Level Change. In: Climate Change 2013: The Physical Science Basis. Contribution of Working Group I to the Fifth Assessment Report of the Intergovernmental Panel on Climate Change [Stocker, T. F., D. Qin, G.-K. Plattner, M. Tignor, S. K. Allen, J. Boschung, A. Nauels, Y. Xia, V. Bex and P. M. Midgley (eds.)]. Cambridge University Press, Cambridge, UK, and New York, NY, USA, pp. 1137–1216. https://doi.org/10.1017/CBO978 1107415324.026)

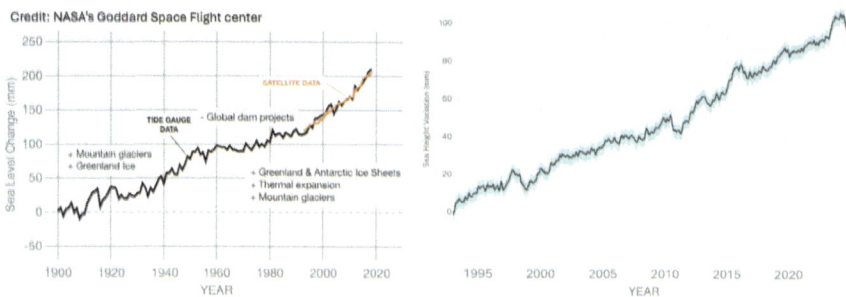

Fig. 6.12 Modern era global sea level rise estimates made from globally distributed tide gauges from 1900 through the present (left-hand panel) and satellite-borne altimeters from 1993 through the present (right-hand panel) (both panels are courtesy of NASA Goddard Space Flight Center, https://climate.nasa.gov)

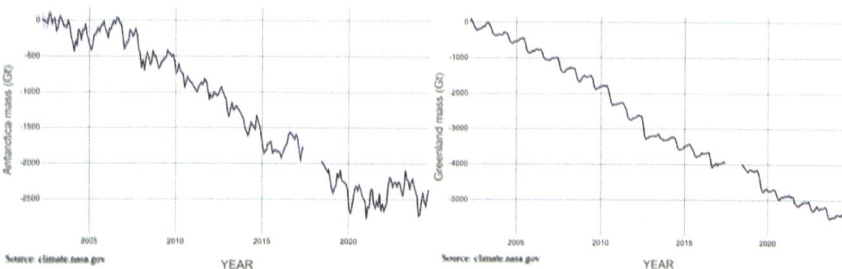

Fig. 6.13 Mass of Antarctic (left-hand panel) and Greenland (right-hand panel) ice sheets, as observed since 2002 by the NASA GRACE mission (courtesy of NASA, https://climate.nasa.gov). The average rates of loss from these two ice sheets are estimated to be 142 and 269 billion metric tons per year, respectively (where a metric ton is about 1.1 tons, or 2204.62 lbs)

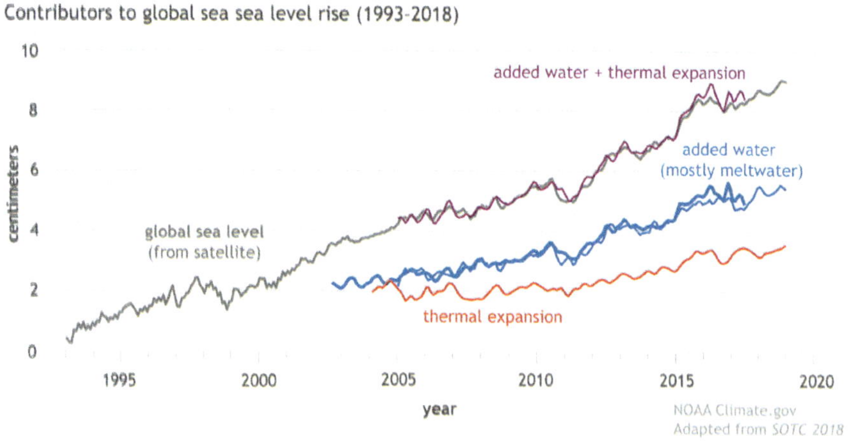

Fig. 6.14 Demonstration of how well the addition of meltwater (primarily from Greenland and the Antarctic, when combined with thermal expansion, can account for the rate of rise of globally averaged sea level (courtesy of NOAA Climate.gov, https://www.climate.gov/media/12868)

Contribution	Scenario	2100	2200	2300	2400	2500
Thermal expansion	Low	0.07 to 0.31 m	0.08 to 0.41 m	0.08 to 0.47 m	0.09 to 0.52 m	0.09 to 0.57 m
Glaciers	Low	0.15 to 0.18 m	0.19 to 0.23 m	0.22 to 0.26 m	0.22 to 0.26 m[b]	0.22 to 0.26 m[b]
Greenland ice sheet	Low	0.05 m[a]	0.10 m[a]	0.15 m[a]	0.21 m[a]	0.26 m[a]
Antarctic ice sheet	Low	−0.01 m[a]	−0.02 m[a]	−0.03 m[a]	−0.05 m[a]	−0.07 m[a]
Total	Low	0.26 to 0.53 m	0.35 to 0.72 m	0.41 to 0.85 m	0.46 to 0.94 m	0.50 to 1.02 m
Thermal expansion	Medium	0.09 to 0.39 m	0.17 to 0.62 m	0.20 to 0.81 m	0.22 to 0.98 m	0.24 to 1.13 m
Glaciers	Medium	0.15 to 0.19 m	0.21 to 0.25 m	0.25 to 0.29 m	0.25 to 0.29 m[b]	0.25 to 0.29 m[b]
Greenland ice sheet	Medium	0.02 to 0.09 m	0.05 to 0.24 m	0.08 to 0.44 m	0.11 to 0.65 m	0.14 to 0.91 m
Antarctic ice sheet	Medium	−0.07 to −0.01 m	−0.17 to −0.02 m	−0.25 to −0.03 m	−0.36 to −0.02 m	−0.45 to −0.01 m
Total	Medium	0.19 to 0.66 m	0.26 to 1.09 m	0.27 to 1.51 m	0.21 to 1.90 m	0.18 to 2.32 m
Thermal expansion	High	0.08 to 0.55 m	0.23 to 1.20 m	0.29 to 1.81 m	0.33 to 2.32 m	0.37 to 2.77 m
Glaciers	High	0.17 to 0.19 m	0.25 to 0.32 m	0.30 to 0.40 m	0.30 to 0.40 m[b]	0.30 to 0.40 m[b]
Greenland ice sheet	High	0.02 to 0.09 m	0.13 to 0.50 m	0.31 to 1.19 m	0.51 to 1.94 m	0.73 to 2.57 m
Antarctic ice sheet	High	−0.07 to −0.00 m	−0.04 to 0.01 m	0.02 to 0.19 m	0.06 to 0.51 m	0.11 to 0.88 m
Total	High	0.21 to 0.83 m	0.58 to 2.03 m	0.92 to 3.59 m	1.20 to 5.17 m	1.51 to 6.63 m

Notes:
[a] The value is based on one simulation only.
[b] Owing to lack of available simulations the same interval used as for the year 2300.

Fig. 6.15 Estimates of the contributions to sea level rise in meters by thermal expansion, glacial melt, Greenland ice sheet melt and Antarctic ice sheet melt under low, medium and high scenario probabilities as defined by the IPCC, 5th assessment report (Chap. 13, Table 6.1). Projections are given in 100-year increments out to 2500 (excerpted from: Church, J. A., P. U. Clark, A. Cazenave, J. M. Gregory, S. Jevrejeva, A. Levermann, M. A. Merrifield, G. A. Milne, R. S. Nerem, P. D. Nunn, A. J. Payne, W. T. Pfeffer, D. Stammer and A. S. Unnikrishnan, 2013: Sea Level Change. In: Climate Change 2013: The Physical Science Basis. Contribution of Working Group I to the Fifth Assessment Report of the Intergovernmental Panel on Climate Change [Stocker, T. F., D. Qin, G.-K. Plattner, M. Tignor, S. K. Allen, J. Boschung, A. Nauels, Y. Xia, V. Bex and P. M. Midgley (eds.)]. Cambridge University Press, Cambridge, UK, and New York, NY, USA, pp. 1137–1216. https://doi.org/10.1017/CBO9781107415324.026)

World Meteorological Organization (WMO) and the United Nations Environment Programme (UNEP), and study reports by this body have been published ever since, the most recent one being the Sixth Assessment Report published in 2022. Pooling scientific expertise worldwide, these reports describe the present state of knowledge regarding ocean, atmosphere, cryosphere and land processes as pertaining to the Earth's climate and provide quantitative estimates on the various factors leading to globally averages sea level rise, and based on different assumptions offer projections on future sea level over the next few centuries. Figure 6.15 provides some results. Provided are estimates of sea level rise owing to the thermal expansion of seawater, the melting of continental glaciers and the reductions in the Greenland and the Antarctic ice sheets. Estimates are given based on the consensus regarding low, medium and high probability of occurrence. For our purposes, the projections out to 2100 are the most pertinent. In all scenarios, the thermal expansion of seawater is the largest contributor, followed by land-based glacial melt and lastly the continued melting of the Greenland and Antarctic ice sheets. Under the high scenario, the amount of sea level rise is from 0.21 to 0.83 m, or from 0.7 to 2.7 ft.

There are also other estimates of sea level rise simply based on preset rates of rise and how these may be increasing. For instance at 3 mm/yr, we would expect about 1 ft in 100 years or about 0.7 ft by 2100. Given what now looks to be an acceleration in this rate of rise, estimates of twice this amount are available, consistent with the range of Fig. 6.15. Sea level rise is certainly something for society to be concerned about and to plan for, although some of the gloom and doom prophesies advanced by some media outlets may be overly pessimistic.

Curiously, despite fears of sea level rise and the unequivocal observations that this is indeed occurring, there is no perceptible slow down in the tendency for people to populate the shoreline. In other words, real estate development is not in synchrony with the potential outcomes of global warming. A common sense approach would be to curtail new development in low-elevation regions and, instead, begin to plan for a slow, orderly retreat from such shoreline areas. Given the numbers discussed above and both human lifetimes and the useful lifetimes of most buildings, such an approach would seem to be a feasible one, if there is a will to do so.

Addendum: A Simple Bouillabaisse

Continuing with a seafood theme, let's turn to a fish stew. Julia Childs provides a more authentic recipe, but its preparation is daunting. Here is a simple version that, while maybe not as good, is still quite satisfying.

The type of fish and other seafood to use is up to you. I usually include three different fish, clams, mussels and scallops with fennel, cherry tomatoes, potatoes, onions, garlic and saffron along with stock of some form. I also shop at the freezer compartment where vacuum-packed frozen fish are available such as mahi, grouper and snapper. There one can also find bay scallops and New Zealand green mussels. For clams, I tend to use a can of diced clams because it also provides clam juice for the broth to which I add a bottle of clam juice.

Dice a large onion and several cloves of garlic, plus about half of a fennel bulb cut into about ¾ inch pieces and begin sweating these in some olive oil with salt, black pepper and a pinch of saffron added. Once soft, add the contents of a can of clams with the liquid, plus the bottle of clam juice. For stock, fish stock is best, although you may be able to find a can of lobster stock, or just use a second bottle of clam juice, plus a little dry white wine. Then add a dozen cherry tomatoes, and the fish cut up into ¾-inch cubes. Cutting up the fish while frozen works well. Also, frozen fish comes in individual packets of ~¼ lb each making this convenient for determining how much to use. For potatoes, I like to use canned ones because they are already cooked and just require a short amount of time to heat up. Cut them in half so that they are about the same size as the fish. Cover the pot with a lid and let cook until the fish thaws, but is not quite done. Finally, add the remaining shellfish (scallops and mussels). As soon as these last ingredients are cooked (it does not take long), then the entire dish is done, and as you can imagine this simple bouillabaisse takes little time from start to finish.

Serve in a bowl with garlic bread on the side. To make the garlic bread, cut a baguette in half lengthwise and put under the broiler to lightly toast. Then take a peeled garlic clove and rub this over the toasted baguette. It may take a few cloves to do the entire baguette. Then drizzle olive oil over the baguette and place back under the broiler (or not) for a very short time.

This is an entire meal so all that is needed is a white burgundy or a chardonnay, or any other dry, mineral-driven, acidic white wine that you may like.

Chapter 7
Waves of All Sizes

In the most general sense, waves are nature's way of transmitting energy over large distances while leaving the medium through which the waves travel undisturbed after their passage. Light and sound, as perceived by humans, provide examples, the light being transmitted as the interplay between electric and magnetic fields and sound being transmitted through pressure variations. Visible light is part of the electromagnetic wave spectrum, whereas sound is a mechanical wave. Waves are repetitive in space and time. The repeat interval in time is the wave period (T), and the repeat interval in space is the wavelength (L). Thus, an observer fixed in space would see the waveform repeat itself over and over again at intervals of time T, whereas an observer at a fixed time would look upon the waveform and see a repeat interval at a length L. The simplest waveform representation is a sine wave (Fig. 7.1), that which a point at the tip of a circle's radius would trace if it is rolled along a straight line. The amplitude of the sine wave is the radius of the circle and the wavelength is the circle's circumference.

The peak positive point of the wave is called the wave crest and the peak negative point is called the wave trough. If the wave were to move a distance L in a time T, it would have a wave speed, or celerity, (C) equal to the ratio between the wavelength and the wave period ($C = L/T$). For electromagnetic waves, the interactions between the electric and magnetic fields, each traveling at the speed of light, are governed by Maxwell's equations. Unlike mechanical waves, electromagnetic waves can propagate through the vacuum of space.

For mechanical waves, the relationship between L and T is dependent upon the medium through which the wave may travel, and this relationship is referred to as the dispersion relationship. If dependent only upon the properties of the medium, then the dispersion relationship is intrinsic to the medium and hence is called an intrinsic dispersion relation, whereas if geometric constraints exist, such as boundaries, then the dispersion relationship may become modified by the specific geometry in which the waves may propagate.

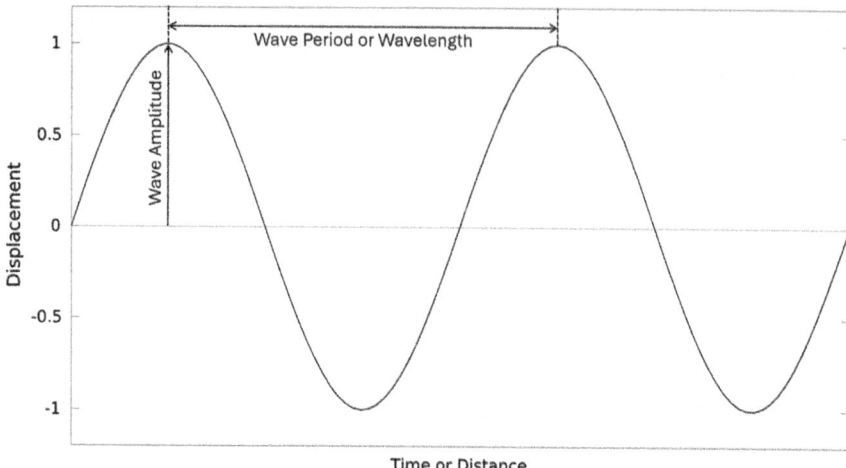

Fig. 7.1 A Sine wave

The energy that the Earth receives from the sun is in the form of electromagnetic energy, and human ingenuity over the past two centuries led to the use of such wave energy for almost all human endeavors. In contrast, for aspects other than heat exchange, the workings of the coupled ocean-atmosphere system are largely influenced by mechanical waves, whose propagation is dependent upon the medium in which they exist, plus the geometry through which they may travel.

A broad spectrum of different mechanical waves, each with their own governing physical balances, is found in both the ocean and the atmosphere. Sound, for instance, propagates by the alternating compression and expansion of solid, liquid or gaseous materials and their differing molecular stiffness and density determines the propagation speed; hence, sound speed slows when going from solids to liquids to gases. A ubiquitous observation of waves in the ocean is the surface gravity waves generated primarily by winds blowing over the ocean surface. Their presence at any location is either by propagation from afar (referred to as ocean swell that may have been generated 100s to 1000s of km from where they are observed), or by local wind generation (referred to as ocean seas). Typical surface gravity wave periods are of several seconds duration, and, being that this is much smaller than the rotational period for the Earth, the Coriolis effect upon surface gravity waves is insignificant.

Similar gravity waves may also occur beneath the surface, generally unseen by human eyes. These are referred to as internal gravity waves, whose repeat intervals can be much longer than those at the surface. At their high frequency (short wave period) limit, the restoring force for internal gravity waves is gravity, but operating on the much smaller density differences existing across the water column (due to variations in temperature, salinity and pressure) than the much larger density difference between air and water as occurs for surface gravity waves. As the internal gravity wave period increases to the point where the rotation of the Earth becomes a factor, these are then referred to as internal inertial-gravity waves, and their orientation (i.e., their direction of wave propagation) rotates to become ever more vertical as their period approaches the Coriolis period, $2\pi/f$. Thus, internal inertial-gravity wave time scales can exist between what is termed the buoyancy period, where these waves propagate horizontally, to the Coriolis period, where these waves propagate vertically. Internal inertial-gravity waves are one mechanism by which mechanical energy input at the surface by winds can penetrate down throughout the water column.

With further increases in wave period and wavelength, to the point where the variation of the Coriolis parameter with latitude becomes a factor, another class of waves called planetary waves (or Rossby waves, named after Carl Gustaf Rossby who first identified these waves in the atmosphere) may be defined. Additionally, the latitudinal changes in angular momentum that define Rossby waves are analogous to angular momentum variations that occur due to changes in bottom topography. Such topographic changes in angular momentum give rise to topographic Rossby waves. There are also other names for waves in either the oceans or the atmosphere depending upon the geometry within which they may travel.

For all of these waves, a dispersion relation follows from the either the force or the angular momentum balances that govern their respective fluid motions. As the simplest example, consider surface gravity waves in deep water observed at the interface between the ocean and the atmosphere. As the waveform is observed to go up and down, the actual fluid particles transcribe circular motions. At the waveform peak, the restoring force by gravity directly opposes the centrifugal force due to the circular water particle motion. Equating these two forces, and recognizing that the fluid particle transcribes a circular loop of circumference $2\pi R$ within a duration of one wave period T and that the circumference also equals the wavelength L, we may arrive at $T = (2\pi L/g)^{1/2}$, and hence $C = L/T = gT/2\pi$. Other waves have more complex force balances so their dispersion relationships do not follow as easily. Nonetheless, the formalism is similar. First, define the balances based on either force or angular momentum considerations to arrive at a governing mathematical relationship, and then seek a wave-like (sinusoidal) solution with periodic behavior in time and space to arrive at a dispersion relationship. The physics defines the force of angular momentum balance, and the mathematics is then applied to specify the waveform and the resultant dispersion relationship.

For surface gravity waves in deep water, the radii of the water particle's circular motions decrease very rapidly (exponentially) with depth such that they are small below a depth of about one-half the wavelength, and they, in essence, become zero below a depth equal to a wavelength. However, as every beach-goer knows, transformations occur once waves propagate into shallow water. As the water particle motions begin to feel the bottom, they can no longer be circular, and in the shallow water limit of water depth of about 1/20th of the wavelength, the particle motions become both horizontal and uniform with depth. Such waves are referred to as shallow water waves, and their dispersion relationship during the transition and at the shallow water limit must contend with the water depth as an added property of the medium through which the waves propagate. In the shallow water limit, the dispersion relationship transitions to $C = (gH)^{1/2}$ where H is the water depth; hence $L = T(gH)^{1/2}$, demonstrating that as the wave speed decreases with decreasing depth, the wavelength must also decrease. To conserve the total wave energy flux (the transmission of wave energy per unit of area and time), the wave amplitude must increase. This increase in amplitude and decrease in wavelength causes the wave to steepen and eventually break, which is one reason why we have surf to enjoy playing in.

The transformations that waves make while propagating through a slowly varying medium give rise to some other important behaviors. Along with wavelength and amplitude changes to conserve energy, once generated and away from the generation region, waves cannot begin or end in the fluid medium. This results in the properties of diffraction, refraction and reflection. Diffraction is the longitudinal (along the wave crest) spreading of wave energy as waves propagate either away from their source and/or into wider regions. As an example, consider what happens as waves propagate past the end of a breakwater. The region on the lee side of the breakwater without wave energy will have to be filled in by some of the wave energy passing by the end of the breakwater, requiring a spread of wave energy to fill that void. With wave energy spreading along the crest, the path in which the wave energy is transmitted, or the wave ray, bends to accommodate this need for wave energy to fill a void. Similarly, light passing through a diffraction grating (e.g., pin holes in an otherwise solid barrier) will spread to illuminate the entire region behind the barrier. Diffraction is therefore the spread of wave energy along the wave crest and the bending of the wave crest and hence the direction of the wave rays to fill what would otherwise be a void. It is for this reason that the spreading of solar energy with distance from the sun follows an inverse square law because instead of being spread along a line, the solar radiation is spread along an ever-increasing spherical surface area. That is why the solar constant for the Earth differs from that of the other planets.

Refraction also entails a bending of the wave rays, but for a different reason. Let's say that a wave train enters a region of varying depth (or anything else that would alter the wave speed). The portion of the wave crest with faster speed will begin to pull away from that with slower speed. But since the wave crests cannot begin or end in the medium, the wave train will have to bend in order to keep the wave crest continuous. As an example, consider surface gravity waves propagating toward a curved beach, and recall that the wave speed slows with decreasing depth. If the beach curve is convex, then the waves will converge upon the beach promontory, whereas if the beach curve is concave, then the waves will diverge from the beach recess. It is for this reason that mature beaches tend to be straight because, were they not, the higher wave energy region at the promontory would erode the sand there to be deposited at the beach recess where the wave energy is less.

Reflection is a simpler concept. If a wave hits a rigid barrier, then all of the wave energy must reflect in order for energy to be conserved. When driving across a bridge causeway, one may notice that the water looks more choppy close to the causeway than farther off in the distance. This is owing to wave reflection.

Reflection gives rise to another related concept, that of the difference between standing and progressive waves. Progressive waves are what we generally see arriving at a beach. The water particle speeds are largest at the wave's crest and trough, being in the direction of wave propagation at the crest and in the opposite direction at the trough. Upon perfect reflection, the amount of wave energy traveling in either direction is zero (because the energy fluxes into and out of the reflecting barrier are the same) resulting in a standing wave. For the case of surface gravity waves, the particle speeds are now zero at the crest and trough whereas they are largest at the mean water level thereby facilitating the flow of water between the crest and trough. Everyday examples of this phenomenon are tides in most estuaries. While not exactly standing waves because the reflections are not perfect and there are also frictional loses, tides in estuaries and embayments tend to be nearly standing waves, with slack water occurring near high and low tide and maximum flood and ebb occurring at the mid-range of the tide. Some embayments, because of their geometry, may also have a natural period at which they tend to oscillate that coincides with the period at which the tides are forced. In such a case, a resonant behavior occurs that amplifies the range of the tide. The Bay of Fundy provides such an example, where the tidal range and certain areas approach about 50 ft.

Waveguides provide examples wherein reflections can steer waves in a certain direction. Organ pipes are examples of waveguides for sound waves. With each pipe having a different dimension, resonant sound waves may be achieved over a range of desired frequencies. Sound waves in the ocean may also be wave guided. Being that sound speed depends on temperature, salinity and pressure, a sound speed minimum tends to occur at the base of the thermocline, resulting in the convergence of sound energy there due to refraction both above and below. This allows sound to travel great distances in the ocean facilitating communication between species such as whales as well as providing a basis for detecting underwater vehicles such as submarines. This is one reason why countries with naval forces engage in oceanographic research. Along with climate, ecology and maritime commerce justifications, there also exists a national defense imperative for the study of oceanography.

For most of us, our experience with waves tends to be limited to sound, light and surface gravity waves. As may now be evident, however, there are multitude of waveforms that affect our lives on Earth. Atmosphere teleconnections, and hence weather, are regularly affected by large-scale planetary (Rossby) waves and smaller-scale (topographic) lee waves associated with winds passing over mountain ranges. Wherever fluid (air and the moisture contained therein) converges in the atmosphere, we tend to get rain and, with a sufficient release of latent heat, these rainfalls may result in major storms. Storms often result from planetary waves becoming unstable and growing to the point where they must release some of the energy contained within them. Instability gives rise to vertical motions that result in condensation, rainfall and latent heat release. Weather reporters typically speak about troughs and high pressure ridges and how these may be moving to provide future weather forecasts. Such movements are associated with planetary waves.

Large-scale waves also move energy throughout the oceans, and topographic features, from beach slopes to continental shelf slopes provide waveguides. Surface gravity waves refracting and reflecting from beaches can result in larger scale gravity waves that are trapped to the vicinity of the shorefront while propagating (being waveguided) alongshore. These are referred to as edge waves. As the edge wave period increases to the point where the Coriolis force becomes important, then edge waves acquire angular momentum that may be altered by virtue of water depth variations. This phenomenon results in what are termed continental shelf waves. These can take on several different forms depending on the bottom slope. Even a vertical wall can support large-scale waveforms when the Earth's rotation becomes a factor, and these take the form of coastal Kelvin waves and Sverdrup waves. Tides propagating along the coast may be described as Sverdrup waves.

Each of these different waveforms may be described mathematically by using what are known as special functions of physics. The concept is referred to as generalized Fourier series, wherein different media geometries and their properties require different mathematical forms to fully account for what may exist within these specific constraints. As an example, variations in time and space that occur within a rectangular geometry with a spatially uniform medium can be expressed as the sum of sines and cosines. Other geometries and medium variations require other functional forms. Applied mathematics is the domain that provides the tools for solving important geophysical problems, thereby gaining insight and describing how the Earth actually works. As previously stated, a firm grounding in mathematics is therefore a fundamental requirement for anyone considering a career in science.

One evolving oceanographic issue is how information is communicated between high northern and high southern latitudes, and conversely. For instance, recall the discussions on the formation of North Atlantic Deep Water and other deep-water bodies in Chap. 3. Given the angular momentum constraints on water traveling between latitudes due to the Earth's rotation and its spherical shape, how do these deep-water masses move between hemispheres? The Sverdrup relation accounts for the gyre scale circulation, but what about these deeper waters?

Waveguiding is one way of setting up a pressure field distribution that can support geostrophic currents whose angular momentum dissipation can facilitate transport like it does near the surface. Planetary waves along sloping topography propagate anticlockwise in the northern hemisphere and clockwise in the southern hemispheres so through a complex series of events a deep-water pressure field can effect southward flows from high northern latitudes along the western boundary of the northern hemisphere that may continue as southward flows along the eastern boundary of the southern hemisphere, and similarly for northward flows from high southern latitudes. With anticlockwise propagation in one hemisphere and clockwise in the other hemisphere, how might these hemispheres actually connect? The answer is that the equator itself serves as a waveguide.

This is a good point to digress upon how waves helped to shape my career. I suppose that it began with playing in the surf during my summers at Rockaway Beach, N.Y. While surfboards were not common at the time, body surfing and raft surfing were among the daily activities. A love of the ocean was instilled in me from an early age. My first cruise as a graduate student also involved waves, but these were not as much fun. What was intended to be a five-day cruise from Narragansett, R.I. to Bermuda, just prior to final exams and winter break, turned out to be an adventure. About a day out of port, a storm evolved that turned into a gale with seas exceeding

30 ft. The precursor waves were fun at first. Traveling slower than the waves we were able to go out on deck prior to the storm approaching and enjoy watching the swell propagate past the ship. I recall being on the aft lower deck in awe of the approaching waves. With the stern in the trough, it was as if one could reach out and touch a wall of water about to hit the ship. I was with another student marveling at such sights when one of these walls decided to partially break, inundating the deck. Fortunately, there was a stanchion that I was able to straddle before being hit. I thought that my classmate was lost, but he had sense enough to go inside prior to this happening. With the winds and waves then increasing, we spent the next four days tacking in the trough, simply to stay afloat, rather than making any headway to Bermuda. The seas were too large to go downwind for fear of broaching, and when heading too closely into them, the bow would submerge causing the ship to shutter before the bow would reemerge. Thus, several days were spent with everything lashed down, with no one permitted on deck and with little sleep as the ship incessantly rolled, with one roll recorded at 58°. There was obviously no studying for finals exams, and instead of having a few days to enjoy Bermuda, I had but a few hours to shop for a gift and get to the airport. This wave experience had me swear off ever going to sea in the North Atlantic during winter again. Subsequently, after completing my Ph.D. dissertation on estuary circulation, I became an equatorial oceanographer, at least initially.

The Global Atmosphere Research Program (GARP), Atlantic Tropical Experiment (GATE) was essentially a meteorological endeavor to which a few oceanographers decided to tag along. My major professor at that time, John Knauss, an established equatorial oceanographer with seminal works on the Equatorial Undercurrent and the North Equatorial Countercurrent of the Pacific Ocean, encouraged another graduate student and me to participate. With John's advice, we crafted a proposal to the National Science Foundation and were successful in being granted GATE Program funding. My portion of the project was to deploy a set of fixed moorings with instruments recording the currents on and about the equator, and after considerable planning and preparations, the field effort commenced in summer, 1974.

A new theory specifying the existence of equatorially trapped waves was published eight years prior to this, and some attempts to observe these using tide gauge records were undertaken in the Pacific Ocean. There were also a few limited velocity records obtained in the Pacific Ocean, but insufficient to be of a definitive nature. So GATE provided an exciting opportunity to plan for a set of observations that might actually observe and describe these waves. My approach was to deploy a set of subsurface moorings with instruments recording water velocity at a few discrete depths. The array consisted of four such moorings with two along the equator and another two straddling the equator at latitudes of 1.5° north and south. Equatorial currents tend to be both strong and varying with depth with a South Equatorial Current (SEC) flowing westward at the surface atop an Equatorial Undercurrent (EUC) flowing eastward beneath the SEC and extending down to about 150–200 m depth. Mooring technology at that time precluded deep ocean observations in high-speed regimes so I opted to aim for a float depth of 200 m in a total water depth of about 4000 m. Taking great care both to measure the rope to be used and to test it for both stretch and creep under load (i.e., the tension to be applied by the buoyancy of the subsurface float), I set out to deploy these moorings as chief scientist on a research cruise from San Miguel, Azores to Recife, Brazil.

Mooring design and instrumentation preparations are but one aspect of deep-sea deployments. Knowing the water depth and how it varies over a several nautical mile region is required because to deploy such a mooring you must first string out the equipment along the surface over about a three-mile distance before dropping the anchor and hoping that it lands at the correct depth. This entails a bottom survey using an acoustic fathometer to first find a suitable location and then to map the bottom. This was challenging because the locations were along the mid-Atlantic Ridge with its undersea mountains and valleys so finding a large enough flat bottom area using very limited satellite navigation was difficult, especially prior to GPS. Using an experimental, Transit Satellite Navigation System with a very limited satellite constellation that offered fixes only at several hour intervals, dead reckoning and luck were also in play. Nonetheless, suitable locations were found and mapped, and in the end, and almost miraculously, we got to within 20 m of the desired float depth, and some two months later, when these moorings were recovered, they yielded the first set of equatorial velocity time series simultaneously both on and across the equator.

Upon analyses, these data demonstrated the polarization and symmetry properties as predicted by equatorial wave theory, and they also showed a convergence of fluid upon the equator with the thermocline, as required by existing Equatorial Undercurrent theory. This accomplishment established me as being worthy of future National Science Foundation support, which allowed me to continue my independent work as a research scientist. It also introduced me to an array of colleagues with whom I continued to interact with throughout my career.

Thus, my first sea-going surface gravity wave encounter led me away from the North Atlantic to the equator where, for nearly three decades, the bulk of my studies were pursued, both in the Atlantic and Pacific Oceans, looking at the large-scale, planetary waves that exist sight unseen beneath the surface. The GATE Program success led in my first stand-alone NSF proposal in further pursuit of equatorial trapped waves, but a little farther east in the equatorial Atlantic's Gulf of Guinea. This work included a collaboration with French scientists stationed in Abidjan, Ivory Coast, whom I met during GATE. Again, using an array of subsurface moorings, we unequivocally identified a packet of Equatorially Trapped, Rossby-gravity waves that were generated near the surface farther west by instabilities observed between the EUC and the SEC. Such instabilities, also seen during GATE, are the ocean's equivalent of hurricanes in that they occur when the SEC and EUC accelerate in response to the seasonal increase in the southeast trade winds, as occurs each boreal (northern hemisphere) spring season. As these currents accelerate, their kinetic energy increases along with the internal (heat) and potential energy that accumulates within a convergent ridge that forms between the SEC and the adjacent North Equatorial Countercurrent (NECC). The tropical instability waves (TIWs) act to reduce these energy (kinetic, potential and heat) excesses owing to the wind-driven seasonally varying current systems. Being that the time (period of about 30 days) and space (wavelength of about 1000 km) scales of the TIWs are close to the period and wavelength of the Rossby-gravity waves, some of their energy goes into forcing these waves at depth.

7 Waves of All Sizes

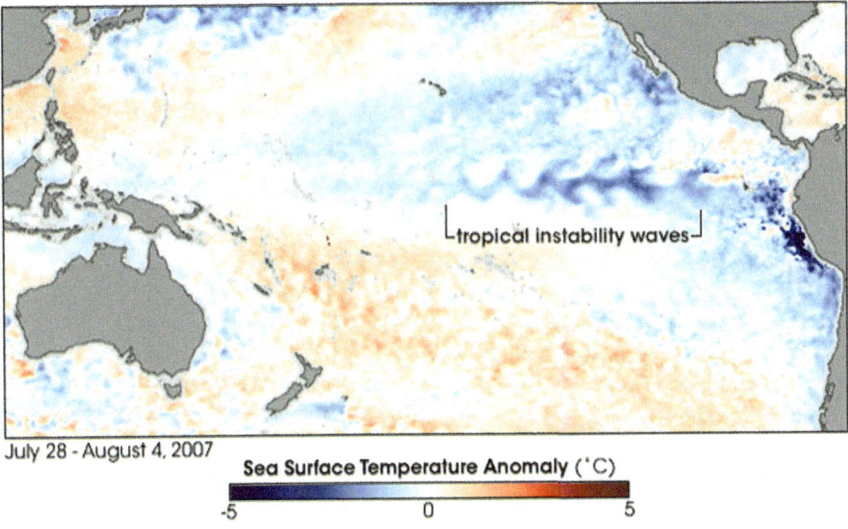

Fig. 7.2 Pacific Ocean SST anomaly image showing a train of tropical instability waves positioned just north of the equator between the South Equatorial Current and the North Equatorial Countercurrent (courtesy of NASA, https://earthobservatory.nasa.gov/images7941/pacific-sea-surface-temperature)

The TIWs are readily observable in both Atlantic and Pacific Ocean satellite sea surface temperature images and an example for the Pacific Ocean is provided by Fig. 7.2. In the Atlantic Ocean, they tend to last for about three months as compared to about nine months in the Pacific Ocean. We learned that the duration difference was due to the different widths of these two oceans. The onset of increasing Trade winds generates several different types of equatorially trapped (Kelvin and Rossby) waves that eventually lead to the adjustment of an east–west oriented pressure gradient force to balance the stress of the wind upon the sea surface. Once this pressure force adjustment occurs, the currents themselves slow to the point where they are no longer unstable, thereby terminating the generation of TIWs. Since the Pacific Ocean is so much wider than the Atlantic Ocean, it takes longer for the Kelvin and Rossby wave adjustments to occur there, which explains the longer duration of the TIWs in the Pacific, than in the Atlantic.

The above TIW behavior details had to await another sea-going expedition that occurred some seven years later. For the Gulf of Guinea expedition that began in 1977, we were more simply concerned with whether or not equatorially trapped waves actually existed. GATE and other work suggested so, but we were still lacking observations of a more definitive nature. Consistent with the scientific method, an experiment was designed to test the existing theory. The design, in essence, was an antenna capable of determining the wave period and wavelength, plus the spatial structure of the velocity field, as predicted by theory. The antenna consisted of velocity recording

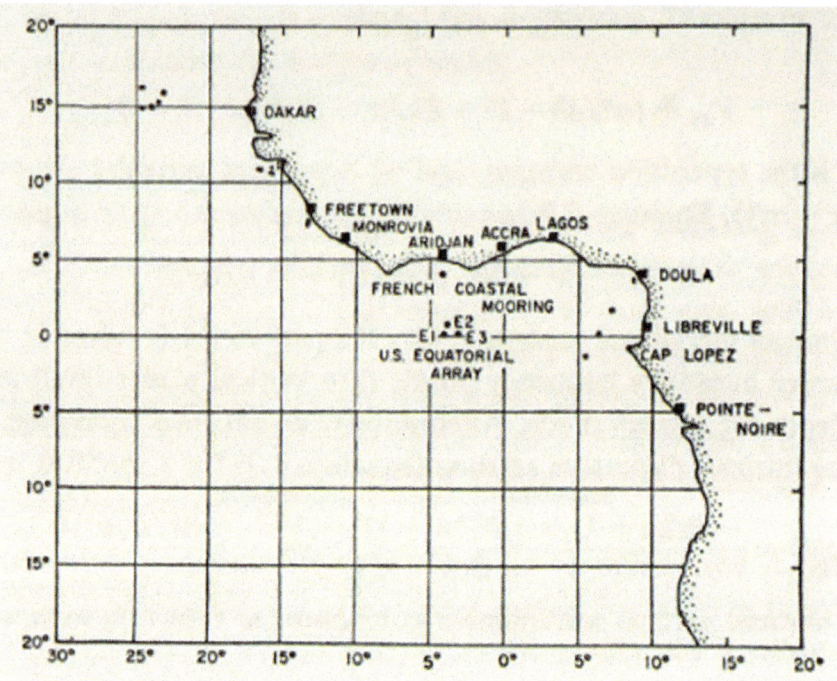

Fig. 7.3 Gulf of Guinea moored array consisting of three subsurface deep-sea mooring with two along the equator and another at 0.5° north of the equator (adapted from Weisberg, R. H., A Horigan, and C. Colin (1979). Equatorially trapped Rossby-gravity wave propagation in the Gulf of Guinea, *Jour. Mar. Res.*, 37, 67–86)

instruments (current meters) positioned vertically on three separate deep-sea moorings, two along the equator to determine both zonal (east–west) and vertical length scales and another positioned to the north of the equator to determine the spatial structure of the possible waveform. More instrumentation would have been nice, but there is always a trade-off between what one would like to deploy, versus what the funds available are capable of supporting.

The mooring array is shown in Fig. 7.3. The duration was January 1977 through June 1978, in which time we mounted three deployment and recovery cruises. Some data losses occurred in the first and third deployments, but we were fortunate to retrieve a complete dataset for the middle deployment, the one that actually coincided with the TIW season.

Examples of the velocity records that were obtained are shown in Fig. 7.4 for the interval July 15, 1977, to January 22, 1978, coinciding with second deployment that yielded a complete dataset. Observed are roughly four cycles of a wave-like phenomenon that transited with antenna array in a very coherent manner. Upon analysis, it was found that the period and wavelength matched the predicted Rossby-gravity wave dispersion relation, and the differences in how the water particles

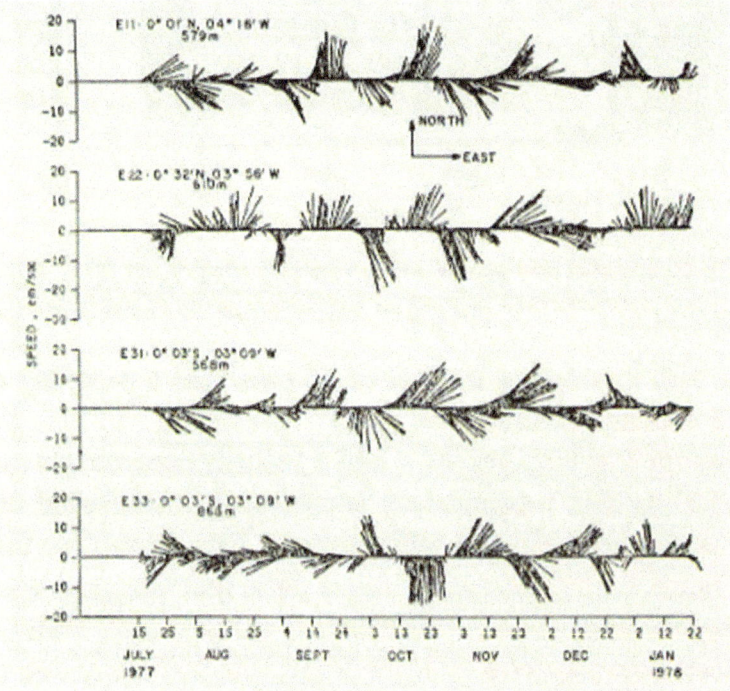

Fig. 7.4 Velocity records from two locations along the equator and another roughly 0.5° north of the equator. Each stick is a velocity vector spaced one day apart. The length is the speed and the orientation is the direction (north being up and east to the right) (adapted from Weisberg, R. H., A Horigan, and C. Colin (1979). Equatorially trapped Rossby-gravity wave propagation in the Gulf of Guinea, Jour. Mar. Res., 37, 67–86)

moved off the equator, versus directly on the equator matched the spatial structure as predicted for a Rossby-gravity wave. The second of these two findings is shown in Fig. 7.5, a planar view of the spatial structure for a Rossby-gravity wave.

The aforementioned fact that a wave propagating away from its source cannot begin or end in the fluid requires that the phase differences (or time delays between locations), when summed around any closed circuit, must be zero, a test that indeed worked for this example, which demonstrated that we actually observed a wave, versus some other phenomenon. Finally, an energetics analysis determined that the observed wave train was consistent with these waves being generated at the surface to the west of the antenna array at the time when TIWs were formed by surface current instabilities. The Rossby-gravity waves at depth were therefore a consequence of the TIWs and served as a mechanism by which the TIW energy could be transmitted away from the source region.

This was arguably the first unequivocal determination of a planetary wave train in the ocean in that the observations withstood tests regarding wave kinematics (description), dynamics (physical force balances) and energetics (source origin). The

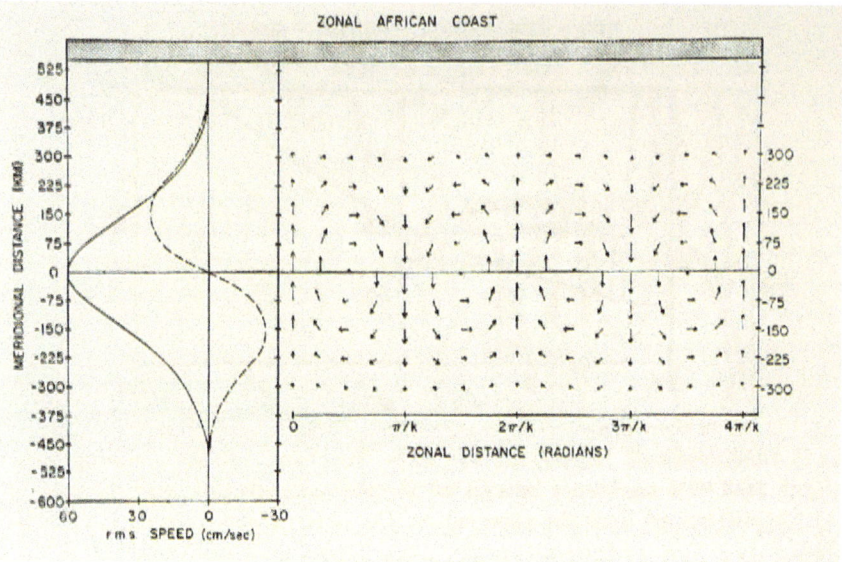

Fig. 7.5 Rossby-gravity wave spatial structure implied from the Gulf of Guinea array observations. The left-hand panel shows the meridional distributions for the north and east components of velocity. The right-hand panel shows a planar view of the horizontal velocity vector distributions over two complete wave cycles (adapted from Weisberg, R. H., A Horigan, and C. Colin (1979). Equatorially trapped Rossby-gravity wave propagation in the Gulf of Guinea, Jour. Mar. Res., 37, 67–86)

scientific method begins with observations, continues with explanations in theory and culminates with experiments and analyses to provide tests of theory. The evolution of physics has always entailed the interplay between theorists and experimentalists. Being a theoretician generally entails a certain level of brilliance that is not innate to all of us. But that does not preclude a career in science. There remains lots of room for smart, dedicated experimentalists. Hypotheses are just that; they require testing, and as an experimental, sea-going, physical oceanographer, the interactions with theoreticians and the actual planning and execution of field experiments and subsequent analyses to help advance the science proved to be quite an enjoyable and stimulating career.

I subsequently led several field campaigns (along with colleagues) in the Atlantic and Pacific Oceans, all entailing how these large-scale equatorially trapped waves and tropical instability waves account for not only the seasonal variations of the equatorial and tropical current systems in response to wind forcing, but also the interannual variations of the coupled ocean-atmosphere system that includes the El Nino—Southern Oscillation (ENSO). This was an exciting period of discovery that may not have happened had I not found myself straddling that stanchion as a wave crashed over the stern of the R/V Trident during my first year as a graduate student.

Addendum: Ahi Poke

I was introduced to this on one of my equatorial Pacific research cruises out of Hawaii. The chefs were from the Philippines, and on their leisure time, they would trawl for fish (it was a several day steam from Hawaii to the equator). Out of the galley on day came this amazingly good appetizer that I was not familiar with, Ahi Poke, a simple mixture of cubed fresh tuna with some soy sauce, scallions, a wee bit of garlic, sesame oil and sesame seeds.

It is as simple as it sounds, but you do need a high-quality fresh piece of sushi-grade tuna (with no white sinew running through it). ½ to ¾ inch cubes are about right. Put these in a bowl and add enough soy sauce to make these appear brown (just enough to coat the fish; you don't want the cubes swimming in soy sauce). Finely slice a few scallions and maybe a small clove or less of garlic and add this to the bowl, plus some sesame oil to dress everything. Sesame seeds lightly toasted work well if you have them and toss all together and let sit for an hour or so in the fridge before serving chilled. A small variation that I like is to add some fresh avocado cut up like the tuna.

This is a wonderful appetizer served on sesame crackers, or make it part of a larger salad bowl with whatever you feel like cutting up.

Chapter 8
Sea Level Extremes by Tsunamis

Recall what happens when a stone is tossed into a lake, as all of us have done—if not, try it. First, there is a depression where the stone falls and an accompanying splash. Then, there are a series of surface gravity waves that radiate out from the center of splash site. The stone imparts both kinetic and potential energy upon the sea surface and the waves provide a way by which this energy is transmitted away from the impact site. At some point in time, the impact site will settle down, and a packet of waves will be seen propagating away in a circular fashion with longer wavelength waves leading the pack. Being that wave energy is spread along the crests in ever-growing circles, the entire event will slowly peter out as these waves get farther from the source. So not only do waves travel at what we term the celerity, or phase speed, i.e., the ratio of wavelength to wave period; they also transmit energy at what is referred to as the group velocity, which is the speed at which the wave packet travels. For deep-water surface gravity waves, the group velocity is ½ the phase speed.

That waves come in packets is readily observed at the beach while playing in the surf. Intervals of larger waves will be seen followed by smaller ones, then larger ones again and so on. This is attributed to waves of differing wavelength adding either constructively, or destructively with one another. That is one reason why waves observed in deep water when offshore boating look rather random: they all travel at different speeds and hence there is little regularity in how they may add constructively or destructively. In contrast with deep-water waves, shallow water waves have equal phase and group speeds so the packet shape does not change other than for the waves getting steeper until they eventually white cap and break. Steepening is a consequence of conserving the number of waves that propagate past an observer, which means that the individual wave periods must not change. With constant wave period and decreasing wave speed $(gH)^{1/2}$ due to decreasing depth H, the wavelength must decrease; hence steepening as the distance between wave crests gets smaller with decreasing water depth. Wave breaking occurs when the steepness reaches the critical value, $2a/L = 1/7$, where a is the wave amplitude and L is the wavelength.

With the wave energy equal to $\frac{1}{2}\rho g a^2$, (where ρ is the water density and g is the acceleration of gravity) and with this energy traveling at the group velocity, the energy flux for shallow water waves (or the energy passing by a given cross-sectional area per unit of time, e.g., energy per m^2 per second), which is also a conserved quantity, equals $(\frac{1}{2}\rho g a^2)(gH)^{1/2}$. From this, we see that as H decreases, the wave amplitude must increase to keep the energy flux a constant.

These concepts may now apply to tsunamis. Analogous to the tossing of a stone in a lake, consider what happens when we have a large displacement of the sea floor, such as by an earthquake, a violent underwater volcanic eruption or a massive landslide into a water body. In each of these scenarios, there occurs a displacement of the sea surface whose energy must subsequently propagate away as a train of surface gravity waves. The distribution of wavelengths (its wave spectrum) and the amplitude of these waves comprising the wave spectrum depend upon the vertical displacement of the sea surface caused by the vertical displacement at depth and the horizontal extent of this disturbance. If the earthquake or volcanic eruption occurs at the sea floor, then the pressure disturbance will extend across the entire water column nearly instantaneously because pressure disturbances travel at the speed of sound (about 1500 ms^{-1} in seawater). The resulting waves will therefore be shallow water waves, even if they are formed in the deepest portions of the ocean.

Anything that displaces water over a sizable horizontal extent will cause a tsunami. The same mathematics (generalized Fourier series) that allows us to represent some functional form within a given medium (in this case water) and geometry may be used for tsunamis such that each different wavelength produced by the disturbance will have its own amplitude. Being that the waves are shallow water waves, once the wavelength is known, then the wave period may be calculated from the dispersion relation $L/T = (gH)^{1/2}$. Empirical evidence from prior tsunamis is that the wave packet emanating from a deep-sea disturbance may have a central period of several minutes to tens of minutes, and since the packet consists of multiple waves, there will be tsunami precursors followed by subsequent waves of varying amplitude.

A photograph from the 2004 Sumatra–Andaman earthquake, which sent tsunamis into Phuket, Thailand (Fig. 8.1), Banda Aceh, Indonesia as well as throughout the entire Indian Ocean provides a demonstration. Seen in this photograph is a precursor wave, after which the most intensive damage occurred by waves estimated to be as high as 12 m. Note that the precursor wave in Fig. 8.1 has already broken before nearing the beach. Wave breaking occurs once a critical wave slope ($2a/L = 1/7$) is exceeded, and this happens at some distance offshore. Why this occurs is as follows. From the conservation of wave energy flux, we see that the square of the wave amplitude times the wave group velocity must remain constant until the wave energy is dissipated by friction. In deep water (4000 m as an example), the group velocity ($c = (gH)^{1/2}$) may be around 200 ms^{-1}, whereas in shallow water (consider 10 m as an example), it may be about 10 ms^{-1}. Therefore, with the square of the amplitude being about 20 times higher, the amplitude is about 4.5 times larger. So a 1 m amplitude wave in deep water would have a 4.5 m amplitude or a 9 m height (trough to crest) at 10 m depth. Theory also demonstrates that the maximum wave height is 0.79 times the water depth, and rarely if ever is this reached. Hence, the tsunami waves

8 Sea Level Extremes by Tsunamis

Fig. 8.1 Photograph capturing the precursor waves of the Sumatra–Andaman earthquake in 2004 arriving at a Phuket Thailand beach on December 26. The initial response was a draw down of water, beckoning tourists to walk out over dry land (courtesy of NOAA, https://www.noaa.gov/jetstream/tsunamis/tsunami-inundation)

generally approach a beach having already broken, and as we know from playing in the surf, the force of a broken wave is not something to toy with. It is further noted that wave theory, based on certain simplifying assumptions, breaks down under extreme conditions. What we normally refer to as surface gravity waves under normal conditions transform to another form of waves referred to as solitary waves with their own, more complex dispersion relationship. So while we can make certain estimations of what tsunami heights may be, these are just estimates. At Banda Aceh, a maximum height of 12 m was estimated and later as these waves propagated west to Sri Lanka the heights estimated there were 7–9 m. These concepts are illustrated in Fig. 8.2. Whereas relatively small amplitude waves with very long wavelengths may form due to a disturbance occurring in the deep ocean, these waves will steepen rapidly as they propagate into shallow water. In essence, as the wavelength shortens, the amplitude must grow in order for the wave energy flux to be conserved.

As should seem obvious to the reader now, the devastation at Banda Aceh Sumatra and other locations hit by this Sumatra–Andaman earthquake tsunami was profound. The USGS was one of the agencies that sent teams to survey the damage. Figure 8.3 provides an example of just how powerful the advancing whitewater of a broken tsunami wave may be. Here we see an entire beach front at the Lho Nga region just to the southwest of Banda Aceh wiped clean of structures and foliage and replaced by a large steel coal barge and a smaller tugboat. The height to which the tsunami made its impact (estimated at 31 m) may be seen on the adjacent hillside and distant island. Figure 8.4 shows the devastation at Banda Aceh, with standing water and broken structures remaining at distances quite far inland. The impacted structures were of various construction types, including frame, block and concrete. Thus, it would be difficult to impossible to engineer residential structures capable of withstanding the forces associated with breaking tsunami waves. One more such photograph (Fig. 8.5)

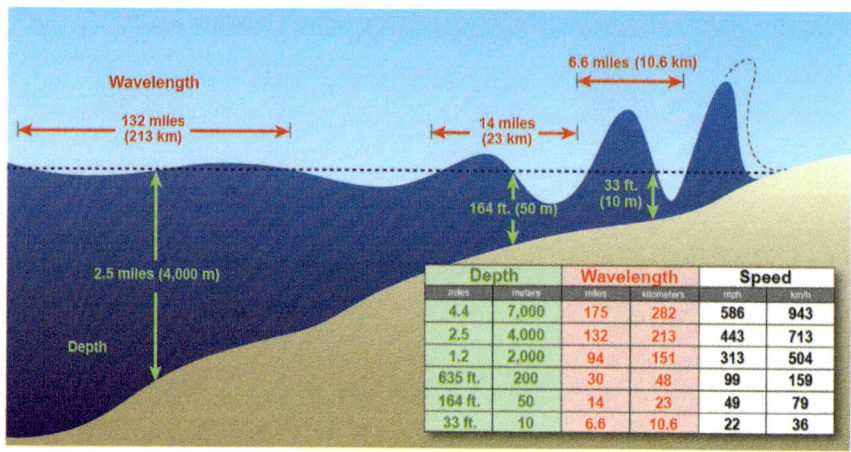

Fig. 8.2 Diagram of wave amplitude and wavelength as a shallow water wave propagates from the deep ocean toward the shoreline. At some point, the wave gets so steep that it breaks (courtesy of NOAA, https://www.noaa.gov/jetstream/tsunamis/tsunami-propagation)

adds to the possible fallacy of engineering structures to withstand large wave forces. Here, we see a former structure made of steel and concrete that was destroyed by the broken wave's whitewater forces. Being that the steel pilings were bolted to their bases, these could not break away; instead, they were forced to bend when they were hit by surging water. This photograph serves as a precursor to Chap. 9 where the forces associated with hurricane storm surge and accompanying waves are discussed and quantified.

To summarize, tsunamis formed in deep water travel very quickly (~200 m/s or ~400 kts or > 500 miles/hr). As they approach shallow water, their wavelength decreases and, consequently, their amplitude increases. So even if the wave amplitude may be imperceptible to an observer in deep water, once they shoal, their amplitude can grow to devastating heights before they break and come ashore as powerful walls of water.

Tsunamis are a regular occurrence, although most are not as devastating as the example just given. Tsunamis may also occur on land if a landslide impacts a large lake, as happened when Mt St. Helens exploded in 1982. Any region prone to tectonic activity with a nearby water body may experience a tsunami, and for this reason, there is a tsunami warning system deployed consisting of buoys with sensitive pressure sensors, and regions where tsunamis have occurred may also have sirens deployed to alert residents to seek high ground if a tsunami is detected. Even the Gulf of Mexico, generally thought to be tectonically quiescent may have its moments. I personally recall a tremor in the early 2000's when, standing at my kitchen counter, the entire house vibrated. I quickly did some arithmetic in my head and waited, knowing that it would be foolish to get in my car and try to drive away. Fortunately, nothing happened. There indeed was a tremor in the deep Gulf of Mexico, but not one that displaced water and hence there was no tsunami.

Fig. 8.3 Beach front area near Lho Nga Sumatra located to the southwest of Banda Aceh (courtesy of USGS, Guy Gelfenbaum, personal communication)

Fig. 8.4 A photograph showing destruction at a Banda Aceh, Indonesia following the Sumatra–Andaman earthquake in 2004 (courtesy of USGS, Guy Gelfenbaum, personal communication)

Fig. 8.5 Photograph showing just how powerful the surging water of a broken tsunami wave may be. Note how the steel girders were bent horizontal because they could not break free of their support bolts (courtesy of USGS, Guy Gelfenbaum, personal communication)

Finally, a tsunami, while potentially devastating locally, can actually propagate its effects worldwide, as the Sumatra–Andaman Sea, 2004 event demonstrated. As the tsunami wave energy flux spreads out, the wave amplitudes decrease becoming nearly imperceptible in deep water, but with nominal frictional dissipation, these waves may be detected throughout the world's oceans, as shown in Fig. 8.6. In the same way that major surfing locales owe their existence to waves generated by storms that may be quite far away, tsunami waves adhere to the same surface gravity wave propagation rules.

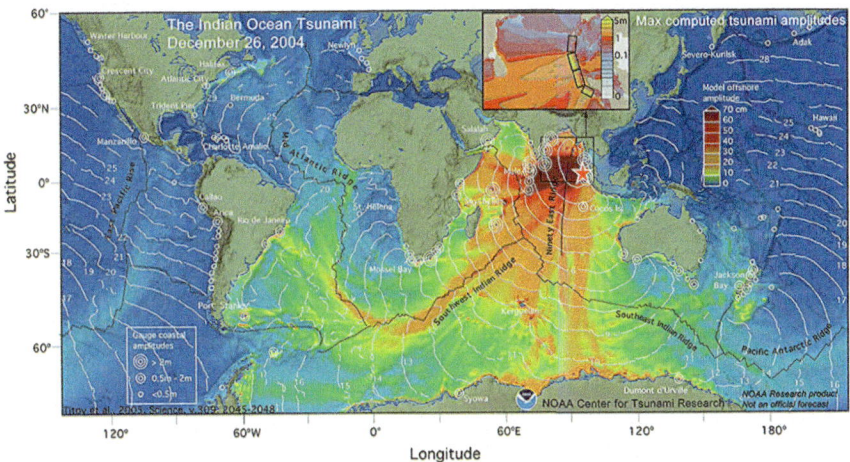

Fig. 8.6 Worldwide propagation of the Sumatra–Andaman Sea tsunami. The contours are in hours showing that within a little more than a day, the effects were seen throughout the Earth's oceans (courtesy of NOAA, https://www.noaa.gov/jetstream/tsunamis)

Addendum: Fish Cakes

My local seafood butcher is also a wholesale purveyor for many of the local restaurants. As such, they prepare serving size portions of fish, leaving trimmings for sale at a reduced price per pound. These are perfect for fish cakes. The standard idea is to mix fish with mashed potatoes at equal proportions and then add other stuff. I prefer to use more fish than potato, maybe two-thirds fish to one-third potato. You can just boil the fish (and freeze the stock for use in bouillabaisse).

It is the stuff that you add to the fish and potatoes that make the dish. Scallions, garlic, fennel, bell peppers, jalapenos, salt and black pepper work well. Dice the vegetables and saute until tender with salt and pepper added. Mix everything together: flaked fish, mashed potatoes, sauteed vegetables, seasoning, plus an egg to bind it all together. Form into ½ lb hamburger-sized patties, roll in corn meal and pan fry until browned on both sides. Serve atop a lettuce salad with some garlic aioli or eat between a bun like a burger.

Chapter 9
Sea Level Extremes by Hurricane Storm Surge

9.1 Hurricane Storm Surge Overview

Public hurricane advisements are generally delivered through commercial media outlets. Whereas some have trained meteorologists to report on weather, others use broadcast personalities to deliver reports provided by contracted services. Entertainment is also a factor, including stalwart bodies who brave the elements by standing on a beach as a storm develops. Rarely are hurricane storm surge advisories provided by individuals who are trained in ocean physics, and because of this, misconceptions abound. For instance, hurricane storm surges are neither a wall of water, nor are they a mound of water beneath the low pressure center that is dragged ashore by the storm, and the Saffir–Simpson 1–5 category scale of storm intensity (1 being a minimal hurricane, with sustained winds of 74–95 mph, or 64–82 kts, and 5 being a very pronounced hurricane, with winds in excess of 157 mph, or 137 kts) is not a reliable determinant of how large a storm surge may be at any given location. Additionally, flooding by storm surge is not the primary public safety factor; instead, the waves that may accompany the storm surge are what destroys structures and takes lives. This chapter will endeavor to explain what a storm surge is and (by using Hurricane Katrina as an example) provide specifics on how damage and loss of life may ensue. It will then explore the potential vulnerability of the Tampa Bay region to landfall by a major hurricane.

Storm surge refers to the sea level rise that accompanies a severe storm. Surges, or, conversely, sea level depressions, may result from both tropical and extra-tropical storms. Storm surges are generally associated with tropical storms since these have the strongest winds, causing the largest surges. For example, the storm surge caused by Hurricane Katrina exceeded 25 ft in some locations.

Tropical storms are large-scale vortices that extend up to the tropopause (about 12 km above the ocean surface). They are nature's way of reducing the heat energy that accumulates in the tropics during summer through fall months. Tropical storms are fueled by the ocean to atmosphere latent heat flux, as water evaporates from the ocean surface, rises aloft in the storm to be released to the atmosphere after the vapor condenses into rain. Once a vortex forms, the convergence of heat into the storm's low pressure core, plus the rotational rigidity of the vortex itself (by its angular momentum), makes it very difficult to dissipate the storm until it is transported over land, where the ocean heat source ceases and the dissipative effect of friction increases.

In contrast to tropical storms, extra-tropical storms are not closed vortices. They are also nature's way of ridding excess energy within the atmosphere itself. These occur when the large-scale wind systems become unstable, producing synoptic-scale weather fronts, resulting in eddy-like motions with updrafts that lift moisture to higher, cooler levels where the moisture condenses to form rain, thereby releasing its latent heat to fuel the storm. Unlike a tropical storm, the latent heat flux, which is also largely of ocean origin, is transported over land and far from the source so once the rainfall is spent, the extra-tropical storm generally dissipates unless it is again transported over water. For instance, storms that may travel east onto the continental United States from the Pacific Ocean often receive a second helping of moisture from the Gulf of Mexico, and if its related low pressure system remains organized as the storm passes into the Atlantic Ocean, then a third helping of moisture may occur. Such an event sequence may result in nor'easters over the New England states, sometimes referred to as bomb cyclones.

Such extra-tropical storms, which are predominant in fall through spring months, when synoptic-scale weather fronts form and intensify at mid-latitudes (owing to the increase in temperature contrasts between the polar and subtropical regions) cause sea level to vary by several factors that must first be explained before the more severe instances of hurricane storm surge may be understood.

As discussed in prior chapters, sea level variations are caused by tides, seasonal steric (density) effects, atmospheric pressure and winds. Other deep ocean influences may also be manifest locally through the propagation of planetary waves, or by long, shallow water gravity waves, such as tsunamis. Here, we concern ourselves with tides, seasonal steric effects, atmospheric pressure and winds, and we will focus upon the Gulf of Mexico, making use of observations and model simulations for actual or hypothesized storms.

The astronomical tides for the Gulf of Mexico are very well represented by existing tide models, and these tides tend to vary with location, mainly due to continental shelf width. Tidal ranges tend to increase with shelf width. Accordingly, the tidal range at Bay St. Louis, MS, is roughly 2.5 ft, whereas at St. Petersburg, FL, it is roughly 3.5 ft. Steric effects are due to seasonal heating and cooling, which alters seawater density and hence water volume, and from Fig. 6.7, this effect amounts to sea level being roughly 0.5 ft higher in summer and 0.5 ft lower in winter than the annual average.

Also, as previously discussed, atmospheric pressure affects sea level as an "inverted barometer" amounting to 1 cm per mbar of pressure variation. Thus, a local decrease in pressure will cause sea level to rise. Typical pressure fluctuations associated with the passage of weather fronts are less than about 10 mbar, resulting in sea level variations of less than about 10 cm (or ~ 4 in.), and under the most

9.1 Hurricane Storm Surge Overview

extreme of hurricanes, a 100 mbar atmospheric pressure drop may cause a sea level rise of roughly 100 cm or about 3 ft. So atmospheric pressure alone cannot account for observed hurricane storm surges and as previously stated to correct a misnomer, hurricanes do not simply drag a mound of water onto the shore.

Wind effects arise by both the along-shelf and the across-shelf components of wind stress, with their relative importance depending on the water depth and the duration over which the wind blows. As explained in Chap. 4, in deep water, and due to the Earth's rotation, the net transport of water by the wind occurs at a 90° angle to the right of the wind (in the northern hemisphere). Coastal sea level will change if this transport impinges on a coast. Thus, an along-shelf component of wind stress will cause a storm-induced set up (set down) if the coast is to the right (left) of the wind. Since these wind-normal transports are a consequence of the Earth's rotation (via the Coriolis force), it takes about a pendulum day ($2\pi/f = 12$ h/sin ϕ) for the response to be fully engaged, or about 1 day along the Gulf Coast (where the latitude ϕ is about 30°). The amount by which sea level may rise depends on the magnitude of the along-shelf current produced by the along-shelf winds. The relevant force balance is between the pressure gradient force due to the across-shelf slope of the sea surface and the Coriolis force due to the along-shelf current. Even under strong, synoptic-scale winds, it is rare for the along-shelf current to exceed about 2 kts because of bottom friction, and, consequently, if the slope begins about 80 km offshore (Fig. 4.5) the mathematics (that we are skipping here) shows that the sea level set up (or down) at the coast may be about 2 ft. Moreover, the winds must blow for about a day for this to be fully realized. So while the along-shelf component of wind stress accounts for the sea level surges of a few feet (in response to synoptic-scale weather fronts), it can only account for a portion of the surge by tropical storms. We refer to this initial set up by the along-shelf component of wind stress as the storm surge forerunner. Having now dismissed the combined effects of tides, atmospheric pressure, seasonal steric variations and the along-shelf component of wind stress as the cause of hurricane storm surge, we must recognize that the real culprit is the across-shelf component of wind stress.

The across-shelf component of wind stress increases in importance as the water depth decreases because (see Fig. 4.5) the bottom stress diminishes the Coriolis force tendency for the transport to be at a right angle from the wind direction. Hence, winds blowing onshore over shallow water will pile water up along the coast, and conversely. The force balance is between the pressure gradient force due to the across-shelf slope of the sea surface and the vertical distribution throughout the water column of the stress as imposed by the wind upon the sea surface. By averaging over the water column, the mathematics (that we are again skipping) shows that the across-shelf slope of the sea surface times the water depth is proportional to the difference between the wind stress acting on the sea surface and the stress of the current rubbing against the bottom. Thus, while the across-shelf slope of the sea surface is directly proportional to stress difference between the surface and bottom, it is inversely proportional to the water depth. This means that the shallower the water, the larger the surface slope, and the farther upslope (inland along the surface slope), the larger the surge. This explains why broad, shallow continental shelves and long estuaries are more prone to large surge than narrow, deep continental shelves. Since hurricane winds vary slowly relative to the surface slope response time, hurricane storm surge tends to build up concurrently (with only a small lag in time) with the winds. As the storm approaches landfall, the surge, which is the surface slope required to balance the difference between the surface wind and the bottom stresses, develops at about the same rate. Rapidly moving storms have more rapidly developing storm surge than slowly moving storms; but, to correct another misnomer, never is there a "wall of water." As an example, during Hurricane Katrina, sea level rose slowly over the course of hours by more than 25 ft in some places, as will be shown later.

In summary, sea level variations owing to storms come about by atmospheric pressure, along-shelf wind stress and across-shelf wind stress, plus the added effects of tides and seasonal steric height variations. For extra-tropical storms (the regular passage of weather fronts), the first two are generally the most important, whereas for tropical storms and hurricanes it is the across-shelf component of wind stress that accounts for the bulk of the storm surge. By virtue of hurricanes occurring in summer months, seasonal heating makes the surge somewhat higher (by about 0.5 ft), and how the surge evolves in relation to high and low tide can either increase or decrease it by another few feet. Not mentioned yet (although we will return to this later) are the wind-induced surface gravity waves whose effects further add to the storm surge.

With hurricane storm surge entailing several factors, each being space and time dependent, its magnitude at any location cannot be generalized. In other words, equating surge heights to the Saffir–Simpson scale may be quite misleading. Three elements are required to simulate a hurricane storm surge: (1) a physics-based, numerical ocean circulation model with provision for the flooding and drying of the land, (2) a supporting dataset on water depths and land elevations, and (3) accurate wind and surface pressure fields to drive the model. Elevations and water depths alone are insufficient because sea level does not rise and fall uniformly. Surge is the highly localized impact of surface slopes by the factors discussed above, and since surge is mostly in response to wind, the winds must be sufficiently accurate to properly drive the model.

We will consider simulations of hurricane storm surges for several different storms beginning with Hurricane Katrina. Because of the severity of Katrina damage, federally sponsored programs were implemented to assess what caused the various levee failures that resulted in much of New Orleans being flooded. Specifically, an Interagency Performance Evaluation Team (IPET) was established in October, 2005, by the U.S. Army Corps of Engineers to evaluate the performance of the New Orleans hurricane protection system during Hurricane Katrina. The National Academy of Engineering/National Research Council (NAE/NRC) then convened a Committee on the New Orleans Hurricane Protection Projects in December, 2005, to provide independent, expert advice to the IPET by reviewing the IPET draft reports. I was selected to serve on this NAE/NRC IPET review Committee.

The IPET marshaled what were the most accurate of the coupled hurricane storm surge and wave models at that time and commissioned reanalysis studies to compose the most accurate wind and atmospheric pressure forcing fields to drive these models. Chosen by the IPET were the Advanced Circulation (ADCIRC) model developed at the University of Notre Dame and the NOAA Hurricane Research Division (HRD) H*Winds with analyses of the peripheral winds, those outside of the H*Winds analysis domain, supplied by Ocean Weather Inc. (OWI).

My involvement on the NAE/NRC IPET review Committee was an enlightening experience for several reasons. The teams (both the IPET and the NAE/NRC IPET review Committee) were composed of outstanding, dedicated individuals who, in the end, did apprise FEMA on why the levees failed. Katrina was a severe storm, but the levees may have held had there not been structural and maintenance issues, as were identified. These flaws resulted in part because there were too many local, state and federal agencies involved with inadequate communications and oversight between them. Recommendations were made, but working within the constraints inherent to large bureaucracies, it is never clear whether, or not, the best recommendations were either advanced or eventually implemented, but that subject lies outside the scope of this book.

One thing that became immediately clear to me as I traversed the Katrina landscape is that while the New Orleans devastation was monumental, what happened along the entire coast of Mississippi was even more so. New Orleans was severely damaged by flooding, primarily owing to the failure of the levee system, and this, along with the human hardship, was extensively covered by the media outlets. In comparison, not only was coastal Mississippi flooded, its structures in proximity to open water (including the Gulf of Mexico, Biloxi Bay, Bay St. Louis and their adjacent waterways) were obliterated by the savaging effects of waves. With New Orleans being so well documented, this chapter will concentrate its Katrina discussions on coastal Mississippi. Along with Katrina, we will also look at the potential for damage throughout the Tampa Bay region, plus provide examples of the actual surges that set water levels either up, or down by some recent storms that affected Florida's west coast.

9.2 The Accompanying Waves

Surface gravity waves enter into the realm of hurricane storm surge in three ways. The first is by raising the surge above what it would be without the consideration of waves. This occurs because the waves themselves have a certain amount of momentum so as the waves propagate into shallow water, steepen to cause white-capping and eventual breaking, this momentum is transferred to the currents. The transfer of wave momentum to the currents is referred to as a radiation stress, which adds to the actual wind stress, thereby requiring an increase in the sea surface slope to balance the total stress by the winds and the waves. Relative to the wind stress, the wave radiation stress is small, and hence, the increase in storm surge height is similarly small, amounting to maybe another foot or less.

The second way is the added height of the wave itself, i.e., the distance between the wave's trough and crest. Whereas waves generated by hurricanes at sea can be enormous (over 100 ft high), there are physical limitations to wave heights in shallow water. These limitations are the same as those that account for wave breaking on a beach under normal weather conditions. A wave will break if the speed of the actual water particles approaches the wave propagation speed, or if the acceleration of the water particles approaches the acceleration of gravity. In shallow water, the wave propagation speed generally equals $(gh)^{1/2}$, where g is the acceleration of gravity and h is the water depth. Given this constraint and the water particle's speed and acceleration criteria, a critical wave slope (the ratio of wave height to wavelength) is reached beyond which the wave breaks, as we see regularly occurring on any beach. In theory, the limit for wave breaking is: $H = 0.79$ h, where H is the wave height (the distance between the wave's trough and crest) and h is the water depth. Thus, as storm surge inundates the land, the height of the waves riding atop the surge cannot exceed this limitation. Given the underlying land variations, the debris and the wind tending to shear off the wave tops, it is unlikely that the theoretical limit is ever reached so smaller values of a wave breaking criteria are employed when attempting to model wave effects. For Hurricane Katrina, the IPET team employed the Simulating Waves Nearshore (SWAN) model coupled with ADCIRC and found that a value of $H = 0.4$ h provided a conservative rule of thumb in the absence of actual wave observations at specific locations. Prior to wave breaking, the total depth of water and hence the flood inundation level equals the background storm surge height, plus the periodic movements up and down of the water level by the waves. Whereas the background surge height increases slowly over the course of hours as previously explained, the added wave variations atop the surge are rapid, with the up and down motions repeating every few seconds.

Lastly, the most important of these three wave effects is the force imparted by the waves as they contact structures. These forces, whether acting upon horizontal bridge roadbeds or the vertical walls of either commercial buildings or residential houses, are enormous. As we will see and quantify later, unlike the inundation by water alone, the waves that ride atop the surge are what destroys structures and leads to the tragic loss of life.

9.3 A Conceptual Overview on the Relative Effects of Winds and Water

Destruction by hurricanes may result from a combination of forces by wind and water. For either of these fluid media, the forces derive primarily from pressure, which may be static or dynamic. Static pressure is simply due to the weight of the fluid per unit area being contacted. Dynamic pressure is owing to the fluid motion. By Bernoulli's theorem, dynamic fluid pressure forces are proportional to the square of the incident velocity component multiplied by the fluid density. Since water density is almost 1000 times that of air, the impact of wind at a given speed is only about 3.5% of the impact of water at the same speed as the wind. To put this in perspective, it would take a wind of about 90 kts to exert the same pressure force per unit area on a structure as a current of only 3.1 kts. Waves also exert pressure forces on structures by reflecting off of the structure, even in the absence of wave breaking. Breaking greatly increases the wave forces because the water particles are then traveling at the wave speed. For instance, for a surge height of 4 m, the wave speed would exceed 6 m/s, and based on water density, the dynamic pressure force would be equivalent to that of a 337 kt wind. More representative quantitative comparisons will be shown later, but the point here is that the weight of water makes its destructive potential much larger than that of air. Local building ordinances in many areas do account for the forces of winds on structures by imposing certain construction standards. For instance, after Hurricane Andrew cut a swath of destruction south of Miami Florida in 1992, building codes were revised to withstand the winds of a category 5 hurricane. However, not all regions adopt such building codes, as was found in 2018 when Hurricane Michael impacted Mexico Beach, FL. Whereas building codes may exist for winds (because insurance companies underwrite wind policies), there is no practical way to design structures to avoid damage by waves.

9.4 A Hurricane Katrina Case Study

9.4.1 An Overview of the Storm

Attention is now drawn to Hurricane Katrina and the damage that Katrina wrought upon coastal Mississippi. Katrina transited the Gulf of Mexico from August 26 to August 29, 2005, making an initial landfall at Buras, LA, and then again at the LA/MS border. The storm reached the Saffir–Simpson scale category 5 before weakening to a category 3 hurricane at landfall. In addition to the flooding of New Orleans, its large size and storm track resulted in devastation along the entire Mississippi coastline. Figure 9.1 shows the track line, Fig. 9.2 shows the National Hurricane Center (NHC) H*Wind product associated with the LA/MS border landfall at 1500 UTC (1000 local time), and Fig. 9.3 shows the maximum sustained wind swath analysis derived from the NOAA post-storm H*Wind analysis. The maximum wind contour over coastal Mississippi in this swath analysis is 102 mph, as observed in the vicinity of Bay St. Louis.

9.4.2 The Various Damage Mechanisms by Hurricane Katrina

Winds, storm surge and waves, along with wind and water-borne debris, are all potentially destructive. We begin a survey of such destruction with conceptual discussions on some of the possible mechanisms by which damage may occur, along with illustrative examples. The potentially damaging factors to be considered are the:

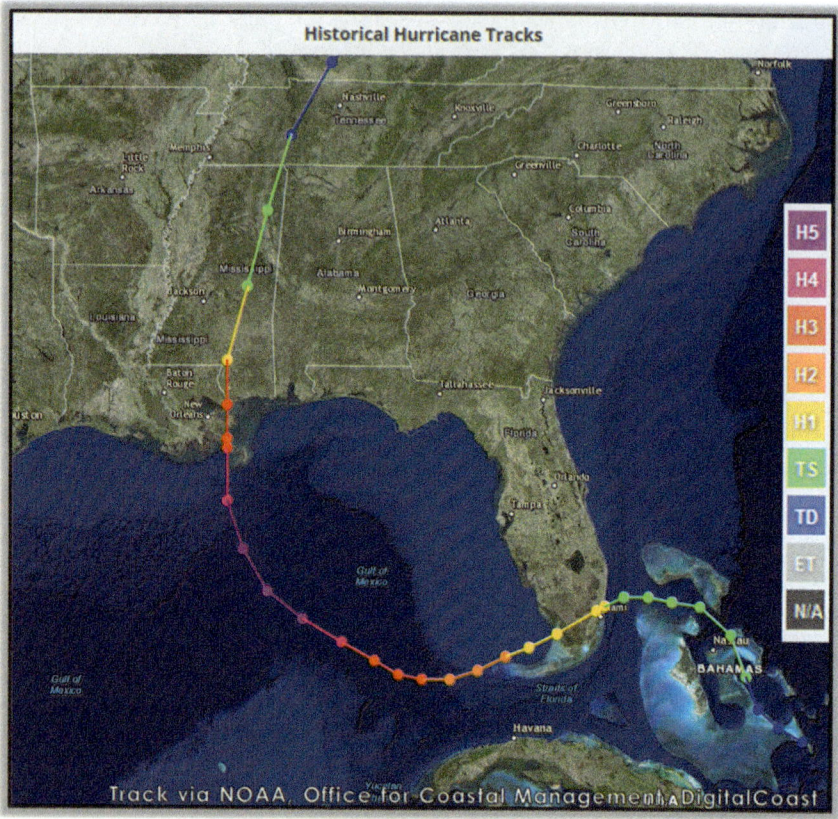

Fig. 9.1 Hurricane Katrina track line (courtesy of NOAA)

- pressure forces exerted by wind,
- pressure forces exerted by water due to the flow past a structure,
- pressure forces exerted by water due to static loading on vertical or horizontal surfaces,
- pressure forces exerted by waves acting on vertical surfaces,
- pressure forces exerted by waves acting on horizontal surfaces.

The evidence left in the wake of Hurricane Katrina suggests that the water-related forces and mechanisms are much more destructive than the wind-related ones so these will be given particular attention. Moreover, of the water-related factors, those by waves appear to have been the dominant ones in many cases.

The pressure forces by wind can topple structures by breaking windows and doors, lifting roofs and collapsing walls. Evidence post-Katrina, either from aerial photographs, or visual inspections, shows that many roofs suffered some degree of shingle loss. Interestingly, however, it is the observation that there are many examples of roofs without any damage at all, versus very few examples of roofs that were

9.4 A Hurricane Katrina Case Study

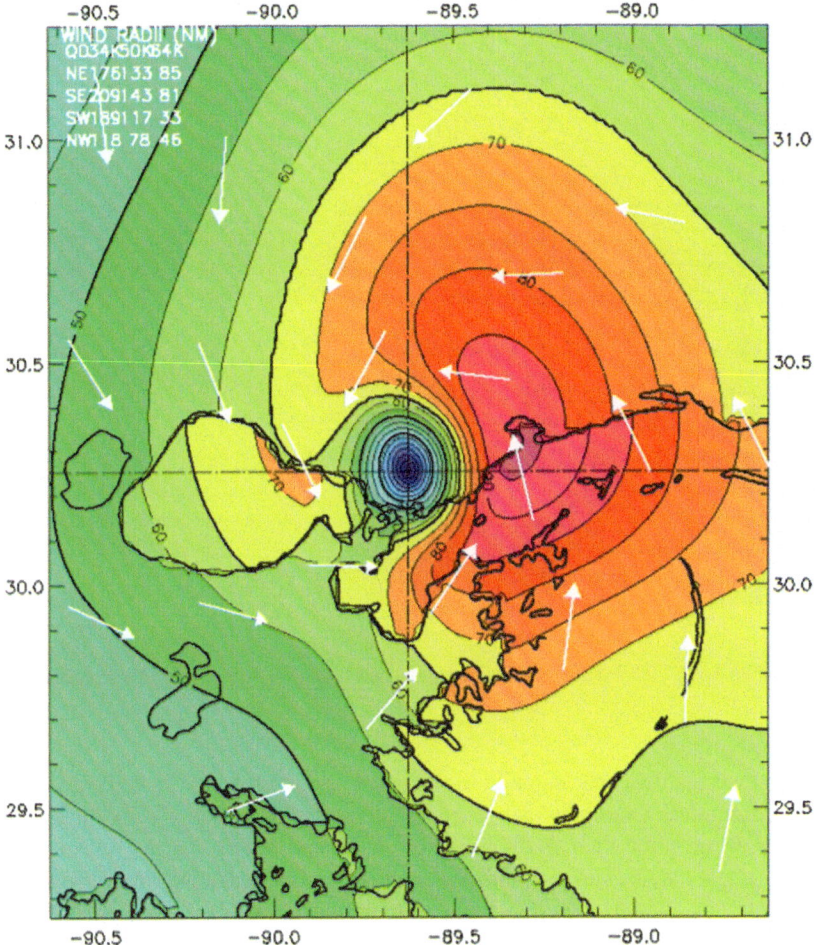

Fig. 9.2 National Hurricane Center H*Wind product (with units in knots) at 1500 UTC (1000 CDT), August 29, 2005 (courtesy of NOAA)

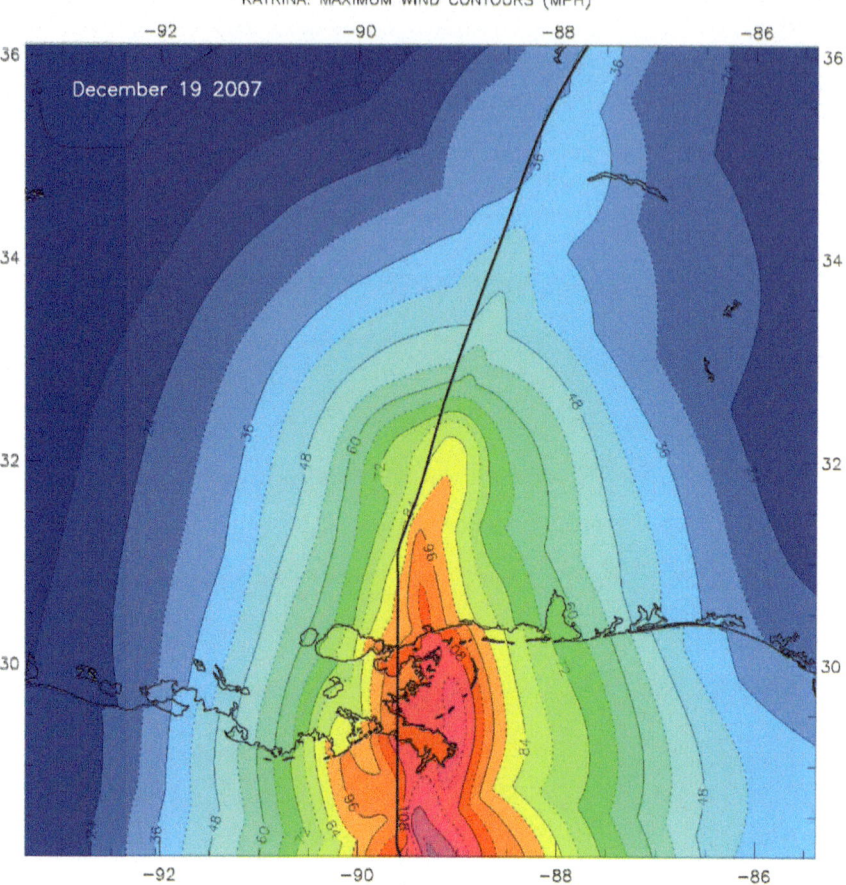

Fig. 9.3 NOAA H*Wind maximum sustained wind swath analysis with units in mph (courtesy of NOAA)

seriously damaged without damage to the underlying structures. In stark contrast, there are countless examples of houses all along the Mississippi coast that were gutted by water while their roofs remained intact and hardly marred. From these observations, it may be surmised that the damage by Hurricane Katrina was largely due to water, as contrasted with wind, and quantitative analyses to be shown later support this assertion.

In the same way, that wind exerts a pressure force when flowing past a structure so does water, and since the water weight is nearly a thousand times that of air, a much slower current can affect a pressure force similar to that of a strong wind. Recall that a 3.1 kt current is equivalent in force to a 90 kt wind when acting on an equivalent area.

9.4 A Hurricane Katrina Case Study

Fig. 9.4 Zoomed view from NOAA aerial photograph 24,331,963, oriented with south (north) to right (left) and west (east) on bottom (top) (courtesy of NOAA, https://geodesy.noaa.gov/storm_archive/storms/katrina/index.html)

The rising surge water can also exert a buoyancy force if the water does not penetrate the structure, thereby equilibrating the pressure both inside and outside the structure. In that event, and with water weighing about 64 lbsft^{-3}, a one-foot height differential between the inside and outside of a structure will result in a buoyant lifting force of 64 lbs for each 1 ft^2 of area. Thus for a 2000 ft^2 house, a one-foot water level differential will result in 64 tons of lift. In the same way, that "a rising tide lifts all boats," if there is not a rapid enough water penetration into a house, then the house could float free of its foundation piers, or buckle in the process of trying to. The evidence shows that many of the houses destroyed by Hurricane Katrina simply floated off their foundation piers or pilings before slamming into other structures or trees. An illustrative example of houses that floated free of their foundations is shown in Fig. 9.4, an aerial photograph of a Pass Christian, MS, neighborhood located inland from the water-borne debris wrack line. While inundated by surge, wave effects were reduced by the wave attenuation by the wrack line. Observed in Fig. 9.4 are numerous vacated foundations and houses that floated free of these to be deposited elsewhere, but otherwise intact. Since the roofs are generally unmarred and the houses themselves are intact, there can be no plausible explanation for this occurrence other than floating free from their foundation piers. Had winds sheared these buildings from their foundations, then they would have been more severely damaged, or completely destroyed. Moreover, had these houses not floated off their foundations then the crushing effect of the water weight would have caused their collapse (if the water height did not equilibrate quickly enough inside and out). For instance, accounts by a photographer stationed at the Gulfport Holiday Inn were that as the water level increased outside of a glass door prior to substantial water penetration inside, the door quickly shattered under the force of the water.

Added to the effects of the buoyant lifting and/or the horizontal pressure forces by hydrostatic loading, or to the dynamic loading of either water flowing, or wind blowing past a structure, are the effects of waves, which may act in several ways. A wave incident on a vertical structural barrier (such as a wall) will reflect, effectively doubling the amplitude of the wave at the barrier. With wave periodicities of only a few seconds, a structure is repeatedly loaded by wave reflection for the duration of the surge, which may last for several hours. Such wave loading is by the dynamical pressure forces exerted by the waves on the structure (which is in addition to the hydrostatic loading discussed above). Since pressure acts omnidirectionally, wave loading acts normal to all surfaces. If a wave reflects off a vertical surface, the loading is by the integrated (or summed) effect of wave pressure on that surface. If a wave attempts to propagate beneath a surface, then the loading is by the pressure force acting upward against that surface. For instance, the failures of the Route 90 bridges that spanned both Bay St. Louis and Biloxi Bay were by the combined effects of wave loading upward on the underside of the bridge spans and horizontally on the vertical sides of the bridge spans. The upward loading jostled the concrete bridge spans off from their support piers, and the horizontal forces pushed the spans in the down-wave direction until they were displaced off of their supports. In the presence of waves, the repeated pounding at periodicities (repeat intervals) of only a few seconds is perhaps the most destructive aspect of the combined effects of wind and water. For instance, the FEMA flood insurance program designates a velocity zone, or V-zone, as that region for which waves are anticipated to be above 3 ft in height. Moreover, if any of these waves are sufficiently steep (by combination of short wavelength and large wave height) as to break at the structure then the full momentum flux of the water between the trough and crest ($\rho a c^2$, in units of force per unit crest width, where ρ is the water density, a is the wave amplitude and c is the wave speed) is imposed on the structure. Residential structures cannot be expected to endure such wave breaking.

Illustrative examples of destruction caused by waves are provided by the Route 90 bridges, a high-and-dry marina, a gutted house and a commercial building. Figure 9.5 shows a section of Route 90 across Bay St. Louis sampled from a NOAA aerial photograph. Note that all of the bridge spans were knocked off of their supports and displaced toward the north. The waves from Hurricane Katrina approached this bridge from the south (from the bottom of the photograph), and the effects of the wave dynamical pressure, both lifting upward on the underside of the spans and pushing northward on the vertical side of the spans, knocked the spans from their supports and deposited them to the north of their supports. Whereas only a small portion of the Route 90 bridge is shown here, it is noted that not a single span was left intact across all of Bay St. Louis after the passage of Hurricane Katrina. The Route 90 bridge over Biloxi Bay suffered a similar fate. There, however, the center portion was sufficiently high as to not be impacted by the waves. This center portion remained intact while all of the spans below a certain elevation were destroyed. The only plausible explanation for this is by the wave forces. Had winds been the cause of

9.4 A Hurricane Katrina Case Study

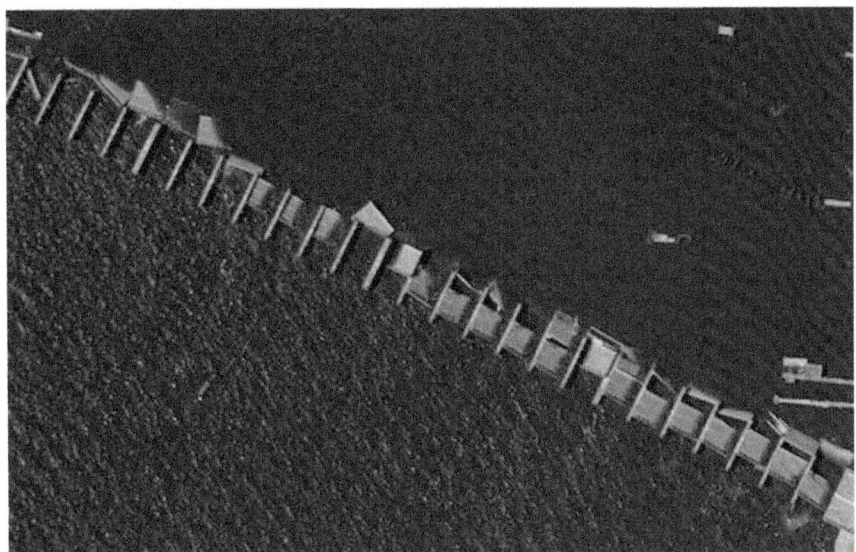

Fig. 9.5 A section of Rte. 90 bridge across Bay St. Louis sampled from NOAA aerial photograph 24,331,945. North (south) is up (down); east (west) is to right (left) (courtesy of NOAA, https://geodesy.noaa.gov/storm_archive/storms/katrina/index.html)

failure, then the highest portion of the Biloxi Bay bridge would have been destroyed, not the lowest portion.

If repeated wave loading can knock out a reinforced concrete bridge, imagine what it could do to less substantial structures. Three additional examples serve as demonstrations. Figure 9.6 shows a high-and-dry marina situated just southeast of the Route 90 bridge ramp of Fig. 9.5. Constructed of steel, it withstood the wave onslaught, but was gutted below a certain level. Just looking down in an aerial image (Fig. 9.7), there would be no indication of damage since the roof was unmarred. Hence, wind was not the cause of this damage; again it was the repeated (every few seconds) pounding by the waves.

The next example is a house located along the Waveland, MS coastline. Unique in its construction, this house used steel girders as vertical supports. This enabled it to remain standing, the only house that did so along this span of beach. As shown in Fig. 9.8, note how the house is completely gutted below a certain level, whereas above that level there is relatively minor damage to the roof, the windows and the siding. Clearly, this damage could not have been wind-induced for if so the topmost portions of the house would be destroyed instead of the lowermost portions. A closer inspection also reveals that the steel pilings are slightly tilted in the down-wave direction (the shoreline is to the right in this photograph). This house took an incredible beating by waves; wind effects, in comparison, were relatively minor.

Many other examples like these may be found in the collection of NOAA aerial photographs or USGS oblique aerial photographs taken directly after Hurricane

Fig. 9.6 A high-and-dry marina situated on western shore of Pass Christian just southeast of Rte. 90 bridge

Fig. 9.7 Figure 1.5 high-and-dry marina from a different perspective (from USGS oblique aerial photograph) (courtesy of USGS, https://coastal.er.usgs.gov/hurricanes/oblique/images/2005/0831/2005_0831)

9.4 A Hurricane Katrina Case Study

Fig. 9.8 A house situated along Waveland MS shoreline

Katrina. A particularly informative one (Fig. 9.9) shows a section of Long Beach, MS, showing a distinctive wrack line defined as the terminus of all of the debris from houses and other structures that were destroyed seaward of the wrack line. The storm surge was estimated to be about 20 ft in this vicinity, with waves of about 6 ft. The wrack line settled in at an elevation of about 16 ft. Evident in this figure are that almost all of the structures seaward of the wrack line were totally destroyed, leaving either foundation slabs that were washed clean or foundation pilings with nothing above them. The exception is the commercial building in the lower left corner of the photograph that will be considered further in Fig. 9.10.

In contrast with what is seen seaward from the wrack line, the houses just landward of it are generally intact. Inspection on the ground showed that these were flooded by the storm surge, but that they were otherwise structurally sound. Even two adjacent houses, one seaward and the other landward of the wrack line suffered drastically different fates, with the seaward one no longer in existence while the landward one, having been flooded, was otherwise intact. The wrack line, in essence, served as a barrier to further wave penetration. Flood waters could penetrate the wrack line but not the waves. It was the waves, riding atop the surge, that destroyed the structures seaward of the wrack line.

The commercial building in Fig. 9.10 further supports this point regarding waves, water and winds. This particular building was the remains of a Walmart. Located just seaward of the wrack line, this building was completely gutted by water-up to

Fig. 9.9 Zoomed view of Long Beach, MS (from NOAA aerial photograph) showing the wrack line (the debris from destroyed structures) and the relative degrees of damage or total destruction that occurred either landward or seaward from the wrack line (courtesy of NOAA, https://geodesy.noaa.gov/storm_archive/storms/katrina/index.html)

Fig. 9.10 Walmart located on the Pass Christian/Long Beach shoreline, cropped from USGS oblique aerial photograph DSC 0923 (courtesy of USGS, https://coastal.er.usgs.gov/hurricanes/tools/oblique.php?mission=Katrina2005#nav)

9.4 A Hurricane Katrina Case Study

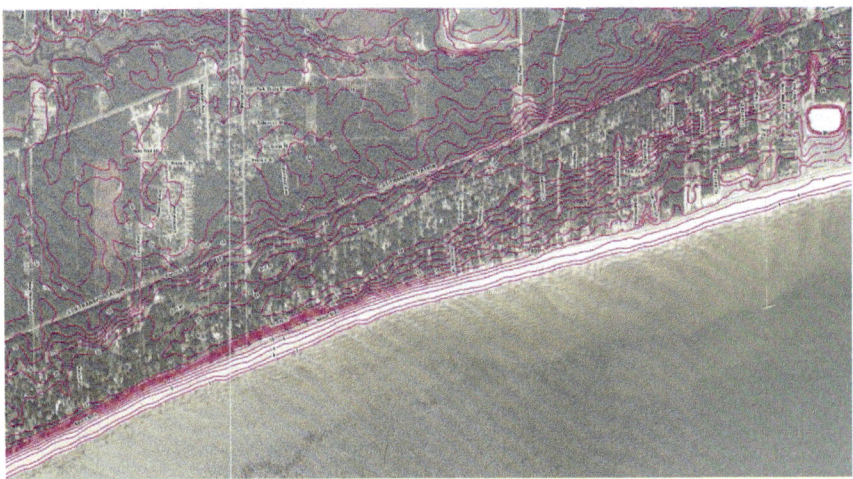

Fig. 9.11 A FEMA topographic map spanning a portion of beachfront from Pass Christian to Long Beach. Note the convergence of land elevation contours west of Long Beach (courtesy of FEMA, Provisional Topographic Contour Maps, Harrison Co., MS)

a certain level, but otherwise intact. The roof appears to be undamaged, along with all of the HVAC equipment positioned on the roof; the signage is largely intact; and even the awning positioned on the left front is largely undamaged. This building was subsequently demolished because it was sufficiently damaged by water. Given the nature of the damage to this commercial structure relative to the surroundings, it is difficult to imagine any scenario other than the pounding by waves and the flooding by water that could have produced this damage. Had the forces by wind been responsible, then the first indication of that would likely have been damage to the roof-based HVAC equipment, but there is no indication of such damage.

These various examples of water-related damage are representative of the entire Mississippi coastline. Oddities existed, however. For instance, the beachfront section spanning a portion of Pass Christian to Long Beach shows surprisingly less damage than adjacent regions, despite it being near the maximum wind speed zone as indicated by the NOAA post-storm H*Wind swath analysis (Fig. 9.3). The explanation is simple, once one considers the land elevation contours. Figure 9.11 is a FEMA topographic map for the region between Long Beach and Pass Christian, MS. Elevation contours are spaced far apart in the vicinity of the Walmart (Fig. 9.10), leaving the entire region between the beach and the railroad tracks vulnerable to surge and waves. These elevation contours converge toward the west, resulting in relatively high land elevations right up to the beachfront and a steep ascent from the beach to the location of the houses. The elevation contours do not diverge again until west of the Pass Christian downtown area, and west of which damage was again severe. This simple demonstration of damage correlation with elevation and open water exposure, especially where the damage is a minimum despite the winds being a

Fig. 9.12 Wooden pilings remaining from a former house in Diamond Head, MS

maximum, again demonstrates the destructive power of waves and water relative to the winds for Hurricane Katrina.

Unlike Hurricane Andrew, which came ashore with category 5 winds at a location where deep water abuts a very narrow continental shelf, Hurricane Katrina, with nominal category 3 winds, came ashore over a fairly board, shallow continental shelf. Hence, the much larger winds and much smaller surge explains why the Andrew damage was mainly by wind, whereas the Katrina damage was mainly by water, with the water causing flooding damage up to a certain level and the waves causing nearly total destruction seaward of the ensuing wrack line.

Within the region of total destruction by waves, there were two distinctly different forms of evidence, one being ground level slabs for houses that were built slab-on-grade and the other being pilings remaining while the houses built upon pilings were no longer there. Two examples from the Diamond Head neighborhood at the northern end of Bay St. Louis are shown in Figs. 9.11 and 9.12. The first of these shows wooden pilings, and the second shows concrete pilings. In both cases, the houses were washed away by water, but note how the wooden pilings are intact, whereas the concrete pilings are broken or toppled over. For the wooden construction, the floor joists are bolted through wood making it easy for the wave forces to break these wooden junctions, allowing the house to be carried away without breaking the pilings. For concrete construction, steel flanges connected the house to the pilings. Given the increased difficulty to break the steel-on-steel connections, the pilings

9.4 A Hurricane Katrina Case Study

Fig. 9.13 Bent and broken concrete pilings remaining from a former house in Diamond Head, MS

themselves, in many observed cases, had to break as the house was being lifted and sheared off of these cement pilings, as evidenced by Fig. 9.13.

Both of these findings for houses built atop pilings dispel the misconception that putting a house on stilts is a remedy for damage by hurricane storm surge. If the surge is high enough such that the waves can act on the underside of the house, then the house will fail regardless of construction type. As a last such example, consider a former house from Ocean City, MS, located just east from Biloxi, MS (Fig. 9.14). In this example, not only were the pilings made of reinforced concrete, but the entire first floor level was constructed as being integral with the pilings by using reinforced concrete. Given that integral construction, the first floor level could not be lifted from its pilings. Thus, the entire structure was forced to break under the influence of waves. All that remained was a jumble of broken concrete sections and rebar. It is not possible, using standard construction techniques, to build a house that will withstand the repeated pounding by ocean waves, either for hurricanes or for tsunamis (e.g., Fig. 8.5).

Fig. 9.14 Remains of a house fronting Gulf of Mexico in Ocean City MS whose first floor level and pilings were constructed integrally using reinforced concrete

9.4.3 A Specific Case Study of How Katrina Destroyed a Particular House

We will now consider a case study for a specific house that was destroyed by Hurricane Katrina. Rather a Gulf of Mexico fronting property, where the cause of the destruction may seem to be more obvious, we will venture to the north side of Biloxi Bay, where proximity to water was still a factor, but where the width of the inland waterway was only about 1.5 miles. In other words, directly fronting the ocean was not a prerequisite for substantial water and wave damage. The general vicinity of the subject property, located on Bayside Drive, is shown (encircled) in Fig. 9.15. During the Katrina storm surge peak, when the winds were southeasterly, this vicinity was exposed to the rising water and the waves that either propagated in from the Gulf of Mexico, or were generated locally.

A closer, zoomed view (Fig. 9.16) provides additional location details. The (circled) lot fronting on the bay sloped up gradually to the slab-on-grade finished floor level of 14.5 ft. Post-storm surveys by FEMA estimated surge heights at 19–21 ft, consistent with nearby high-water marks of 18.9 and 20.4 ft, and the FEMA estimated wave heights were above 3 ft. Based upon this house and adjacent property damage, this entire section of Biloxi provides examples of what was found throughout

9.4 A Hurricane Katrina Case Study

Fig. 9.15 Location of subject property (encircled) sampled from NOAA aerial photographs (courtesy of NOAA, https://geodesy.noaa.gov/storm_archive/storms/katrina/index.html)

coastal Mississippi, that structural damage was correlated with open water exposure and land elevation.

Prior to the house remnants being demolished, the damage was extensively photographed, thereby aiding in causal factor identification. The house was a single-story, wood framed ranch, with brick veneer. It was built slab-on-grade (Fig. 9.17a, b)

Fig. 9.16 Zoomed view of the lot location and the subject house (encircled) vicinity sampled from a NOAA aerial photograph (courtesy of NOAA, https://geodesy.noaa.gov/storm_archive/storms/katrina/index.html)

with an approximate area of 3,200 ft^2, and it had a segmented gable roof. Two immediately noteworthy observations from the photographic record are that: (1) damage to the front of the house was much less than to its rear, bay-fronting side and (2) minimal, if any, damage is seen to the trees in the background.

These observations are consistent with waves hitting the house on its bay-fronting side, and they are inconsistent with wind damage because the peak southeasterly winds, if strong enough, would have damaged some of the trees.

Next consider damage to the east-facing sides of this house and to an adjacent house located just to the west (Fig. 9.18a, b). Both show minimal, if any, roof damage, in contrast with the walls, whose cladding is missing despite the wall studs remaining intact. In other words, these walls did not collapse inward in their entirety. Something destroyed the cladding, but with the winds evolving from easterly to southeasterly as the storm peaked in intensity, these observations of east-facing walls argue against wind pressure being a cause of damage.

Fig. 9.17 Left-hand panel **a** is of the front of the subject house looking south. Right-hand panel **b** is of the front of the subject house looking southeast

Fig. 9.18 Left-hand panel **a** shows the east side of the subject house. Right-hand panel **b** looks west at the house immediately to the west of the subject house

9.4 A Hurricane Katrina Case Study

Fig. 9.19 Left-hand panel **a** shows the west side of the subject house. Right-hand panel **b** shows the east side.

Comparing the damage to walls facing either east or west is illuminating. From Fig. 9.19a, b, it is difficult to tell the difference in damage between these two oppositely facing walls. Both show intact studs and brick veneer that fell outward. Had wind been the cause of such damage then the east side would have shown brick veneer (along with the studs) displaced inward by the force of the wind, versus outward on the west side. On the other hand, once water caused a separation between the studs, the cladding and the brick veneer, gravity would then allow the veneer to fall outward away from the wall cladding and studs. Regardless of wall orientation, a common observations for brick veneer houses throughout coastal Mississippi is that, where these failed, the brick veneer fell outward, consistent with water damage causation, versus wind damage causation.

Perhaps the most telling of the post-Katrina damage photographs taken of the house are the observations that damage occurred from the bottom-up, versus from the top-down. This is evident in Fig. 9.20a, b, where we see views of the house looking north from the south side that fronted on Biloxi Bay. Note the lack of damage above a certain level, versus the destruction below a certain level. An exception is the cladding that is missing on the western section of the roof, i.e., the vertical section that faces eastward. This cladding was damaged by wind since the water did not exceed that level. Note, however, that the damage is limited to exterior cladding; the studs and interior cladding remained intact. The roof section seen in between the east and west sides shows minimal shingle damage. That roof section collapsed from below once the supporting elements were damaged.

This theme of bottom-up, versus top-down damage continues with photographs of the house interior, as shown in Fig. 9.21a–d. The upper left-hand panel (a) is a view looking at the south facing wall, which (as with the east and west facing walls shown previously) also fell outward, not inward; this despite the fact that the studs on the south facing wall are missing. Obviously, the brick veneer failed prior to the studs. In other words, the south wall did not collapse inward under the force of the winds. If so, then the brick veneer would have fallen inward.

Fig. 9.20 Left-hand panel **a** shows the west side of the subject house looking north from the south, Biloxi Bay fronting side. Right-hand panel **b** shows the east side of the house also looking north

Fig. 9.21 Clockwise from upper left are: **a** a view from the south side of the subject house looking northeast, **b** a view from garage looking south, **c** and **d** two interior views looking northeast from the rear of the house

Now consider the interior of the house. We see that the wallboard is intact above a certain level and that the ceiling wallboard is largely intact, as is a ceiling fan. Panel (b) looking south from the garage area also shows a decrease in damage with height. Whereas the ceiling shows damage, the cladding is generally intact. Where the ceiling did drop, we see that the insulation stayed in place. Had wind been the

9.4 A Hurricane Katrina Case Study 141

cause of damage then we would have expected to see more destruction aloft and more of the attic insulation blown out. Panels (c) and (d) reinforce these findings. Taken from the northeast section of the house, that was not as severely gutted by water, we again see the decrease in damage with height.

To summarize this section, the evidence on the ground clearly points to water, versus wind as the causal damage agent. Additionally, of the water-related factors, waves appear to be the primary culprit. Now let us see how the winds, surge and waves evolved in the vicinity of the subject house and what the actual forces may have been.

9.4.4 The Winds, Surge and Waves at the Subject Location

There are several sources of information on the Hurricane Katrina winds, surge and waves. The IPET study used the NHC H*Wind and the modified Ocean Weather Incorporated (OWI) products, along with the ADCIRC (Advanced Circulation Model for Coastal Oceans, Inlets, Rivers and Floodplains) and SWAN (Simulating Waves Nearshore) models for surge and waves. FEMA studies continued to use ADCIRC and SWAN. A separate study by the NIST (National Institute of Standards and Technology) used another commercial vender for winds (ARA, Applied Research Associates) and the NOAA SLOSH model. All of these wind, water and wave models and products provided compatible results. Here, we employ OWI, 10 min averaged winds and ADCIRC surge and SWAN wave model results from a combined National Oceanographic Partnership Program and a FEMA Flood Insurance Rate Map Revision Program for the State of Mississippi [courtesy of Dr. D. Slinn, University of Florida].

The versions of ADCIRC and SWAN employed have different grid resolutions. As with any numerical model, the grid is overlain on bathymetric and elevation data so the results for a given property location can only be sampled at the set of closest grid points. Given the scale of the storm and the resultant wind, surge and waves there is not much difference between adjacent grid points, as will be demonstrated. Figure 9.22 shows developments over time of the water elevation (surge), wind speed, water speed and significant wave height for one of the model grid nodes nearest the subject house. The horizontal line signifies the model land elevation at this particular grid node, which is about 300 ft north-northeast from the subject house.

Consider the wind speed time series first. The wind speed gradually increased, peaking at about hour 39.5 (~ 1530 UTC on 8/29 or ~ 1030 AM local time) in this plot at a speed of 42.0 m/s or about 81.5 kts (94 mph). The application of gustiness factors can be used to elevate these sustained winds to account for higher wind gusts. For instance, gustiness factors of 1.2–1.4, when applied to sustained 10 min winds, would imply peak gusts of 98 kts (112 mph) to 114 kts (131 mph), consistent with the maximum 3 s gust analyses by NOAA and ARA. It is emphasized that these wind speed estimates derive from analyses, versus actual observations at the site. The analyses were refined by the originators (OWI in this case), and the values used

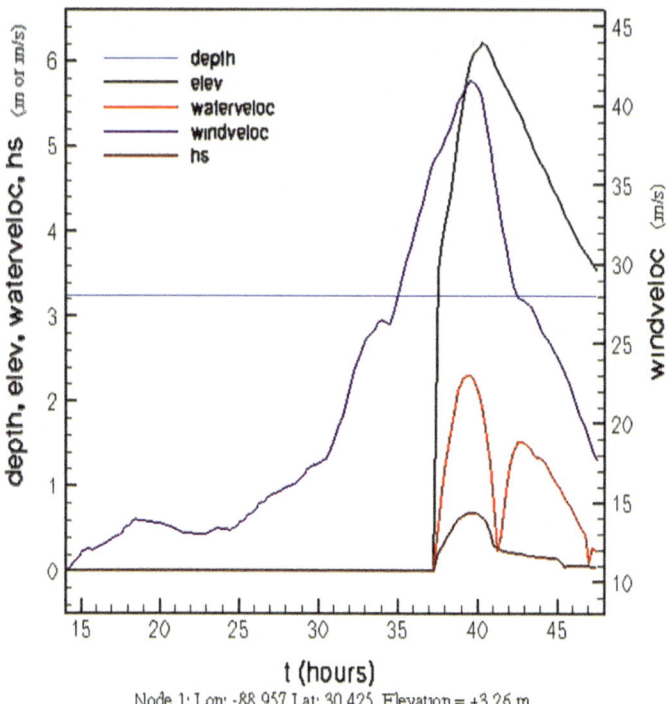

Fig. 9.22 Time series of water elevation (surge), wind speed, water speed and significant wave height for grids nearest subject property. The horizontal line signifies the land elevation at the surge model grid node sampled (model simulation courtesy of D. Slinn, University of Florida, personal communication)

herein are those that were used by the IPET in their final report. The gustiness factors of 1.2–1.4 are a rule of thumb employed by the National Weather Service.

From the storm surge time series, we see that wetting of this land point occurred once the water level exceeded the grid node elevation, and the surge then built up rapidly over the next hour or so with the winds. Surge in this simulation peaked at 6.2 m (20.3 ft) just after hour 40 (~ 1615 UTC or ~ 1115 AM local time). Given the grade elevation estimated at the subject property of ~ 4.3 m (about 14 ft), the implication is ~ 1.9 m (~ 6.3 ft) of water inundated the land at the subject property. On-site inspection, IPET simulations, plus FEMA surge height mapping provide consistent findings for this location. The IPET report (Vol 4, Fig. 74) gives a surge height of about 18–20 ft, and one of the FEMA maps (FEMA Figure B6 for Harrison Co.) provides similar estimates (18.9 and 20.4 ft are the two nearest outdoor high-water marks), both provided as Figs. 9.23 and 9.24, respectively.

As water was flooding the property, the water speed estimated by the surge model reached ~ 2.3 m/s, (~ 4.4 kts). The significant wave heights peaked at ~ 0.7 m (~ 2.3 ft). The water current speed peaked with the wind, and the wave height peaked in between the wind and the surge at this model location. Hence, all of these factors

9.4 A Hurricane Katrina Case Study

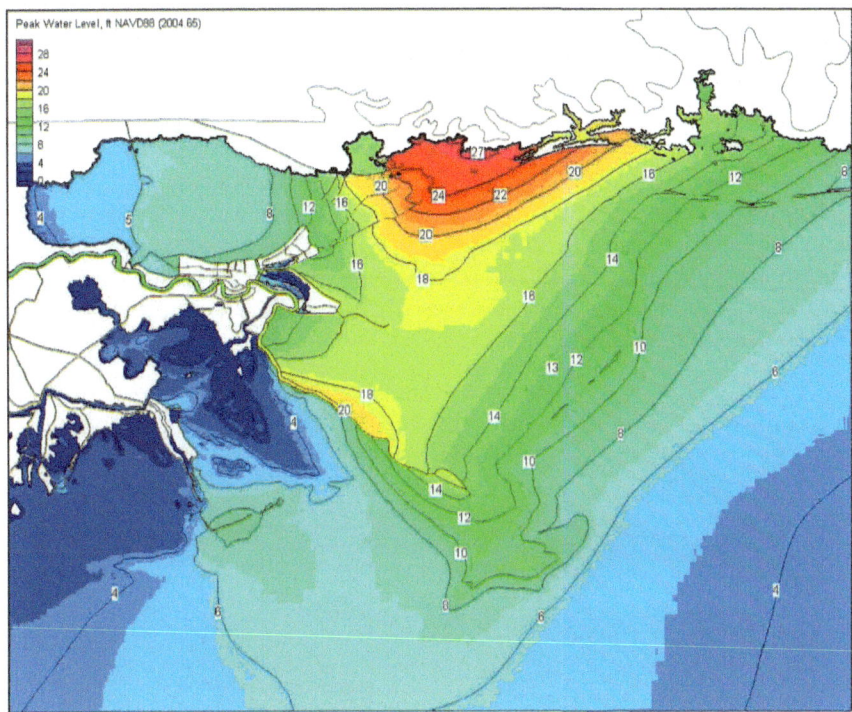

Fig. 9.23 Maximum surge heights (from IPET draft final report, Courtesy of USACE)

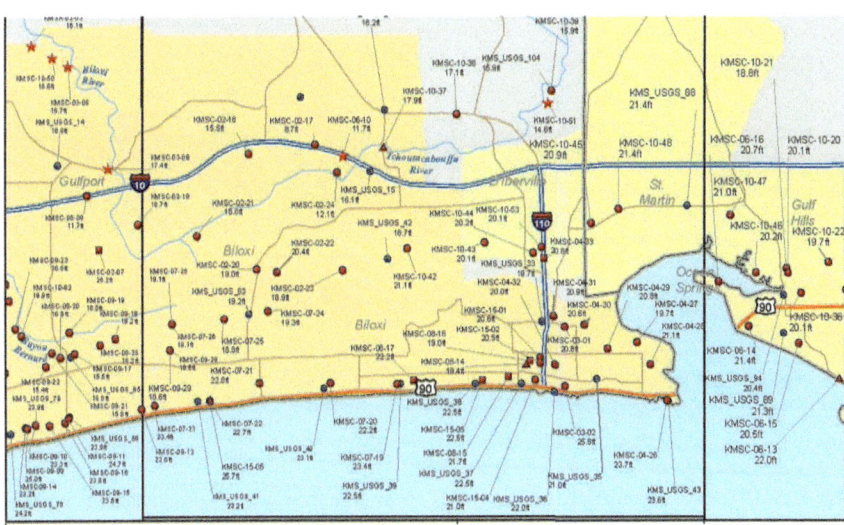

Fig. 9.24 Observed surge heights compiled by FEMA (courtesy of FEMA)

(winds, current, surge and waves) peaked within ~ 45 min of one another. With the subject house having been built slab-on-grade, the finished floor level was ~ 14.5 ft (~ 4.4 m), the house would have been inundated by 1.8 m (5.8 ft) based on the model simulation. It is likely that the house was most heavily damaged as the surge exceeded the grade level and the forces by waves and currents began to work on the house exterior. This would have occurred prior to maximum winds.

How representative these single point time series are of the overall fields of winds, surge, currents and waves is shown in Fig. 9.25a–d, all sampled from Fig. 9.21 simulation at either hour 39 (1500 UTC on August 29th, 2005) for surge height and currents, or hour 39.5 for winds and waves. The approximate location of the subject house is denoted by the purple dot.

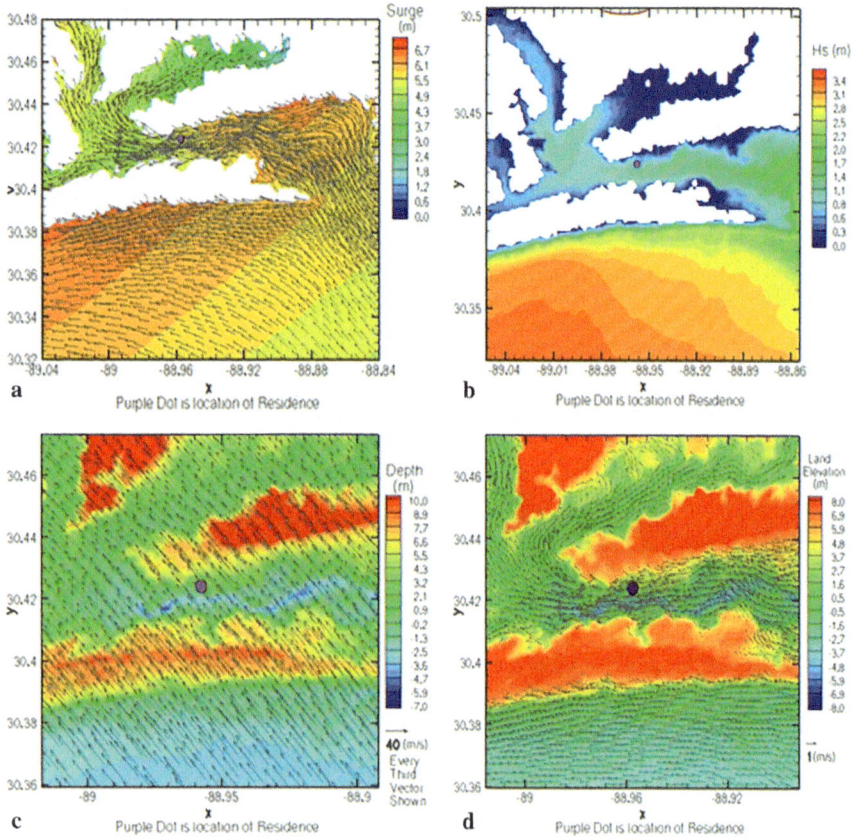

Fig. 9.25 Clockwise from upper left are fields of **a** surge, **b** waves, **c** currents and **d** winds all sampled from Fig. 3.1 simulation at either hour 39 (1500 UTC on August 29, 2005) for surge and currents or 39.5 for winds and waves (courtesy of D. Slinn, University of Florida, personal communication). The purple dot is the location of the subject house

9.4 A Hurricane Katrina Case Study

The waves were large in the adjacent Biloxi Bay, and these diminished rapidly inland. We see that the neighborhood was fully inundated. With the surge still increasing, the currents by the house at this time were flowing toward the west, sweeping debris with it in that direction. The subject house was sufficiently close to the bay to be subjected to waves, and these results are consistent with (albeit a little less than) the 3 ft wave height estimate by FEMA. Thus, Fig. 9.25a–d show that the time series sampled near the subject property are representative of the larger fields of winds, surge, currents and waves in the vicinity of this Biloxi, MS neighborhood. A caveat is that the waves at the house were likely larger than the waves in Fig. 9.22 because the grid point sampled in Fig. 9.22 is farther inland than the house. The closest gird point located at the bay showed significant wave heights of 1.3 m (4 ft).

The evolution of the wind, surge and waves from the simulations relative to the grade and finished floor levels of the subject house are illustrated in Fig. 9.26. Note that the waves add to the surge height. The color coding is such that the green region represents the wave amplitude (1/2 the significant wave height). [I digress to note that the use of 1/2 the significant wave height is a very conservative estimate of amplitude for two reasons. First, the wave peak for steep waves is typically 0.7 times the wave height. Second, with significant wave height defined as the average of the highest 1/3 of the waves, waves can be twice as high as the significant wave height].

Results from the model are plotted from about 0830 AM onward. Surge waters were at the grade level by ~ 0900 AM and at the finished floor level by ~ 0915 AM.

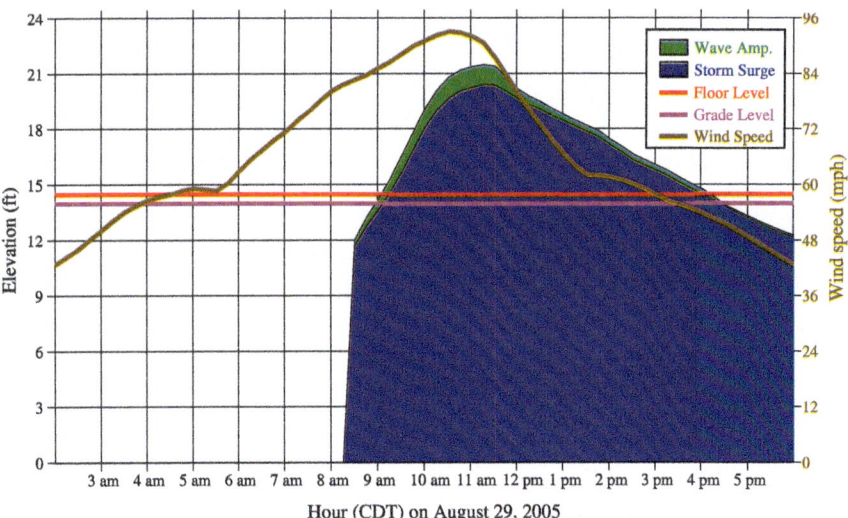

Fig. 9.26 Wind, surge and wave evolutions relative to grade and finished floor elevations of subject house. The purple line is the grade level and the red line is the finished floor elevation. Blue denotes surge height; green the added water level by the waves (taken here to be significant wave height divided by two); brown is the (10 min sustained) wind speed. All times are local, and all units are ft and mph (model storm surge and wave amplitude simulation courtesy of D. Slinn, University of Florida, personal communication).

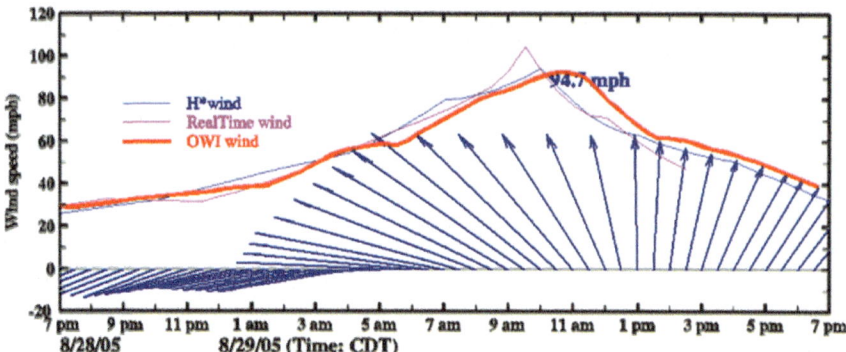

Fig. 9.27 NOAA post-storm analyzed H*Wind vectors and a comparison between NOAA H*Winds (blue) and OWI 10 min sustained winds (red) sampled in vicinity of the subject house. The pink line represents the NOAA "real-time" winds, which are a part of the H*Wind analysis. Units are mph and time is local CDT

It was within this narrow window of time that waves started affecting the house, and damage would have followed very rapidly thereafter. Note that the wind speed did not peak until ~ 1030 AM, 1.5 h later. At 1000 AM and still in advance of maximum winds, the surge level would have been at 18 ft, with the waves adding additional height to this. At the time of maximum winds, the combined surge and waves would have exceeded 21 ft. Thus, substantive damage by water would have occurred prior to maximum winds. Once the surge reached and exceeded the finished floor level, the house would have had to contend with: (1) the hydrostatic pressure differential between the exterior and interior of the house owing to the rising water, (2) the forces exerted by currents flowing past the house, (3) the wave forces acting on the house and (4) water-borne debris striking the house. Even if the house did immediately fill with water, it would still have had to contend with the currents and waves. The current forces would have been larger per unit area than the wind forces, given the speeds (reaching 4.4 kts) estimated by the surge model, and carried with the currents was debris from houses already destroyed. As large as these forces may have been relative to the wind forces, they were still smaller than the wave forces, as we will see. So once the surge level reached the finished floor level well in advance of maximum winds level and the water kept rising, there was little chance for this house to escape severe damage. At the time of maximum winds, with the surge at ~ 20 ft and the surge plus waves at ~ 21 ft there would have been some 5.5–6.5 ft of water above the finished floor level. Whereas the house may have been designed to withstand the wind forces for the speeds that were estimated, it could not withstand the various forces by water and waves.

As a further check upon the wind estimations, Fig. 9.27 compares the 10 min sustained OWI winds with the H*Winds from which they were derived. The red line is the OWI wind speed, the blue line is the post-storm H*Wind speed (the arrows are the post-storm H*Wind vectors), and the pink line is the NOAA HRD "real-time" wind analysis that used for the post-storm H*Wind determination. The three

9.4 A Hurricane Katrina Case Study

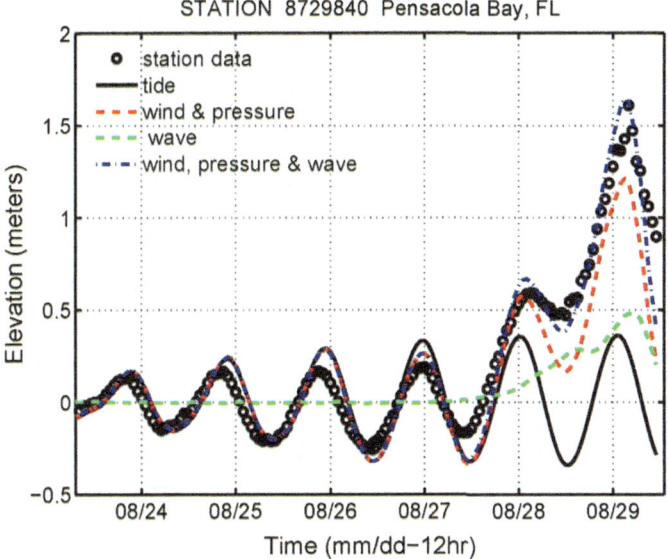

Fig. 9.28 A comparison between sea levels observed and modeled at Pensacola, FL, for the period 8/24-8/29/05, inclusive of Hurricane Katrina. The model is the same as used before, forced by the same OWI winds and pressure, but only including these incrementally to distinguish the relative effects of tides, waves, winds and pressure (model simulation courtesy of D. Slinn, University of Florida, personal communication)

different wind analyses are similar to one another. Each ramps up about the same with the OWI winds and the H*Winds peaking at about the same time, at values less than the "real-time" winds. The OWI winds then persist at higher speeds longer than either the "real-time" winds or the H*Winds. These discrepancies point out the errors inherent to any objective analysis of data from different sources, each subject to certain assumptions and errors. Both the post-storm H*Winds and the post-storm OWI winds are analyzed in a manner that is intended to minimize the errors in these post-storm analyses. Hence, they are considered to be more accurate than the "real-time" winds.

Given the discrepancies between wind products, can these be reconciled on the basis of other information? The validity of the OWI wind phase relative to the H*Wind phase may be addressed on the basis of how well the model simulations of the Hurricane Katrina storm surge agree with observations from tide gauges that were not destroyed. Figure 9.28 provides an example for Pensacola, FL, tide gauge. Shown are the astronomical tides (solid black line), the observed sea level (circles), and the model simulations of sea level based on: (1) tides, winds and atmospheric pressure (red); (2) waves (green); and (3) the sum of tides, winds, atmospheric pressure and waves (blue). As previously explained, the winds are the primary determinant of the storm surge, pressure next and lastly waves. Whereas Pensacola, FL, was distant from the storm center, there was nevertheless a storm surge, with contributions made by all of these factors. The only factor not included in the surge values used in this chapter are the astronomical tides. The fact that the Hurricane Katrina storm surge peaked just after high tide caused the surge to be about 1 ft higher than it otherwise might have been. Thus, the modeled values shown may be biased lower than the actual values. The importance of Fig. 9.28 is that it demonstrates that the modeled surge, when using the OWI wind analyses, not only estimated the magnitude approximately correct; it also estimated the phase correct. Correct phase implies that the time history of the OWI winds is correct.

9.4.5 The Relative Forces by Wind and Water

Force determinations require appropriately derived mathematical formulae, but in keeping with the other chapters of this book, the mathematics are omitted and an attempt is made to explain the relevant physics conceptually. The following subsections explain the force calculations, and these are followed by figures summarizing the relative magnitudes of the forces experienced by the subject house under the two scenarios given above (i.e., as modeled and as forensically estimated, independent from the models). The modeled inputs are as shown in the previous section; the forensic inputs are from the FEMA maps and the IPET report. To estimate the significant wave height when using the forensically determined surge heights, a conservative rule of thumb is employed specifying wave heights to be 0.4 times the water depth. These two determinations, while a little different from one another, provide mutually supportive results leading to the same conclusions.

9.4.5.1 The Wind and Current Forces

The forces exerted by winds blowing perpendicular to a wall derived from the Bernoulli relationship, which says that the pressure exerted by a fluid, as it goes from its approach value to zero at the wall, is proportional to the fluid density times the square of the approach value. The approach value is the wind speed, which also varies with height above the ground level due to friction. Knowing the height at which the winds are estimated and the logarithmic manner in which the wind speed scales vertically, we may estimate the vertical distribution of wind speed from the base to the top of the wall, calculate the pressure and then sum up the total force per unit wall width and the total force on the entire wall.

The water forces follow similarly, using the water density instead of the air density and noting that the height of the water varies in time as the surge rises against the wall. Given the ratio of water density to air density, a current flowing over a similar wall height will equal that of the wind if the water speed is only about 1/29th the wind speed. Thus, for the subject house, the modeled currents would have produced a force per unit area that was larger than that of the winds. However, to the extent

9.4 A Hurricane Katrina Case Study

that waves existed locally, the forces by the waves would have been larger than either the forces by the currents, or by the winds so we will now consider the waves.

9.4.5.2 Wave Forces

Two types of wave forces are considered, the force of a wave on a vertical surface, such as a wall, and the force of a wave on a horizontal surface, such as the underside of a house that is either set atop foundation piers or pilings, overhanging a foundation wall, or a roof section cantilevered over a porch or carport. The latter of these wave forces is what destroyed the Route 90 bridges across Biloxi Bay and Bay St. Louis. Whereas the forces on horizontal surfaces did not impact the subject, slab-on-grade house they are included here for completeness because of the destruction to the bridges and to so many houses that were built on pilings.

Wave forces on a vertical surface

A wave incident on a wall will reflect such that the amplitude of the combined incident and reflected waves is twice the amplitude of the incident wave. The forces exerted by the water on the wall equals the sum of the hydrostatic pressure due to the water itself in the absence of any waves, plus the added dynamical pressure force due to the oscillatory waves. For a shallow water wave (as is the case here), the dynamical pressure is uniform with depth. The waves are oscillatory so the water goes up and down along the wall, reaching its peak height once every wave period, which, for the waves under consideration, is once every 3–4 s. Thus, the wave dynamical pressure exerts and unrelenting repetitive pounding upon the wall. For the calculation based on a derived formula, a wave height of 0.4 times the water depth is used as a conservative value (recall that the theoretical value is 0.79 times the water depth). The use of a larger value would result in a larger force, and for wave breaking, the force would even be substantially larger. The force of a breaking wave, versus a reflecting wave, may be appreciated by thinking about the last time you stood in the surf zone while at the beach. Whereas the propagating wave may jostle you; the breaking wave may more likely knock you over.

Wave forces on a horizontal surface

Waves propagating toward a house either set atop pilings or cantilevered over a foundation wall, or a bridge span resting on support columns, will exert a lifting force due to the omnidirectional wave dynamical pressure. For shallow water waves of the type considered here, the wave dynamical pressure is uniform with depth (as is the wave particle velocity). Whereas waves cannot propagate beneath a rigid surface because there is no gravitational restoring force, there must nevertheless be a communication of the pressure beneath the rigid surface along with a continuation of the wave particle velocity imposed at the leading edge of the surface. For the structural (houses and bridges) scales and wavelengths pertinent to inundation by Hurricane Katrina, the lifting force on the underside of such a surface is proportional to the wave amplitude (a) times the surface area (A), or $F_{\text{lift force}} = \rho g a A$, where ρ is the water density and g is the acceleration of gravity. This force begins when the surge level approaches the floor level and the wave crests begin slapping the underside of the house, and it is fully achieved when the surge level is at the finished floor level. Assuming that the wave amplitude does not change with increasing surge level then the lifting force will remain unchanged as the surge level increases since the wave dynamical pressure, which is independent of depth, is unchanged. Increasing wave amplitude with surge height (as expected) will increase the lifting force proportionately. These wave-induced lifting forces can be enormous, as is so clearly evident by the reinforced concrete roadbeds having been knocked off the Route 90 bridges. Along with the lift is also a time-dependent moment tending to rock the house from its pilings, plus the horizontal wave forces (discussed above) tending to push it off its pilings.

9.4.6 Results

Figure 9.29 summarizes the relative horizontal force determinations. The hydrostatic forces and the dynamic wave forces are the largest. Additionally, to the extent that water rises against a house the wind force decreases since it acts over a lesser height. The wind forces presented here are therefore overestimates because such reduction in height is not taken into consideration. Consistent with Fig. 9.26, the water forces (either hydrostatic or wave dynamic) largely exceeded the wind forces early in the inundation sequence (by ~ 0915 AM for the modeled wave dynamical forces and by ~ 0930 AM for the other hydrostatic and wave forces). Based on horizontal wall forces, for instance, the maximum wave forces exceeded the maximum wind forces by a multiplicative factor of ~ 10, and this excludes the added effects of hydrostatic pressure (and the forces by the currents alone would also have exceeded the forces by the winds).

Given the evolution of the surge and waves and the magnitude of the forces associated with these, I suspect that the water-related forces damaged the subject house by ~ 0915 AM and that this damage was beyond repair almost immediately thereafter. With the only obvious wind damage to a section of (vertical) roof and without significant wind damage to the roofs of remaining nearby houses (except for some shingle damage), plus all of the pine trees and other shrubbery left undamaged on and near the property, it is unreasonable to attribute the most substantial aspects of house damage to winds. Had the house been on higher ground and sheltered from waves, then the damage would have been similar to the more nominal damage observed just inland. It was the elevation of the house and its open water exposure that led to the damage of sufficient magnitude to require demolition after the fact.

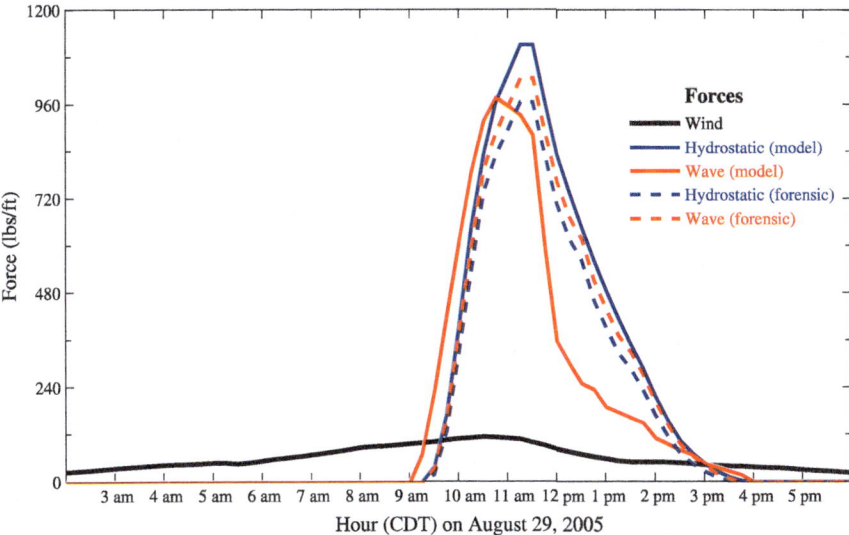

Fig. 9.29 Evolution of the forces acting on a vertical wall as estimated for the subject house based on either the model simulated surge and waves (courtesy of D. Slinn, University of Florida, personal communication) or the forensically estimated surge and a conservative rule of thumb for the waves. The times are local and the force units are in lbs/ft of wall width

9.5 Hurricane Inundation and Damage Potential for Tampa Bay

Given the reality of what happened along the Mississippi coastline by Hurricane Katrina, what might be the potential for inundation and damage for the highly populated Tampa Bay region? My interest in inundation by hurricane storm surge did not begin with Katrina. Long before, as a preteen, I watched the bay meet the ocean at Rockaway Beach, NY where I spent my summers. The water level barely made it over the sidewalks, but it did leave a lasting impression. Early in my Florida residence, a nearby encroachment by Hurricane Elena in 1985 resulted in my first evacuation. The water level barely broached the seawalls, but it did provide a lesson on potential vulnerability. The next several years had nearby misses, notably by a storm headed for Cedar Keys located to the north of Tampa Bay. As with any storm, the television camera crews were at the shore panning the water with their cameras, excitedly advising that sea level would rise as the storm approached landfall. At some point in time, they gave up when the water level, instead of rising, went down. After hours of fear mongering, these storm hardened reporters never bothered to explain why sea level went down. With their behavior being akin to either "Chicken Little," or "The Little Boy Who Cried Wolf," I thought it important to begin informing the Florida west coast public about how storm surge actually works.

Landfall and intensity appear to be the items most discussed by media weather personnel. The path taken by a hurricane, and hence where it may make landfall, is dependent upon the large-scale atmosphere circulation. When forming over the tropical Atlantic Ocean, hurricanes follow the trade winds and eventually curve toward the north. If they do this while still over the Atlantic Ocean, they tend to put the southeastern states of Florida to North Carolina at risk, although they can also travel farther north to New England or even curve further under westerly winds influence to head toward England. Instead of curving north over the Atlantic, if they traverse further to the Caribbean, they may then head into Central America or enter the Gulf of Mexico. Once in the Gulf of Mexico, the landfall bullseye is the region around New Orleans, with landfall occurrences diminishing both to the west and east of New Orleans. Figure 9.30 demonstrates that no United States east coast nor Gulf of Mexico location is immune from hurricane landfall, although some are more prone to this than others (Fig. 9.31). Hurricane occurrence is also seasonal (Fig. 9.32) as is the origin of such storms. The designated hurricane season is June through November, with the months of maximum occurrence being August through October when the Intertropical Convergence Zone, the confluence of the northeast and southeast trade winds, is most developed and tropical Atlantic sea surface temperatures tend to be the warmest. While quite rare, a hurricane may also develop within any month.

Hurricanes striking the Tampa Bay region along paths conducive to incurring major storm surge and wave damage happened in 1848 and 1921. The 1848 event destroyed a military installation located in Tampa, FL, but with no other habitation at that time, the consequences were minimal. The 1921 storm did produce considerable damage, but again, with limited population density at that time, the consequences

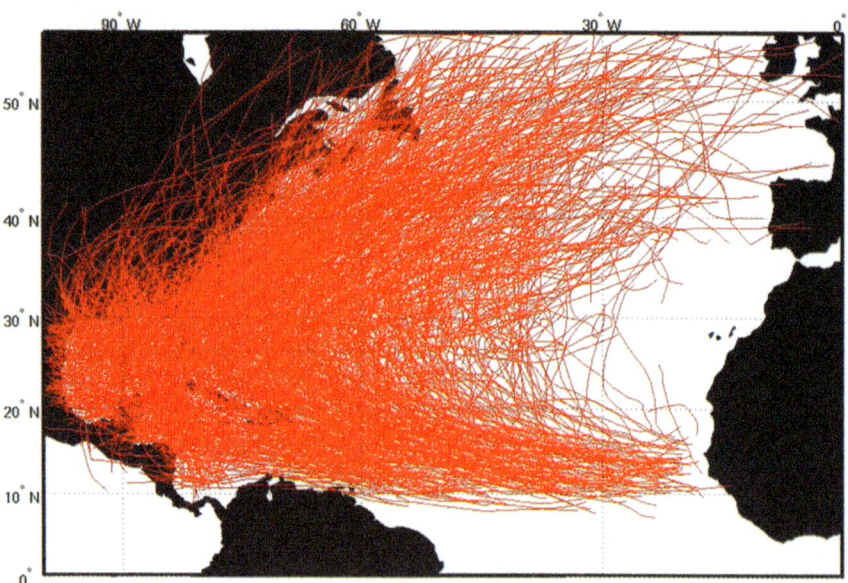

Fig. 9.30 Composite of hurricane tracks from 1851 through 2005 (tracks are courtesy of NOAA, and the graphic is courtesy of J. Virmani, personal communication)

9.5 Hurricane Inundation and Damage Potential for Tampa Bay

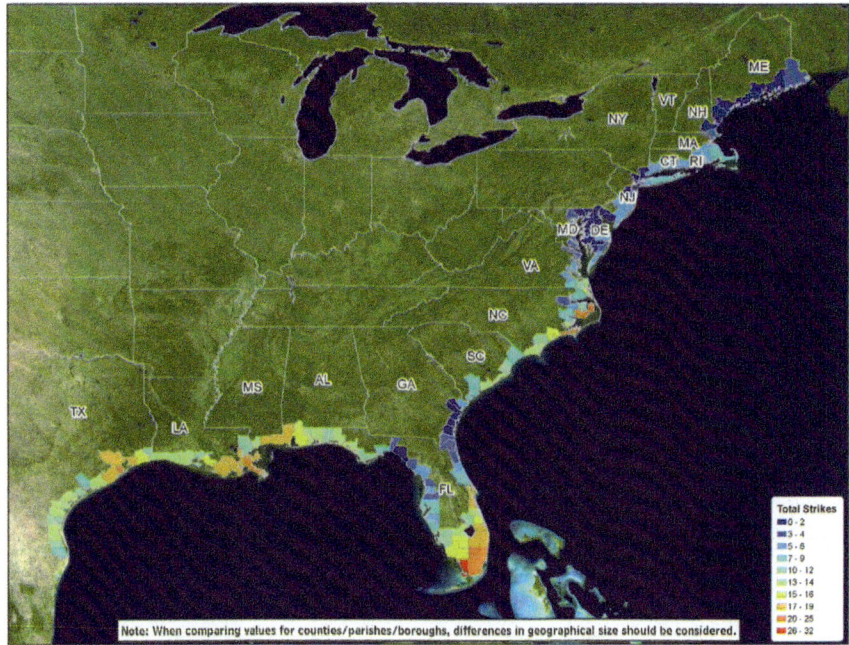

Fig. 9.31 Total number of hurricane landfalls for period 1900–2010 by county (courtesy of NOAA)

were small relative to what they could be today. As rare as a local landfall may be, it is important to understand the risks and plan accordingly.

Unlike many tropical cyclone and hurricane prone nations, where historical practice was to avoid shoreline habitation, such locations are generally sought after in the United States. Consequently, between 1921 and the present time, the Tampa Bay regional shoreline became highly developed, both for residences and tourist destinations. Thus, a 1921 style storm, if occurring today, could be disastrous. It is for this reason that I decided to add hurricane storm surge modeling to my research portfolio. At around that time, I was also fortunate to recruit a new post-doctoral associate trained in numerical circulation modeling and employing a new, unstructured grid technique that allowed for inclusion of shoreline features at resolutions previously unavailable by other modeling techniques.

Our first application addressed the various factors that determine storm surge height at any Tampa Bay vicinity location, including point of landfall, wind intensity, storm dimension (or its radius to maximum winds), direction and speed of approach. All of these factors were found to be important.

Intensity is the easiest to appreciate. Wind stress varies as the square of the wind speed and the sea level slope, which sets up to balance the wind stress, is larger with larger stress so the larger the storm category (e.g., the Saffir–Simpson scale 1 through 5 designation), the higher the storm surge potential. Where the storm makes landfall is

Storm tracks by month

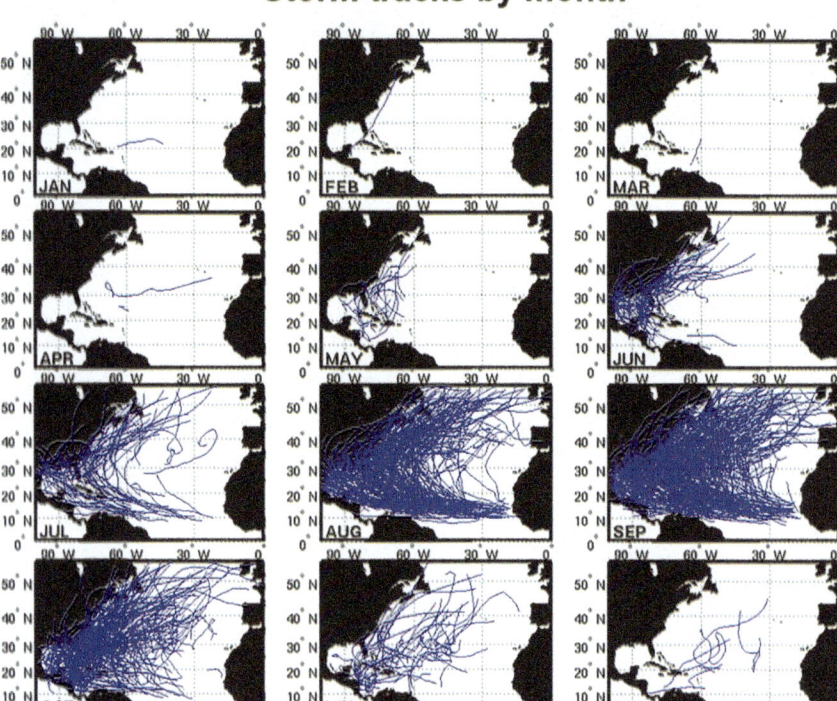

Fig. 9.32 Monthly distribution of hurricanes from 1851 to 2005 (tracks are courtesy of NOAA, and the graphic is courtesy of J. Virmani, personal communication)

equally important. With winds blowing in a cyclonic (anticlockwise) manner around the storm center (or eye), if along a west coast (such as where Tampa Bay is located), then a storm making landfall to the north will have onshore directed winds to the south of the landfall point and offshore directed winds to the north of the landfall point. Hence, sea level will rise to the south while falling to the north. Such was the case in the aforementioned Cedar Keys example, when the storm went ashore to the south of Cedar Keys. With the radius to maximum winds being the distance between the eye of the storm and where the winds are swiftest, locations at that distance from the storm center may be subjected to the largest surges.

The physical size of the storm is also a major concern. The radius to maximum winds is a key factor because it determines the horizontal extend of the storm surge. Larger storms impact larger areas, and conversely. Katrina provides an example of a large storm, as compared to Hurricane Charley, which was smaller in dimension. Charley, while a category 5 storm at landfall, did much less damage than Katrina, a category 3 storm at landfall.

9.5 Hurricane Inundation and Damage Potential for Tampa Bay

The Charley surge was further reduced by its direction and speed of approach. It approached landfall at a nearly shore parallel orientation so that sea level in advance of the storm was first set down so rather than the surge adding to an existing sea level, it added to one that was a little lower to begin with. Charley also traveled quite fast so there was insufficient time for the sea surface slope to fully development. Katrina, on the other hand, approached landfall at an angle that was perpendicular to the shore, and at a slower speed than Charley. Charley caused massive damage where it did hit, but such impact would have been much worse had Charley approached landfall more slowly, from a more shore normal direction and with a larger radius to maximum winds. Hence, the storm category by itself may be quite misleading.

Finally, every individual region has its own geographical uniqueness. Tampa Bay, for instance, is not only adjacent to a broad, gently sloping continental shelf, itself a contributor to a storm surge; it is also a long, shallow water body within which the surge incident at the bay mouth may continually grow larger toward the head of the bay. Thus, here are no simple rules applicable to all water bodies; each has its own characteristics that contribute to a storm surge evolution, distribution and damage potential.

Previous studies regarding Tampa Bay by my associate (Lianyuan Zheng) and I used an idealized hurricane wind distribution. To add a little more realism, we then asked the question: What may have happened here had Hurricane Ivan made landfall near Tampa Bay instead of at the border between Alabama and Florida? Thus, we took the Hurricane Ivan winds and adjusted the track lines to simulate landfalls at various Tampa Bay vicinity locations, as shown in Fig. 9.33.

Before showing results, it is instructive to explore the inundation potential for a storm surge of a certain height. FEMA publishes flood insurance rate maps (FIRMS) that are used for FEMA underwritten flood insurance policies, as required by most mortgage lenders. Prior to the early 1980s, these FIRMS set a finished floor level at 9 ft above mean low water. Such designation was changed to be 11 ft prior to 1984. Figure 9.34 illustrates what this means with regard to storm surge and tides. Even at 11 ft, the finished floor level at high (neap tide) would only be about 9 ft above sea level and at high spring tide in summer months, this would be reduced to only about 8 ft. Additionally, only about a 3 ft surge could allow seawater to broach many seawalls, inundating roadbeds. There is not much room for error in the existing FEMA FIRMS, either for the older, or for the newer versions.

Given these margins of error for flooding, Fig. 9.35 shows the storm surge evolution for several Tampa Bay vicinity locations for the five different landfall point investigated. Several generalizations may be made. First, and as anticipated, for a landfall to the south of the bay mouth, the surges are small and locally dependent upon the wind direction as the storm traverses eastward. Flooding within the bay itself would not be a major factor. Second, whereas the Indian Rocks Beach and Tarpon Springs landfalls yield similar results, landfalls farther away yield ever decreasing surges. Third, and maybe a surprise to some, the surges along the barrier Island beaches are not the largest; instead, the farther up the bay the larger the storm

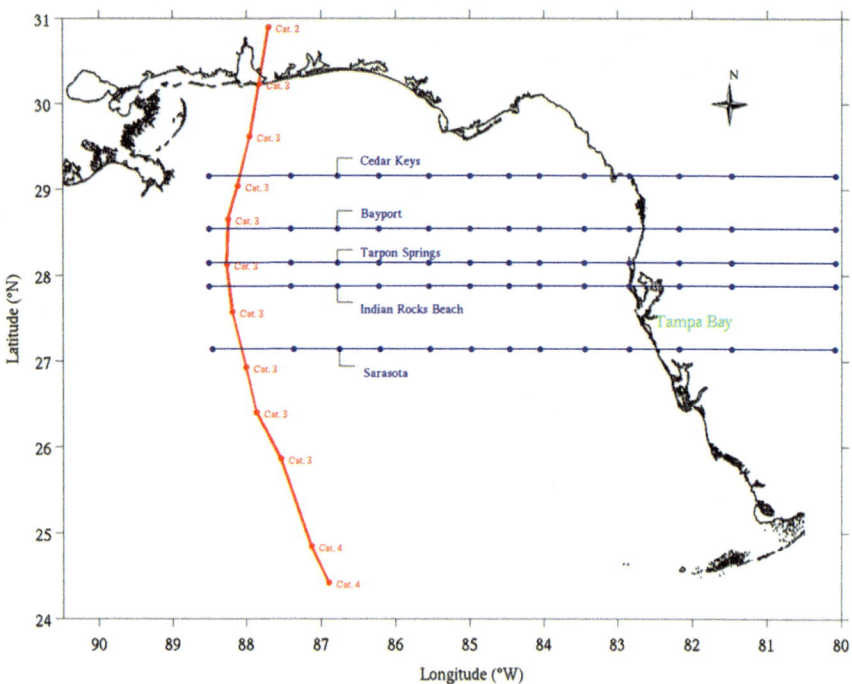

Fig. 9.33 Actual Hurricane Ivan track (in red) and redirected tracks (in blue) used to simulate responses to Ivan winds based upon various points of Tampa Bay vicinity landfall (adapted from Weisberg, R. H., and L. Zheng (2008), Hurricane storm surge simulations comparing three-dimensional with two-dimensional formulations based on an Ivan-like storm over the Tampa Bay, Florida region, *J. Geophys. Res.*, 113, https://doi.org/10.1029/2008JC005115)

surge. This is consistent with storm surge resulting from a sloping sea surface, whose slope actually increases with decreasing water depth as previously explained. Thus, whereas the surge at Egmont Key at the mouth of the bay may be between 3–4 m, the surge at the upper bay bridges and the Port of Tampa may be 5–6 m. The worst scenario for Tampa Bay is a landfall occurring a radius of maximum winds to the north of the bay mouth, thereby focusing the strongest winds on the bay mouth. Moreover, given the orientation of Tampa Bay, such scenario would worsen if the storm direction of approach were to be parallel to the axis of the bay.

The actual surge heights vary throughout the bay owing to the geometry and how different shoreline regions are oriented to the winds as the storm traverses eastward. Figure 9.36 shows samples for different sections of the bay at different times when the storm surge was at its maximum extent. These results are for the Indian Rocks Beach point of landfall. Where land inundation occurs, the surge heights are shown relative to the land elevation. The left-hand panels show the barrier island beaches and the central portion of the peninsular at two different times. A portion of the peninsular mid-section does suffer inundation where the land elevations are low, consistent with the Cross Bayou Canal location. The barrier islands themselves may

9.5 Hurricane Inundation and Damage Potential for Tampa Bay

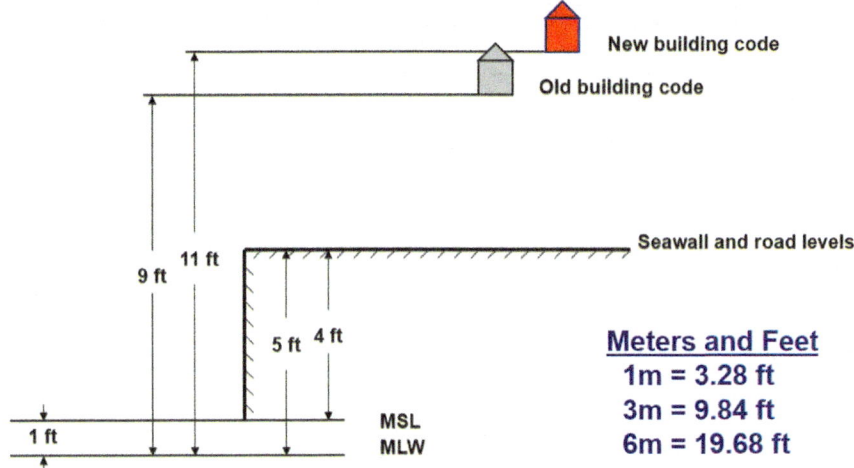

Fig. 9.34 An approximate sketch of residential finished floor levels for older and newer building codes relative to mean low water (MLW), mean sea level (MSL) and both seawall and road levels. With FEMA FIRM building codes being relative to MLW, there is much less room for flooding with high tides in summer months when the 9 ft or 11 ft heights may be reduced by another two to three ft

expect to see 2–3 m of inundation under such a storm scenario, but the worst of the flooding would be in the St. Petersburg Northeast section (the middle panels) where 3–4 m of inundation are seen and on the eastern side of the bay near Apollo Beach (the right hand panels) where some 4–5 m of inundation appear.

Results of simulations, such as these, have been known for almost two decades (the paper in which some of these figures appeared was published in 2008, and seminars showing the results were given in 2006). Despite such findings, some of the most rapid developments within the Tampa Bay region have been in the potentially most vulnerable areas.

A follow-up to this study used the same winds, but with the added effects of waves. As regards surge height, waves increase this by about 0.3 m or 1 ft, but more damaging than the increased flooding is the effects of the wave forces upon structures. Figure 9.37 provides an example of the wave field throughout Tampa Bay including the estimated significant wave heights over the inundated land areas. Whereas 0.5–1.0 m waves may not seem overly alarming to some, recall the damage that similar waves did along the Mississippi coastline. Had Ivan hit the Tampa Bay region, the results may have been devastating.

Rather than conclude this chapter on such a depressing note, we can at least take solace in the fact that other than the 1848 and 1921 storms, the region has escaped such serious hurricane damage. That does not mean that hurricanes have not visited us. Figure 9.38 shows a composite of hurricane track lines that intersected the Tampa Bay vicinity from 1851 to 2006. Several have done so, and subsequently, there have been more. Nonetheless, as this chapter makes clear, how a hurricane may

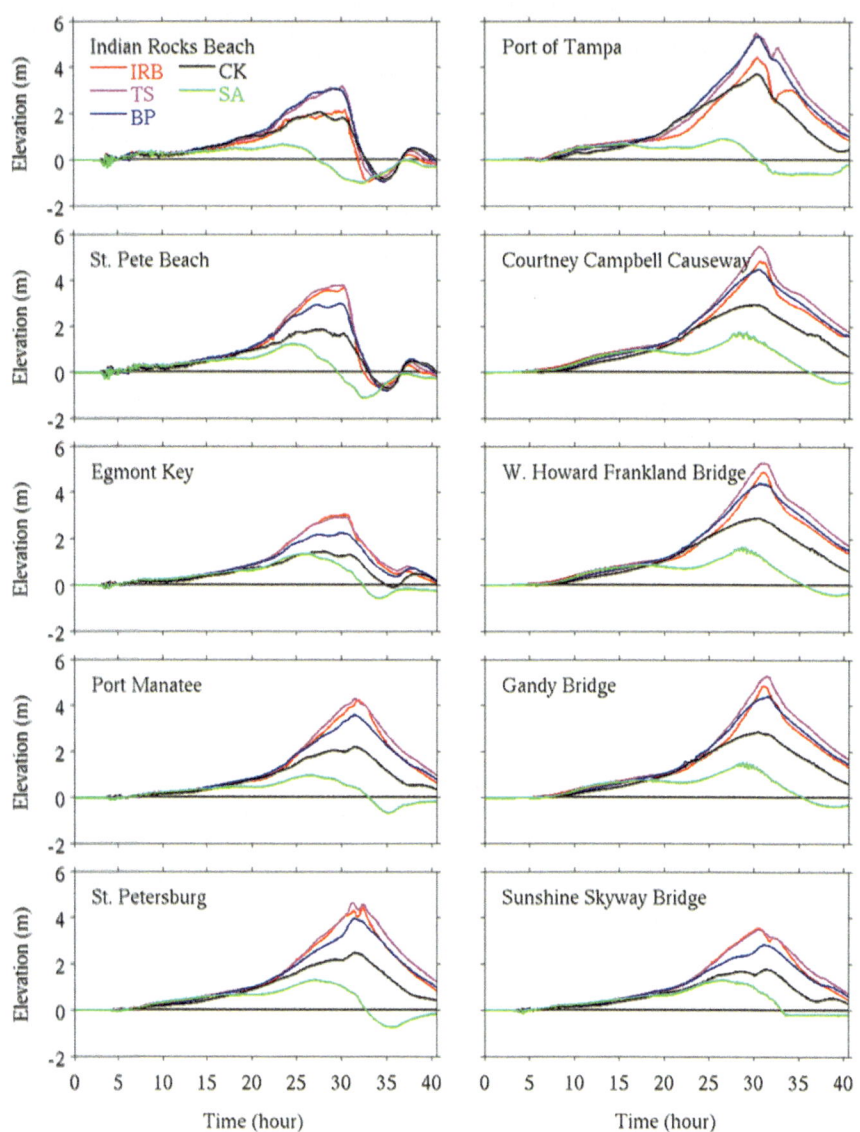

Fig. 9.35 Storm surge height evolution at several Tampa Bay locations for five different landfall points: Indian Rocks Beach (red), Tarpon Springs (pink), Bayport (blue), Cedar Keys (black) and Sarasota (green). Sarasota is south of the bay mouth, whereas all other locations are to the north, Indian Rock Beach being the closest and Cedar Keys being the farthest away (adapted from Weisberg, R. H., and L. Zheng (2008), Hurricane storm surge simulations comparing three-dimensional with two-dimensional formulations based on an Ivan-like storm over the Tampa Bay, Florida region, *J. Geophys. Res.*, 113, https://doi.org/10.1029/2008JC005115)

9.5 Hurricane Inundation and Damage Potential for Tampa Bay

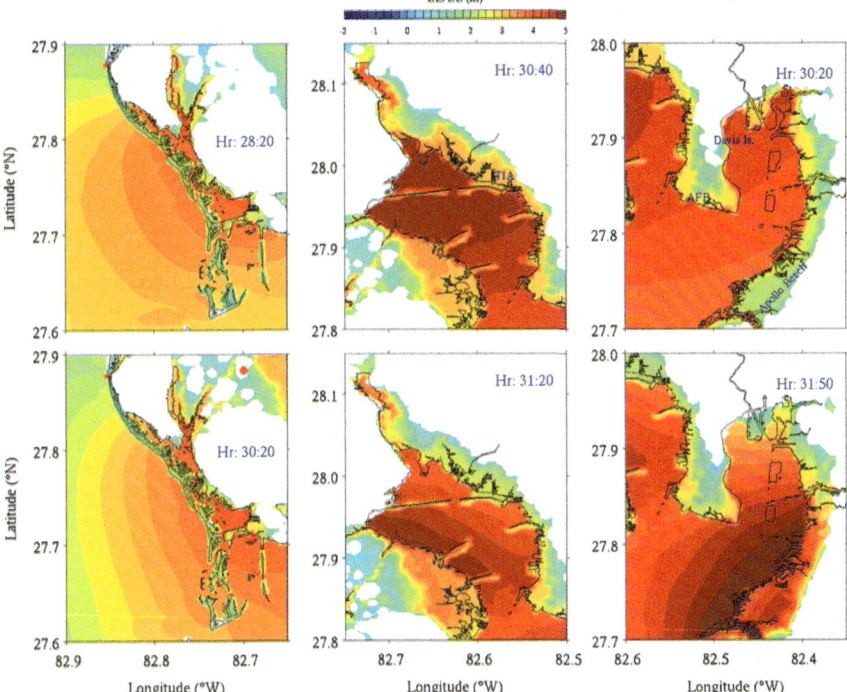

Fig. 9.36 Inundation examples at certain times and locations where these were at their worst. Values on land are shown relative to land elevation. For instance, in the lower left-hand panel, we see inundation of up to 4.5 m above land level, or more than 14 ft above the land level (adapted from Weisberg, R. H., and L. Zheng (2008), Hurricane storm surge simulations comparing three-dimensional with two-dimensional formulations based on an Ivan-like storm over the Tampa Bay, Florida region, *J. Geophys. Res.*, 113, https://doi.org/10.1029/2008JC005115)

impact a given location is critically dependent upon point of landfall, direction and speed of approach, radius to maximum winds and of course the storm intensity. The lesson to all concerned is that we take such threats seriously and that we all heed warnings by emergency managers and personnel. A second lesson, which has not been given enough attention, is that the locations of critical infrastructure, specifically electric power generation facilities and other essential utilities (water and sewer), plus hospitals, bridges and airports all need to be evaluated.

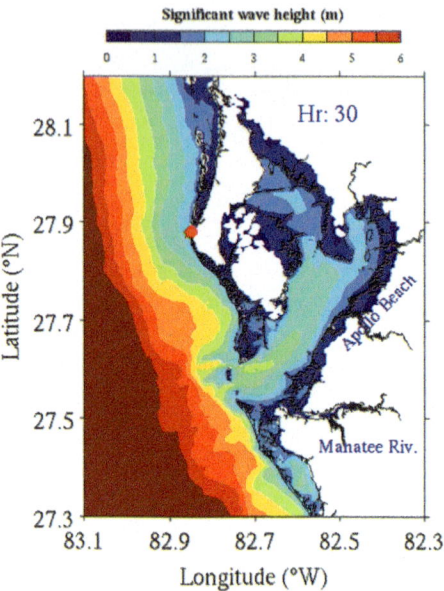

Fig. 9.37 Significant wave heights at the time of landfall for an Ivan-like hurricane making landfall at Indian Rock Beach, FL. Waves atop surge for inundated land areas are modeled to be around 0.5–1.0 m (adapted from Huang, Y., R. H. Weisberg, and L. Zheng (2010). The coupling of surge and waves for an Ivan-like hurricane impacting the Tampa Bay, Florida region, *J. Geophys. Res.*, 115, C12009, https://doi.org/10.1029/2009JC006090)

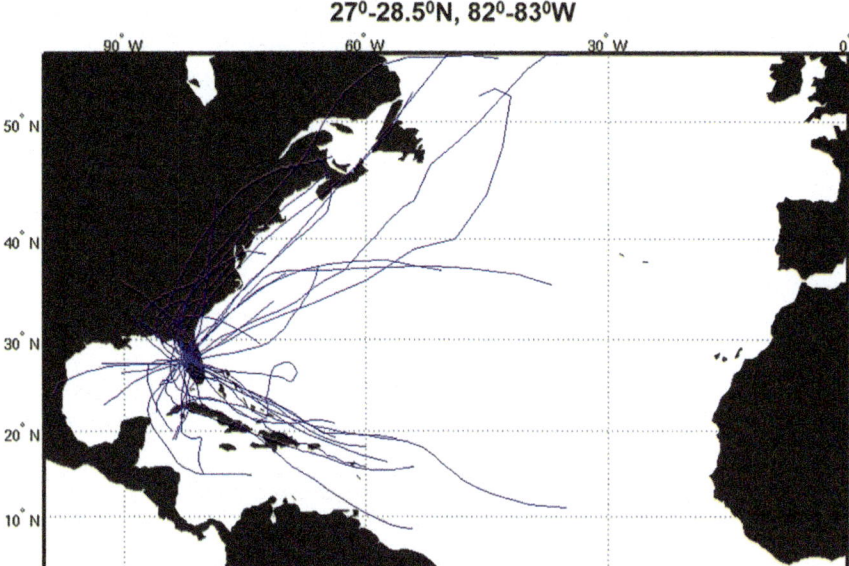

Fig. 9.38 Track lines for all hurricanes that passed through a box spanning 27°–28.5°N, 82°–83°W from 1851 to 2006 (tracks courtesy of NOAA, and the graphic is courtesy of J. Virmani, personal communication)

Addendum: Grilled Halibut

Halibut is like a monster flounder; it is quite mild and tastes like what you cook it with. It is a plentiful fish in Maine so once a year I do the following on a gas or charcoal grill. Get a couple of Halibut steaks about an inch thick and marinate these for an hour or so in lemon juice, garlic and olive oil after sprinkling on some salt and black pepper. A large zip-lock bag works well for this purpose. Put this in the refrigerator and turn now and then so that the steaks marinate evenly. Don't be frugal with the garlic—use several cloves chopped up finely.

This fish goes well with grilled sweet corn so get at least one ear per person. Since the corn is also to be grilled, you need to cut the ends off and shed just a few of the outermost husks. Now that you are down to clean husks and corn with no worms (you'll know by cutting of the ends), let these soak in water for a half hour. When you are ready to cook, grill the corn first. The water will eventually steam off and the husks may begin to burn. Turn the corn a few times to let these cook evenly, maybe 10–15 min of so. Remove the corn and aside, knock any hulks for burnt pieces off of the grill and then use an oiled piece of paper towel to place oil on the grill so that the fish will not stick. Grill the fish sort of like a steak, five minutes on a side. It should turn easily if you properly oil the grill. Fish is finicky; undercook it, and it is lousy;

over-cook it and it is even worse so test it just like you would a steak, and if you are unsure, make and incision to peek inside.

Fish and corn; nothing else is needed, except for a good glass of wine. By now, you've gone through several whites; try a Rose, or a sparkling Rose.

Chapter 10
The Air–Sea Interactions that Determine Water Temperature

Given that the sun is the initiating contributor to the Earth's climate (Chap. 1), we may surmise that with the longest day and highest sun angle occurring at the summer solstice (June 20th or 21st in the northern hemisphere) that this would be the time of maximum ocean heating. However, the rates of ocean heating or cooling, by virtue of the net exchange of heat across the atmosphere–ocean interface, is somewhat more complex. The surface heat flux (or the flow of heat energy per unit area per unit time, as expressed in units of Joulesm^{-2} s^{-1} or Wattsm^{-2}) depends not only on solar radiation, but also on the radiation that emanates from the ocean surface itself, plus that from clouds, greenhouse gases found in the atmosphere and the turbulent exchanges of heat across the atmosphere–ocean interface. This chapter explores these radiative and turbulent heat fluxes and how these lead to a seasonal cycle in the upper ocean's temperature.

Two general forms of radiation enter into the Earths heat budget, one being shortwave radiation, the other being longwave radiation. Both are part of the electromagnetic energy spectrum that comprises a much larger range of components, from the very short wavelength (high frequency) gamma rays, whose length scale approaches that of an atom, to very long wavelength (low frequency) radio waves, whose length scale approaches that of the Earth's radius.

All matter emits electromagnetic radiation at all wavelengths, and Planck's law provides the spectral distribution (how the energy is distributed across all wavelengths) of the emitted energy for what is termed a black body (i.e., a body that absorbs all of the energy incident upon it) at a given absolute temperature measured in degrees Kelvin (°K). Such a radiation spectrum also peaks at a wavelength that is determined by the emitting body's temperature, with this wavelength of peak energy being inversely proportional to the temperature (Wien's law), i.e., the higher the temperature of the emitting body, the shorter the emitted wavelength. Complimenting these two radiation laws is another one embodied in the Stefan–Boltzmann law, which states that the total amount of energy per unit area that is radiated from an emitting body is proportional to the fourth power of the absolute temperature.

What we refer to and measure as shortwave and longwave radiation depends upon the temperature of the body from which this radiation emanates. The origin of what is referred to as shortwave radiation is the sun, which radiates at a temperature of about 5778 °K. The range of wavelengths encompassing this shortwave radiation is between about 295 and 2800 nm (where a nanometer is 10^{-9} m, or a billionth of a meter). The total amount of shortwave radiation that reaches the Earth is about 1360 Wm^{-2}. This amount is the result of what is actually generated at the sun and the spherical spreading of this solar energy as it propagates (travels at the speed of light) from the sun to the top of the Earth's atmosphere. A lesser amount reaches the Earth's surface due to absorption of some of this energy within the atmosphere. The origin of the longwave radiation is the Earth's surface, which radiates at a much lower temperature (around 288 °K), plus radiation from clouds, water vapor and other greenhouse gases, whose temperatures vary from what may be at the surface to that at the height of the cloud tops and beyond. The range of wavelengths encompassing this longwave radiation is about 4–50 μ (where a micron is 10^{-6} m, or a millionth of a meter). The total amount of longwave radiated energy also follows the Stefan–Boltzmann law, and the wavelength of peak energy follows Wien's law. Given these distinctions between shortwave and longwave radiation and their spectral distributions, sensors are designed to measure each of these radiative inputs to the Earth's energy (heating and cooling) budget.

Shortwave radiation incident at the top of the Earth's atmosphere is subject to absorption, reflection and scattering so not all of it reaches ground level. Depending on cloudiness, humidity and other molecules and particulate matter, some 10–30% or more of the incident solar radiation is depleted before reaching the surface and an additional amount is reflected or scattered back to space. In the same way that a black body absorbs all of the radiated energy, a white body will reflect such energy, with the fraction of energy reflected being termed the albedo. Thus, ice has a very high albedo, whereas a surface paved with blacktop has a very low albedo. You can appreciate albedo by performing the following experiment on a sunny day. Find a parking lot with both white and black vehicles. You can place your hand on the white car with impunity, but be very careful when touching the black vehicle, or you may leave some skin on the hood.

Longwave radiation emanating from the Earth's surface is also subject to absorption, reflection and scattering. If all of this radiation returned to space, we would have a much colder planet. The retained amount is what is referred to as the greenhouse effect. How the combination of absorption and reflection and scattering of both shortwave and longwave radiation impacts the Earth's climate is quite complex and beyond the scope of this chapter. As an example, Fig. 10.1 provides a depiction of reflected shortwave radiation and emitted longwave radiation as seen from orbiting satellites using sensors that sum the radiation across the distinctive spectra that encompass the shortwave and longwave radiative energy distributions. Both the spatial and temporal variability for either form of radiation is large and varying with season, landform, land temperature, ocean temperature, ice distribution, cloudiness, water vapor and other greenhouse gas distributions.

10 The Air–Sea Interactions that Determine Water Temperature

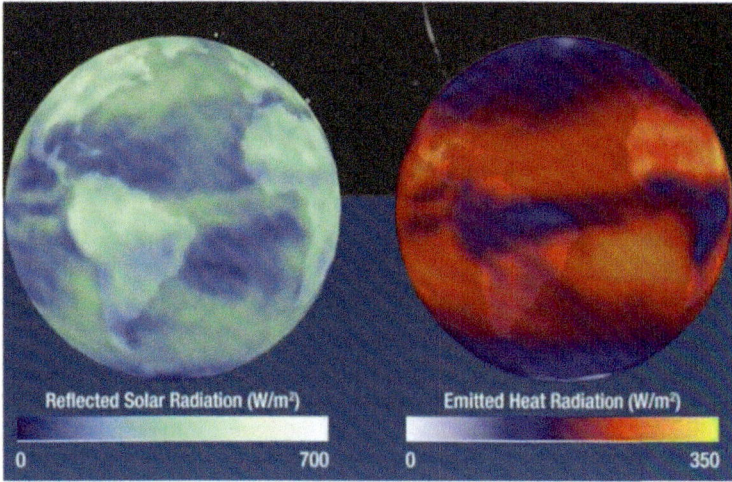

Fig. 10.1 Depictions of reflected solar (shortwave) radiation and emitted longwave radiation as seen from space by satellites equipped with shortwave and longwave radiation sensors (courtesy of NASA, https://science.nasa.gov/ems/13_radiationbudget/)

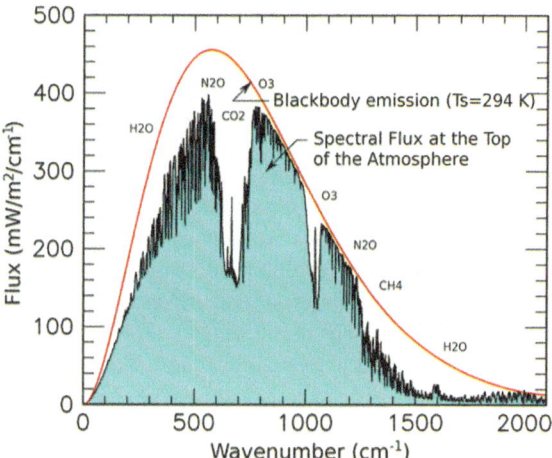

Fig. 10.2 Blackbody emission spectrum at 2940 °K compared with estimated spectrum at top of atmosphere showing wavelengths at which water vapor (H_2O), carbon dioxide (CO_2), ozone (O_3), methane (CH_4) and other greenhouse gases are most absorptive. In order of importance are H_2O, CO_2, O_3 and CH_4 (courtesy of NASA, https://www.giss.nasa.gov/research/briefs/archive/2010_schmidt_05/)

Of the various greenhouse gases, water vapor is the major one. This may be appreciated by looking at the difference between the longwave spectrum for black body emissions at a given temperature (294 °K in this example, or about 70 °F) and what would be observed at the top of the atmosphere given the greenhouse gases that are present (Fig. 10.2). Most of the reduction is due to water vapor, although a very prominent reduction is also seen as a notch associated with CO_2, and others by ozone and methane.

As discussed in Chap. 1, the turbulent heat fluxes are both sensible and latent. Whereas these occur at molecular levels, the air–sea interface is hardly at rest. Winds, waves and ocean currents provide a very turbulent, time and space-varying environment, necessitating a macro-scale statistical approach to these exchanges. The turbulent exchanges themselves can be estimated through measurements of the vertical fluxes of temperature and moisture at the air–sea interface and then relating these turbulent fluxes to more readily observed quantities, such as sea surface temperature, air temperature, humidity and winds at a standard level, typically 10 m above the ocean surface. We refer to such an approach as a parameterization, whereby we equate actual turbulent fluxes that are difficult to observe to quantities more readily observed through a statistically determined parameter. Such a parameter at the molecular equilibrium scale would be a molecular diffusion coefficient, whereas at the turbulent, macro-scale this becomes a much larger turbulent diffusion, or eddy coefficient. As an example, consider the winds blowing over the ocean. Anyone who has sailed a small boat knows that the winds are never constant. Tiller adjustments must continually be made with each puff or lull that results in a header (a tiller adjustment to head higher toward the windward mark) or a footer (a tiller adjustment to head lower toward the windward mark) and with each passing wave that also changes the local relative wind velocity between its crest and trough. Thus, to estimate the stress of the wind acting on the sea surface we must average over both time and space to arrive at what is referred to as a bulk parameterization of the wind stress given the wind velocity measured at a 10 m height.

From the preceding discussion, we find that a parameter is not something that can be observed. Instead, it is something to be estimated statistically and then used with other ocean or atmosphere properties that can be observed. Observables appear in mathematical expressions as state variables that are functions of space and time. For instance, temperature, as a state variable, will depend in some functional way on both space and time, and since we can actually observe temperature, we can test whether or not the functional formulation is a physically reasonable one. Parameters, on the other hand, are not observed; they must be inferred statistically in a way that best fits the actual observations of the state variables that they are used with. A pet complaint of mine is the interchangeable way in which some scientists use the words variables and parameters. They are not the same. Variables are observable; parameters are things that we estimate. These estimates may be good ones, but estimates based on assumptions, as contrasted with actual observables, they remain.

Given the foregoing introduction of both radiative and turbulent heat fluxes, let us now explore how these determine the ocean temperature. It all begins with observations; in fact, the scientific method itself begins with observations. We observe something and we ask why that might be. We then construct a hypothesis, devise an experiment to test that hypothesis and continue with that process until we find something that seems to explain the observation. It was in this manner that all of the conservation laws (e.g., mass, heat, momentum, angular momentum) that we employ in studying the role of the ocean circulation in Earth's climate and ecology were deduced.

Fig. 10.3 CMS-USF Ocean Circulation Lab's west Florida shelf moored array as presently deployed (2023). C10, C12, C13 and C22 are surface moorings; C11, C15 and C19 are bottom-mounted moorings

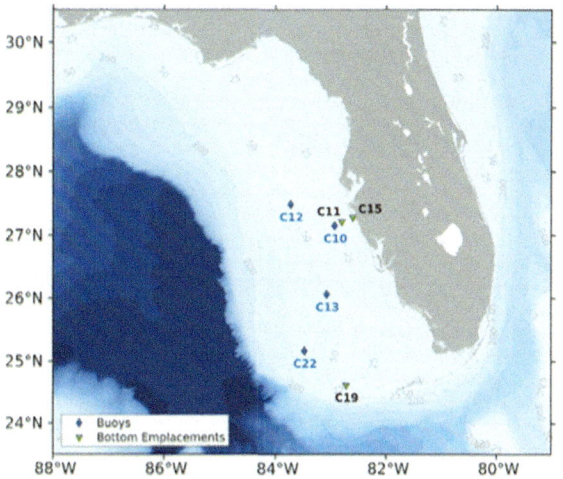

In an effort to describe and understand the working of the coastal ocean off Florida's west coast, the Ocean Circulation Lab housed in the College of Marine Science at the University of South Florida has sustained a set of moorings with both oceanographic and surface meteorological instrumentation since around 1998. The present array is shown in Fig. 10.3, and it consists of two types of moorings, one with surface floats that support both ocean and atmosphere observations and another with bottom-mounted instruments providing only ocean data. One of these, mooring C10, positioned at the 25 m isobath offshore from Sarasota, Florida, has a complete suite of instruments for use in heat flux analyses. Some 25 years of surface meteorological data from C10 show that while both the air temperature (AT) and the sea surface temperature (SST) show very distinctive annual cycles, with maxima in August and minima in February, as contrasted with the solstice months of June and December, all of the other variables are more complex. This may be seen in Fig. 10.5 for calendar year 2005, a year with the fewest data gaps in all of the observed variables.

Whereas the foregoing discussion on shortwave and longwave radiation considered what is incoming from the sun and outgoing from the surface, Fig. 10.4 also shows that there is incoming longwave radiation. This incoming longwave radiation is the greenhouse effect. The Earth's surface does indeed radiate, but some of that longwave radiation gets absorbed by clouds and greenhouse gases, upon which it is reradiated with half of it returning to the surface. The returning longwave radiation (the incoming part) is observed and the outgoing part is estimated using the Stefan–Boltzmann law, whereby the outgoing longwave radiation is proportional to the fourth power of the absolute temperature. Note that the incoming longwave radiation varies considerably throughout the year, particularly in fall through spring months due to clouds and water vapor.

Without developing the detailed formulae in use, Fig. 10.5 shows the radiative heat flux [the net incoming shortwave radiation (after accounting for reflection from the sea surface), plus the incoming minus the outgoing longwave radiation] in the

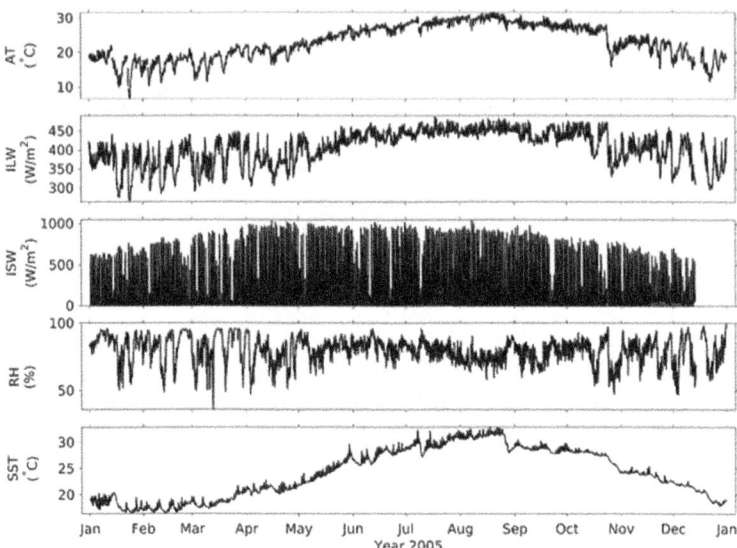

Fig. 10.4 Hourly observations at mooring C10 used in calculating the net surface heat flux for 2005, a year of no significant data gaps

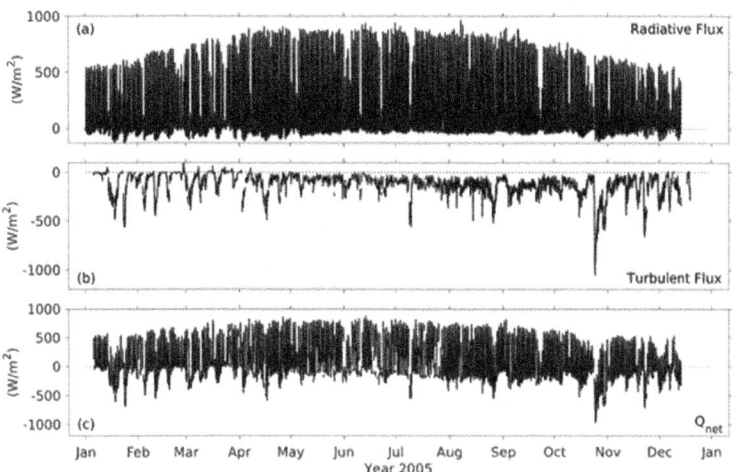

Fig. 10.5 Hourly time series of surface heat flux constituents at mooring C10 during 2005 and net surface heat flux. From top to bottom are: **a** the total radiative flux, **b** the total turbulent flux, and **c** net surface heat flux (Q_{net})

upper panel and the turbulent heat flux (sensible, plus latent) in the middle panel that together comprise the net surface heat flux in the lower panel. Positive values denote warming, and negative values denote cooling. It is difficult to interpret these hourly values, although in general we can say that radiation is generally warming in daytime and cooling in nighttime, whereas turbulence, with few exceptions, provides cooling. As pertains to the seasonal cycle, it is more advantageous to consider daily or even monthly averages.

Nonetheless, there is a reason for looking first at hourly values, as the following mathematics digression explains. The calculation of nearly all of these radiation or turbulence quantities is nonlinear, either by the product of variables multiplied together or a variable raised to a power greater than one. For example, outgoing longwave radiation depends on the fourth power of temperature, and the parameterizations for the turbulent fluxes depend on products of wind speed with other variables. Thus, it is important to make the nonlinear calculations first before averaging these because nonlinear calculations from averaged quantities are much different from the averages from nonlinear calculations. In the language of mathematics, linear operations commute (it does not matter what order you do them in), whereas nonlinear operations do not. As reiterated several times, it cannot be stated forcefully enough that proficiency in mathematics is essential for a career in science. Students adverse to mathematics should choose a different career path.

Data gaps preclude such detailed calculations for all years, but by first calculating all of the hourly values when these are available and then averaging these for each individual day and averaging the same days for each year (for years in which data for a given day exist), we may arrive at what is referred to as a climatologically averaged annual cycle, i.e., the averages for all January days, all February days and so on. Such daily mean climatologies for the heat flux constituents, their sum, plus SST are as shown in Fig. 10.6 for the time span of 1998–2022.

Thus, for the annual cycle, the top panel shows that the net shortwave radiation acts to warm the ocean with values ranging from around $100 \ Wm^{-2}$ in winter to around $300 \ Wm^{-2}$ in summer, whereas the net longwave radiation acts to cool the ocean with values ranging between around $50 \ Wm^{-2}$ in summer and $80 \ Wm^{-2}$ in winter. The smaller summer values are because the outgoing longwave radiation is larger then. The latent heat flux is minimal in spring and maximal in fall and with considerable variations in these two periods owing to the passage of cold fronts. The sensible heat flux is the smallest of these constituents, and it peaks in fall though winter months when the difference between AT and SST is largest. Adding these together yields a net surface heat flux that is positive from February through August and negative from the end of August through early February. Thus, it is not surprising that SST peaks at the end of August, when the net heat flux changes sign from warming to cooling, and it has its coolest values in February, when the net heat flux changes sign from cooling to warming. The reason why the phasing for these transitions is not synchronous with the sun may be explained on the basis of the turbulent contributions to the net surface heat flux, which depends on AT, SST, moisture and wind speed; in other words, climate is complex!

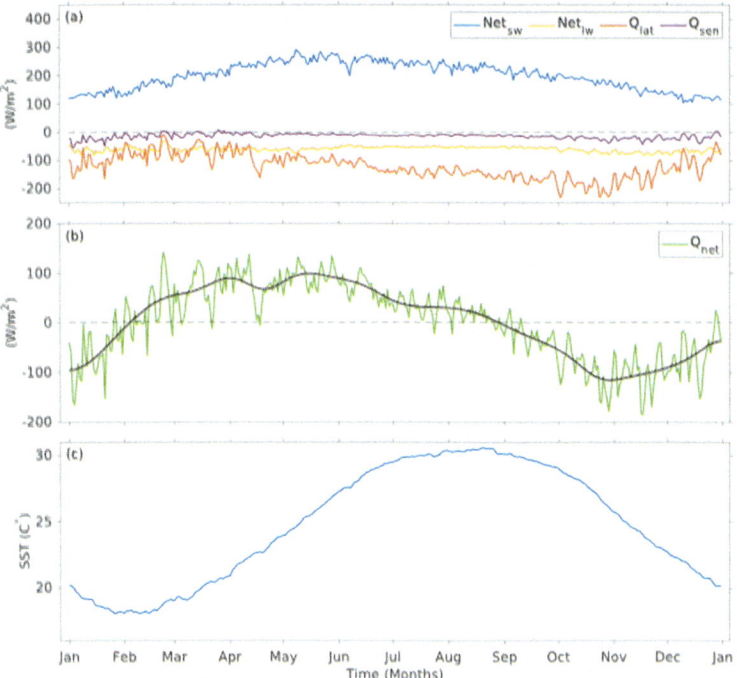

Fig. 10.6 Daily mean climatologies at C10 (by averaging from 1998 to 2022) of net heat flux constituents: **a** net shortwave radiation (Net$_{sw}$), net longwave radiation (Net$_{lw}$), latent heat flux (Q_{lat}) and sensible heat flux (Q_{sen}), **b** Net heat flux (Q_{net}) along with its 36-day low-pass filtered version (solid black line), plus **c** SST (adapted from Sorinas, L., Weisberg, R. H., Liu, Y., Law, J. (2023), Ocean–atmosphere heat exchange seasonal cycle on the West Florida Shelf derived from long-term moored data, *Deep-Sea Research Part II*, 212, https://doi.org/https://doi.org/10.1016/j.dsr2.2023.105341)

It is also edifying to note just how much smaller the climatologically daily averaged net shortwave radiation is when compared with the hourly values. In rough numbers, what might begin as around 1300 Wm^{-2} at the top of the atmosphere at noon in summer gets reduced to around 1000 Wm^{-2} at the sea surface, but with no incoming shortwave radiation at night, the daily average hovers around 300 Wm^{-2} in summer with much smaller values in winter. This is one reason why solar panels that are rated based on 1000 Wm^{-2} generate so much less power than their nameplate capacity would lead a consumer to believe (more on that topic will be addressed in Chap. 13).

The net longwave radiation provides a cooling influence, hovering around 50 Wm^{-2} with considerable interannual variability, which is one reason why some seasons may seem to be either warmer or colder than what the long-term climatology might suggest. Both of the turbulent heat flux contributors vary seasonally with the latent heat flux, hovering around 100–200 Wm^{-2}, being much larger than the sensible heat flux at only around a few 10 s of Wm^{-2}. Thus, the largest contributors to the net surface heat flux, which varies from positive values in summer to negative values in winter, are the net shortwave radiation and the latent heat flux; in other words, and as discussed in Chap. 1, climate begins with the sun and how both heat and moisture are then transferred from the ocean to the atmosphere.

What determines the ocean temperature is also complex. How the surface heat flux at the ocean–atmosphere interface gets distributed within the ocean depends on ocean processes. In shallow water (mooring C10 is at the 25 m isobath), turbulent mixing can effectively result in a uniform temperature from top to bottom. However, this mixing effect can be offset if the ocean currents themselves transport (or advect) either cooler of warmer water to a given location, either near the surface or the bottom.

Examples exist at C10 for years in which ocean processes other than mixing are, or are not important, as provided in Fig. 10.7. As we found in Chap. 4, the water properties of west Florida continental shelf can be profoundly affected by the adjacent deep Gulf of Mexico by virtue of how the Gulf of Mexico Loop Current interacts with the shelf slope region. Calendar year 2000 was one without such interaction, whereas calendar year 2010 saw a very strong and persistent interaction. The result in 2010 was an upwelling of relatively cold water onto and across the continental shelf that manifested in cool near-bottom waters at C10 from spring through summer months [i.e., compare the surface and near-bottom temperatures at C10 for these two years (2000 in the upper panel and 2010 in the lower panel of Fig. 10.7)]. From the top panel, we see that the surface heat flux alone, when distributed vertically across the water column, provides a reasonably accurately depiction what transpired during that year, at least for the portion of the year when observations are available for making the net surface heat flux calculation (note that the yellow line does not cover the entire year owing to data gaps). Calendar year 2010 tells an entirely different story. In that year the near-bottom temperature at 19 m depth is much colder than either the near-surface temperature at 1 m depth or the simulation using the net surface heat flux and the assumption of complete vertical mixing. These differences diminish once fall arrives because once the net surface heat flux changes sign to surface cooling, the colder water at the surface, being more dense than the warmer water beneath, is immediately mixed downward by turbulent convective mixing. So both the ocean–atmosphere exchanges and the internal ocean dynamics themselves are important in determining the water properties. The ocean may derive its properties from the atmosphere, but these properties can then be advected over great distances from where they may have first been derived. This transport of properties from one location to another adds additional complications to understanding how the Earth's climate functions.

The west Florida continental shelf is merely a microcosm of the larger global ocean. Nonetheless, what happens here does impact climate locally. If we compare how SST varies from year to year relative to the 1998–2022 climatology, we see that years when strong upwelling occurs tend to be cooler than normal, and conversely. At the time of this writing, it can be said that 2018 was the last time that a major upwelling event occurred during summer months. From 2019 onward, it would also seem that the summers have been somewhat warmer than normal. Is this a manifestation of

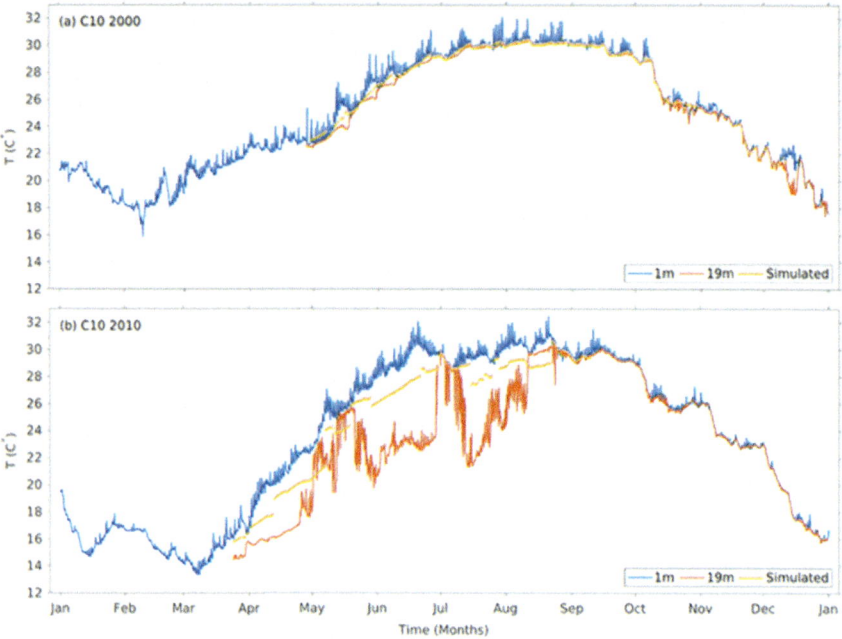

Fig. 10.7 Comparison of water temperatures in **a** 2000 with those in **b** 2010, a year with a strong offshore-induced upwelling. For both years, temperature observations were made at depths of 1 m (the blue line) and 19 m (the red line). These observations are compared with a one-dimensional distribution of the net surface heat flux (the yellow line, assuming no substantive ocean advection). The yellow line shows the simulated depth average temperature using the 1D temperature equation (adapted from Sorinas, L., Weisberg, R. H., Liu, Y., Law, J. (2023), Ocean–atmosphere heat exchange seasonal cycle on the West Florida Shelf derived from long-term moored data, *Deep-Sea Research Part II*, 212, https://doi.org/https://doi.org/10.1016/j.dsr2.2023.105341)

a globally warming climate, or is it simply the result of local ocean–atmosphere interaction, whereby the local SST is somewhat warmer than normal due to a lack of summertime upwelling. Once again, climate is much more complex than some in the media would like for you to believe!

Addendum: Flounder Meuniere

Sole meuniere is a classy dish when plated tableside, but this is just as good. The secret is fresh flounder, not frozen or sitting at the butcher for too long (flounder is delicate). The preparation is a quick saute followed by the making of a brown butter sauce. I like to do a reverse coating of the fish. Instead of flour, egg and whatever comes next, for this preparation I like to do egg, flour, egg. Melt some butter in a non-stick saute pan over low to medium heat. Salt and pepper the fish and then do

Addendum: Flounder Meuniere

egg, flour, egg before putting the fish in the pan. Cook until it begins to lightly brown and then gingerly turn it over to the other side and brown it lightly. The egg exterior should look golden versus dark. This should only take a couple of minutes per side. Remove the fish and then wipe the residual cooking fat and whatever else is there from the pan. Add some new butter and very slowing heat this swirling the pan now and then until the butter just begins to brown ever so slightly. Don't burn it or you will have to start over. Prior to doing this, squeeze a whole lemon into a cup and remove and pits. Also, chop up some fresh parsley. As soon as the butter browns a little, remove the pan from the heat and pour in the lemon juice, swirling the pan as you do so. This is the brown sauce that you can spoon over the fish filets that are already plated. Then sprinkle some parsley over the fish and you are done. I also like to put some capers on top.

I like to serve this with sauteed spinach made with some onions and garlic. A rutabaga mashed is also a nice side dish. To make this boil up some diced rutabaga along with the potatoes. Rutabaga takes longer so cut it into smaller pieces and boil these up first to make sure that everything is soft enough to mash together (about one part Rutabaga and three parts potato).

Chapter 11
Florida Red Tides

Florida residents and vacationers, especially those on the west coast between the Tampa Bay and Charlotte Harbor estuaries, often find themselves beset by the unpleasantries of red tide, a phenomenon that is neither red, nor a tide. Red tide is a euphemism for a harmful algae bloom (HAB) of which there are many different kinds worldwide. Florida red tides are generally associated with a bloom of the toxic dinoflagellate, *Karenia brevis*. *K. brevis* is known to produce toxins, which kill finfish, taint shellfish to the point of being hazardous for human consumption, cause respiratory distress in marine mammals and humans and make shore conditions repulsive to tourism and dining. Like *K. brevis*, other HAB species worldwide have similar deleterious effects, some much worse, and the ocean circulation lessons learned from *K. brevis* are, at least in a general sense, applicable elsewhere.

Before proceeding further, I must confess that the only thing that I really know about biological oceanography is how to filet a fish. Nonetheless, my limited knowledge of this species does derive from collegial interactions and co-authored publications with experts in the field of phytoplankton ecology, including the late Dr. Karen Steidinger for whom *Karenia brevis* is honorifically named and a former colleague at USF, Dr. John Walsh, who retired several years before I did.

As may be guessed from the foregoing chapters, *K. brevis* ecology is not just biology; it also entails all of the elements that are necessary for this phytoplankton species to outcompete other species to form a largely monospecific bloom. Once certain water properties (or lack of) were recognized as being necessary for this to occur, the role of the ocean circulation, being as important as the organism's biology, became readily apparent. This realization did not occur until around 1998, when a truly interdisciplinary study of *K. brevis* was initiated, a National Oceanic and Atmospheric Administration (NOAA) funded, Ecology and Oceanography of Harmful Algal Blooms (ECOHAB) regional field study. Through a series of physical, chemical and biological observations systematically acquired over several years and supported by both circulation and biological models, a more quantitative understanding of *K. brevis* ecology began to emerge. While a complete understanding remains elusive, we at least have reasonable theories on what may facilitate the occurrence of a major *K. brevis* bloom in any given year and, based on these theories, the ability to make seasonal predictions regarding bloom severity. Once a bloom is observed, we also have the ability to track its movement.

Nonetheless, misunderstandings continue to abound, and the public is often treated to oversimplified explanations and remedial suggestions by Non-Governmental Organizations (NGOs) seeking to influence political action. For instance, it has been purported that if we would only stop sprinkling fertilizer on our lawns or curtail the established sugar and cattle industries, then red tide would miraculously disappear. By first laying out the historical facts to see what may be sensible, we may be able to debunk some of the myths surrounding red tide and better understand what may, or may not, be feasible.

Once Florida red tide was identified as being associated with a certain toxic dinoflagellate, its sampling, via the collection of water and the counting of cells using microscopy, systematically ensued around 1953. Shown in Fig. 11.1 is the ensemble of locations where such sampling identified *K. brevis* cells from 1953 to 2019, and several apparent facts are as follows. First, the epicenter for *K. brevis* throughout Florida is the region from just north of Tampa Bay to just south of the Charlotte Harbor–Sanibel estuary complex. A secondary region of occurrence is the Florida Panhandle, especially to the west of Apalachicola Bay, with much less abundance found within the Florida Big Bend region. Lastly, we see that some cells may also be found along the Florida Keys and the east coast, but at a much lesser abundance than in the epicenter region. Explanations regarding the ellipse and arrows shown in this figure will follow later.

Whereas sampling of water and counting of cell abundance began around 1953, there exists anecdotal evidence dating back to 1878, as documented by the Florida Wildlife Commission (Fig. 11.2). The earliest reports were primarily by fishermen whose catches were found to mysteriously die owing to the red tide toxins, or who themselves were subjected to the respiratory irritations caused by these toxins.

Evident from Fig. 11.2 is that red tide occurrences and durations differ from year to year. While red tides tend to occur in almost every year, the severity and duration of such a bloom are quite variable. To generalize, the observations show that a red tide onset occurs in late summer to early fall with subsequent cessation in winter. Despite such generalizations, red tide blooms can exist throughout the year, with 2005 and 2018 providing the two most recent examples. Even prior to the documentation in Fig. 11.2, evidence of red tide may date back to the sixteenth-century logbooks of Fernando DeSoto's Gulf of Mexico exploration and from Native American practices of eschewing shellfish as part of their diets in summer and fall months. By pre-dating Florida's population expansions, its commercially organized agriculture (crops and livestock) and the more than the tenfold increase in the sugar industry over the past half century, it is apparent that Florida red tides are a natural, versus a human-made phenomenon. Whereas simple explanations and remedies are desirable, not only can we not fix something if we do not know how it actually works, we do not know if the consequences of certain remedial actions may actually be worse than the red tide itself.

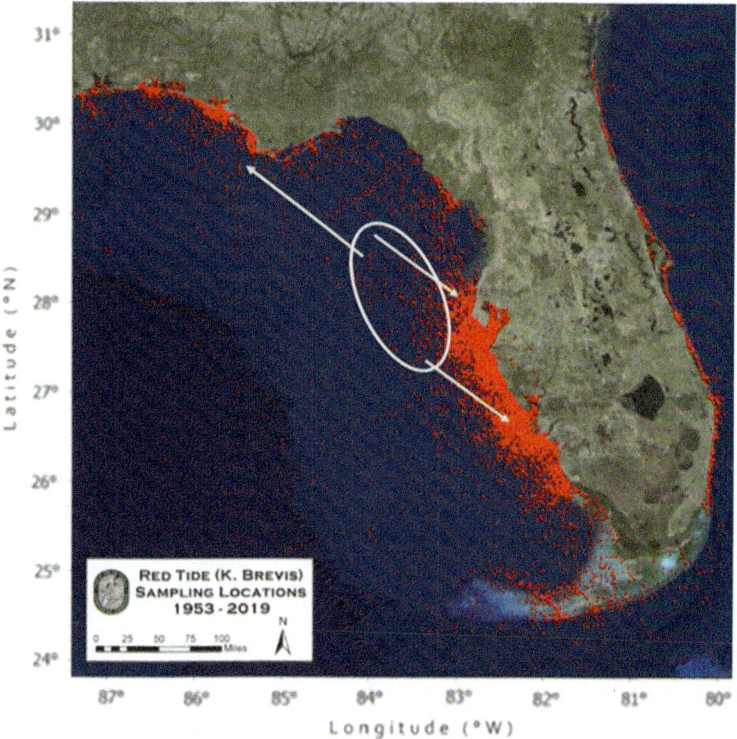

Fig. 11.1 Composite of all observations of *Karenia brevis* made by the Florida Wildlife Research Institute from 1953 through 2019. Note the epicenter region from just north of the Tampa Bay to just south of Charlotte Harbor estuaries. The ellipse denotes the hypothesized, mid-shelf initiation region, and the arrows denote the transport pathways between initiation and manifestation within the epicenter region and along the Florida Panhandle (adapted from Weisberg, R. H., Liu, Y., Lembke, C., Hu, C., Hubbard, K., Garrett, M. (2019), The coastal ocean circulation influence on the 2018 West Florida Shelf K. brevis red tide bloom, *J. Geophys. Res. Oceans*, 124, 2501–2512, https://doi.org/10.1029/2018JC014887)

Phytoplankton ecology is an expansive, complex field, with over 20,000 different species identified as populating the oceans. Among these are the various species of diatoms, dinoflagellates and cyanobacteria. Diatoms are photosynthetic unicellular organisms with cell walls made of silica. Dinoflagellates are also unicellular and photosynthetic, and some are also thought to be mixotrophic, meaning that they can also feed upon prey (analogous to a Venus flytrap). By virtue of flagella, dinoflagellates can also move vertically across the water column under their own locomotion. As an example, flagella are evident in Fig. 11.3, a scanning electron microscope image of *K. brevis*. Unlike diatoms, dinoflagellates do not require silicic acid (silica or silicate combined with water) for their cell structure. Cyanobacteria, also called blue-green algae, like diatoms and dinoflagellates, are unicellular and photosynthetic, and some (known as diazotrophs) have the added ability fix nitrogen compounds from

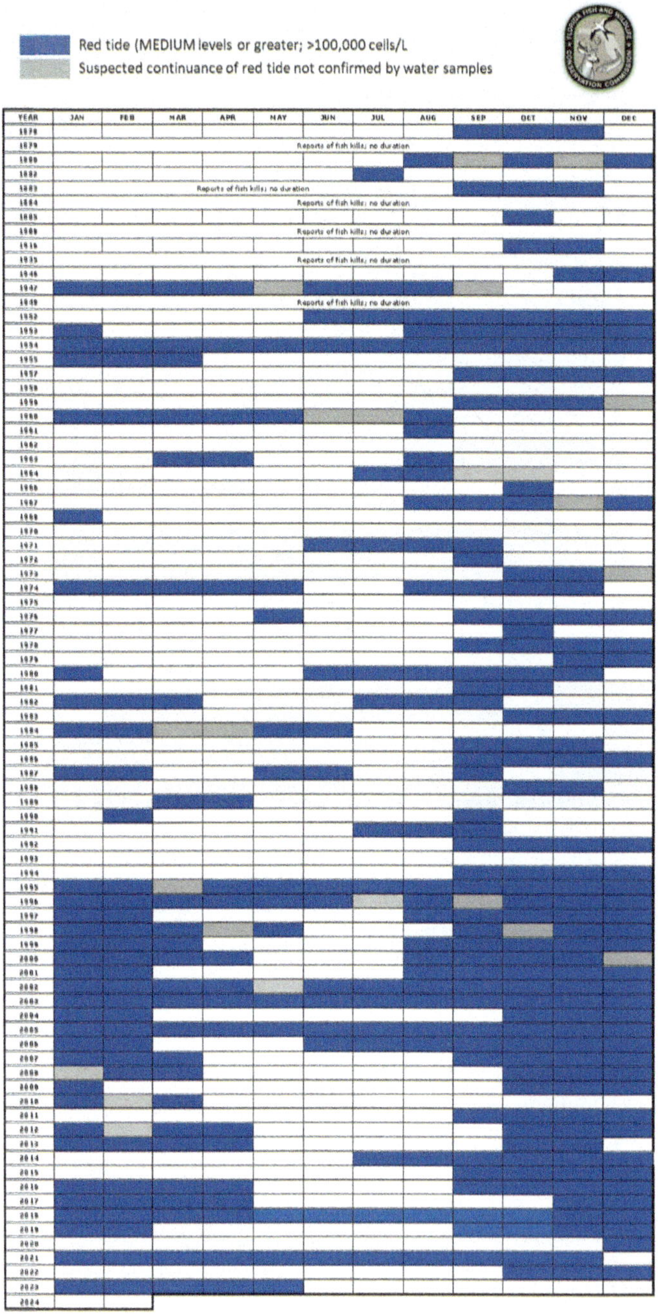

Fig. 11.2 Anecdotal and in situ evidence of red tide occurrence from 1878 to 2024 (courtesy of Florida Fish and Wildlife Conservation Commission, https://myfwc.com/media/drOfc1sm/bloom-historic-database.pdf)

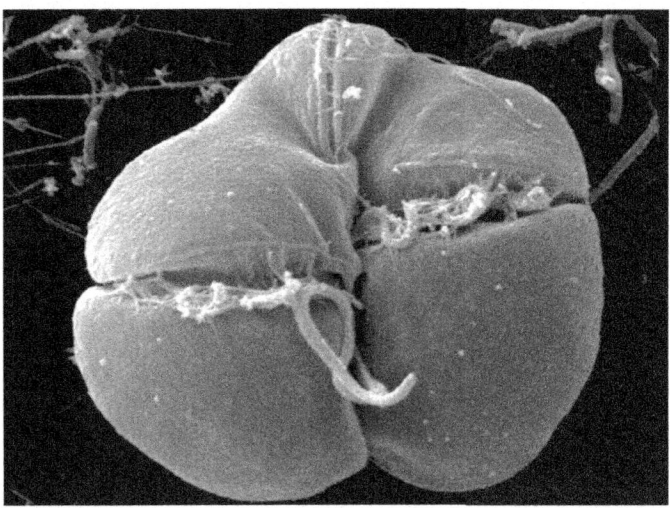

Fig. 11.3 Scanning electron microscope image of a *K. brevis* cell (courtesy of Florida Fish and Wildlife Conservation Commission, https://www.flickr.com/photos/myfwc/albums/72157626901199940/)

the nitrogen gas of the atmosphere. The cyanobacterium, *Trichodesmium* is one such diazotroph. All of these various plant forms, plus others, have their own unique attributes, and these three categories are thought to play central roles in *K. brevis* ecology.

The varying attributes among these three classes of organisms appear to determine whether or not a *K. brevis* bloom may occur in any given year. One additional ingredient is growth rate, or how quickly individual cells may divide to produce new cells. Diatoms tend to grow rapidly, whereas dinoflagellates grow much more slowly. Thus, if all of the water properties are conducive to diatom dominance, then this will occur. However, diatoms require silicic acid for cell wall production, whereas the dinoflagellate, *K. brevis* does not. So despite their slow grow rates, in the absence of sufficient silicic acid, *K. brevis* may be able to dominate over diatoms.

The foregoing biological information, terse as it may be, is sufficient to generate a hypothesis on *K. brevis* bloom formation. The overall scenario, while convoluted, may be simplified as follows. Photosynthesis requires both light and nutrients, nitrogen and phosphorus compounds in particular, plus, in the case of diatoms, the added ingredient of silicate. The west coast of Florida, for paleontological reasons, has an abundance of phosphorous compounds, leaving nitrogen and silica compounds the limiting ones. Recall from the discussions of Chap. 5 that the middle portion of the WFS tends to be nutrient deplete, or oligotrophic, making it difficult for diatoms to grow in the absence of silica. Given adequate phosphorus compounds, what might happen if a source of nitrogen compounds were to materialize? Under such condition, and without much competition, *K. brevis* might be able to gain a foothold as a monospecific phytoplankton bloom at mid-shelf. This is where cyanobacteria come into play. *Trichodesmium* is one such cyanobacterium species that fixes nitrogen, allowing it to persist in otherwise oligotrophic waters. However, *Trichodesmium* requires iron to function, which occurs each year as Saharan dust is transported westward aloft in the atmosphere by the trade winds, particularly in spring and summer months. When all of these ingredients coalesce (a lack of silica, abundant phosphorus and a source of iron to support *Trichodesmium*, which fixes nitrogen), then *K. brevis* can thrive. Once *K. brevis* becomes the dominant species, it can then continue to compete with diatoms if transported to an environment where diatoms may also reside. Conversely, if conditions at mid-shelf are conducive to diatoms thriving there, then *K. brevis* is unable to gain such a foothold.

From the foregoing discussion, we see that the key to *K. brevis* blooming, or not, is the oligotrophic nature of the WFS. A springtime onset coincides with the seasonal increase in light, as required for photosynthesis. Increasing water temperature is also helpful, and the converse occurs in winter when sunlight and temperature are less. Along with these obvious factors, the main controlling element is the suite of nutrients that may be available to support a bloom.

For reasons entirely serendipitous, the 1998 ECOHAB regional field study coincided with a Northeast Gulf of Mexico hydrographic survey funded by the Bureau of Land Management and conducted by oceanographers from the Texas A&M University. This provided another complemental dataset to the north of the ECOHAB sampling domain. 1998 was also a year when the Gulf of Mexico Loop Current set the entire WFS in an upwelling favorable motion by impacting the shelf slope region near the Dry Tortugas. The upwelling circulation lasted long enough to bring new, inorganic nutrients from the deeper Gulf of Mexico onto the WFS and subsequently to transport these shoreward along the bottom (as explained in Chap. 5). The Northeast Gulf of Mexico observations confirmed the elevated nutrient levels, and the ECOHAB observations showed the upwelling circulation and only a muted *K. brevis* bloom in that year. Through a description of the observations and quantitative modeling of both the circulation and the phytoplankton ecology, my colleague, Dr. John Walsh and I, along with our students, published a coordinated pair of scientific papers that demonstrated the simplistic scenario explained above.

Continued observations and model simulations added further credence to these ideas. 2010 was a year with virtually no red tide, and this was explained on the basis of the circulation alone. Upon comparing the red tides of 2012 and 2013, the first of these being relatively pronounced, whereas the second was muted, and the circulation again provided an explanation. During 2012 we also had water column mappings of temperature, salinity, chlorophyll and oxygen using a robotic glider, a torpedo-shaped device with wings that is propelled by buoyancy changes. By going from deeper to shallower waters, we observed the *K. brevis* bloom (as indicated by high chlorophyll values) within a narrow, near-bottom depth range between the 50 m isobath and about the 20 m isobath. Only in water depths less than 20 m did the bloom appear at the surface. This confirmed the offshore origin of the bloom and the shoreward transport pathway within the bottom Ekman layer via an upwelling circulation.

It was previously known that a *K. brevis* bloom could be detected using satellite color imagery, but the aforementioned observations demonstrated that such detection can only occur after the bloom reaches shallow enough water to be mixed to the surface. Thus, these glider observations confirmed that *K. brevis* blooms form offshore along the bottom at mid-shelf, only making their presence known when the get close to shore. In other words, they cannot be detected by satellite color imagery until they actually become a near-shore nuisance bloom. Whereas this had been previously surmised, it took actual observations to find confirmation.

The foregoing discussion brings us back to Fig. 11.1 and the ellipse and arrows shown therein. Based upon the oligotrophic, mid-shelf argument for conditions conducive to *K. brevis* bloom development and the directionality of flow within the bottom Ekman layer for delivery of such a bloom to the nearshore, the ellipse is the hypothesized *K. brevis* bloom formation region and the arrows are the pathways along the bottom to the nearshore. Such configuration further helps to explain why the region from just to the north Tampa Bay to just to the south of the Charlotte Harbor–Sanibel estuary complex is the epicenter region: That is where red tide cells are delivered by the circulation once a bloom is formed offshore.

If, while the offshore bloom is developing, there occurs strong currents directed along isobath toward the north-northwest, then some of the red tide cells may be transported to the Florida Panhandle region, especially to the west of Apalachicola Bay, but rarely would there be conditions that would transport red tide cells to the Big Bend region; hence, the only nominal occurrences observed there. Furthermore, once red tide cells get to the epicenter region, if the upwelling circulation persists and is strong enough, then some of the cells in the nearshore that are mixed to the surface can be transported offshore to eventually become entrained in the Loop Current/Florida Current/Gulf Stream system to facilitate their transport to Florida's east coast. In other words, all of the observations over the years from 1953 through the present time, whether in the epicenter region, the Panhandle coast or the east coast, can be explained by red tides initiating in the hypothesized formative region (within the ellipse shown in Fig. 11.1).

An example that illustrates these findings was provided by the *K. brevis* red tide bloom of 2018. As stated previously, the manifestation of a typical bloom as a nearshore nuisance has its onset in late summer to early fall and its demise in winter. The bloom in 2018 was an example of a very severe bloom, which lasted into the following year, perhaps the most oppressive bloom since 2005. The 2018 bloom actually started in 2017 as a normally occurring event that for reasons not fully understood continued throughout the winter/spring of 2018. The conditions in 2018 were again conducive to the formation of a new, near-bottom, mid-shelf bloom, and once an upwelling circulation set in after the Loop Current made contact with the shelf slope near the Dry Tortugas in August, this new bloom was transported to the near shore, adding to what was already in place from 2017.

The evolution of the bloom concentrations (cells/liter) is shown in Fig. 11.4. The 2017 bloom started as is typical in October when concentrations exceeded 10^6 cells/liter. The black, red and blue dots represent observations in the epicenter, Panhandle and east coast regions, respectively. The red dot observations were subsequent to

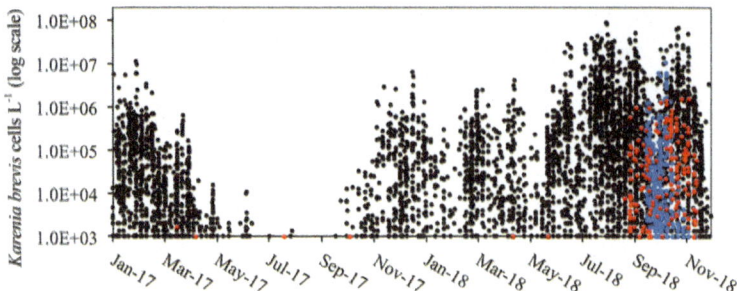

Fig. 11.4 *K. brevis* cell counts plotted using log scale from 1/1/17 through 12/16/18 (from the Florida Fish and Wildlife Conservation Commission—Fish and Wildlife Research Institute HAB monitoring database) show the persistence of the *K. brevis* bloom from 2017 through the spring and early summer of 2018 in the epicenter region (black circles), and the initiation of bloom conditions in the Panhandle (red) and on the east coast (blue). Only samples with $\geq 10^3$ cells per liter are shown (adapted from Weisberg, R. H., Liu, Y., Lembke, C., Hu, C., Hubbard, K., Garrett, M. (2019), The coastal ocean circulation influence on the 2018 West Florida Shelf K. brevis red tide bloom, *J. Geophys. Res. Oceans*, 124, 2501–2512, https://doi.org/10.1029/2018JC014887)

tropical storm Gordon, which imposed strong, wind-forced northward currents for a few days. The blue dot observations were after the *K. brevis* cells were first transported to the nearshore and upwelled to the surface from which they were subsequently transported offshore and to the south to become entrained into the Loop Current/Florida Current/Gulf Stream system. Once entrained, the transit time to around Palm Beach on the east coast is very short, only a few days.

A glider survey conducted from 8/24/18 to 9/17/18 mapped out the location of near-bottom chlorophyll, with highest concentrations observed between the 30 m and 40 m isobaths. Using the West Florida Coastal Ocean Model (WFCOM), a numerical circulation model (see Chap. 5), the trajectories taken by these phytoplankton cells (presumably *K. brevis* cells) were determined, as shown in Fig. 11.5. The trajectories are in accordance with an upwelling flow within the bottom Ekman layer that is directed onshore and southward. These trajectories are also consistent with the epicenter region from just north of Tampa Bay to just south of the Charlotte Harbor–Sanibel estuary complex.

Continued trajectory simulations using WFCOM further showed how some of the *K. brevis* cells, if mixed up closer to the surface, may have been transported to the Panhandle by Tropical Storm Gordon and also how *K. brevis* cells from the epicenter regions may have subsequently been transported to the east coast. A demonstration of the latter transport pathway is provided in Fig. 11.6. Two month-long simulations, one beginning on August 1, the other on September 1, in the top two panels are borne out by the satellite and in situ observations in the bottom two panels. Each of these simulations shows that once the near-bottom transported *K. brevis* cells reach the near shore and are upwelled and mixed to the surface (e.g., Fig. 11.5), they may then be transported offshore and southward at the surface by the same upwelling circulation. When they get far enough south, they may then be entrained into the Loop Current/

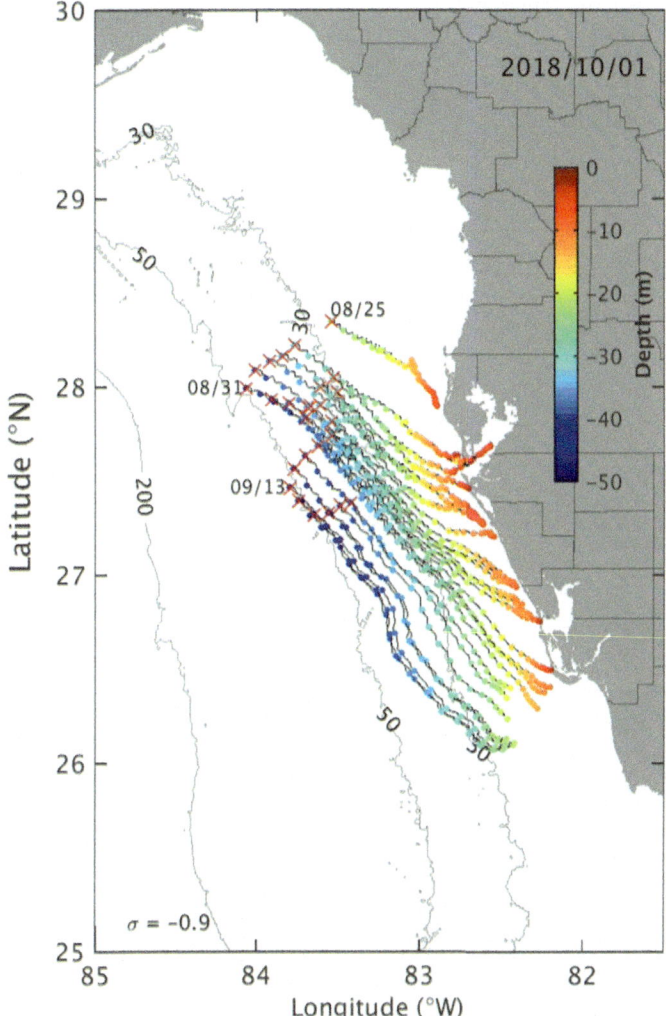

Fig. 11.5 WFCOM estimated trajectories for *K. brevis* cells initialized at the locations and days for which high chlorophyll levels were observed by a glider survey (red X's). Color coding indicates the simulated cell depth (adapted from Weisberg, R. H., Liu, Y., Lembke, C., Hu, C., Hubbard, K., Garrett, M. (2019), The coastal ocean circulation influence on the 2018 West Florida Shelf K. brevis red tide bloom, *J. Geophys. Res. Oceans*, 124, 2501–2512, https://doi.org/10.1029/2018JC014887)

Florida Current/Gulf Stream system as they flow around the Dry Tortugas. Such offshore transport was captured by satellite imagery as well as by in situ sampling and subsequent microscopy.

With the origination zone and transport pathway established for this 2018 *K. brevis* red tide event, one additional question may be addressed by using the same WFCOM circulation model. That question is: What may have caused the termination of this

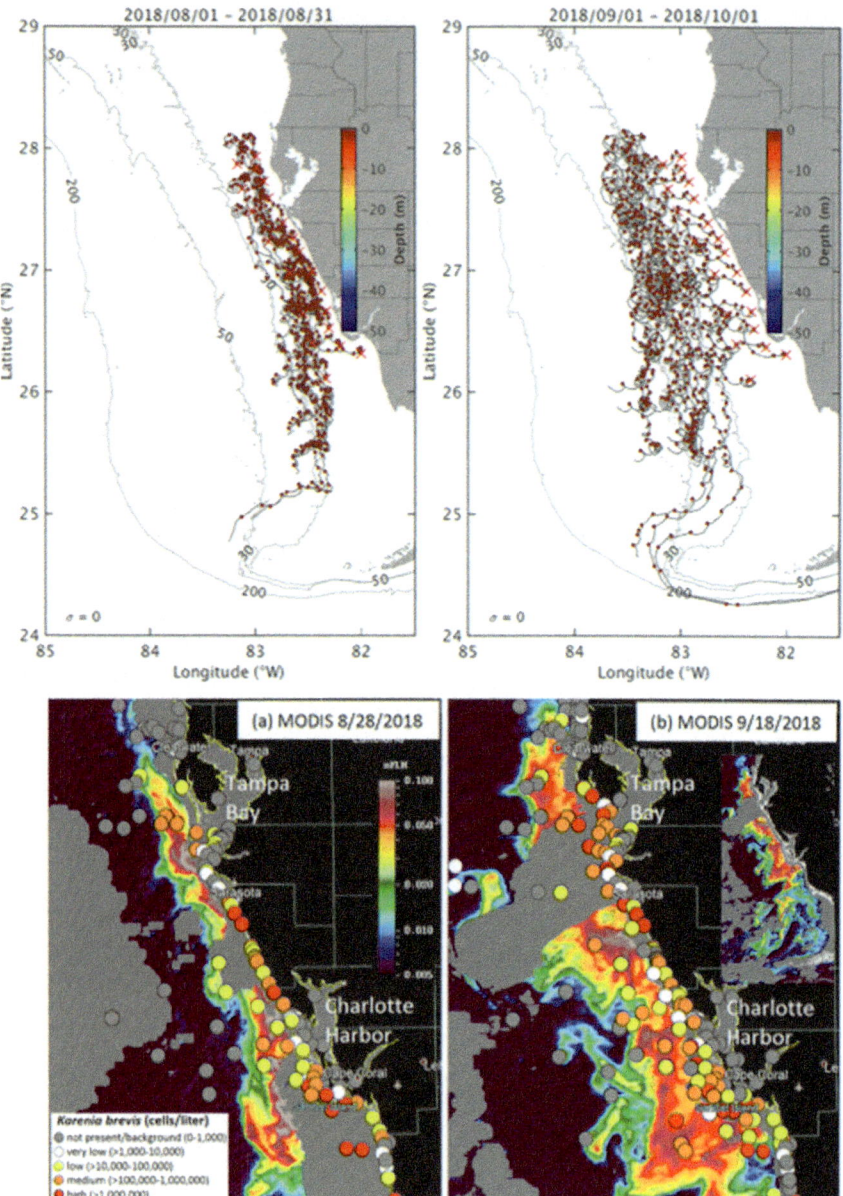

Fig. 11.6 Top panel shows WFCOM simulations of particles initialized at the surface and tracked for one-month intervals starting on 8/1/18 (left) and 9/1/18 (right). The bottom panel shows satellite imagery (C. Hu, personal communication) findings for *K. brevis* and actual cell counts (K. Hubbard, personal communication) for 8/28/18 (left) and 9/18/18 (right)

bloom? Could the circulation, by itself, be responsible for bloom termination? To address this question, we considered a situation when the offshore source of new cells was no longer available, either by depletion, or by the establishment of diatoms in place of *K. brevis* once the nutrient levels were sufficient to support these. In other words, how long would it take for the *K. brevis* bloom, as observed nearshore, to end under the influence of a persistent upwelling circulation, and might this be consistent with observations?

Figure 11.7 shows the spatial distribution of *K. brevis* cell counts for the months of September 2018 through February 2019. September, October and November show cell concentrations at bloom levels, whereas December and January show limited regions with nominal levels, and by February the bloom, in essence, is gone. The simulation in Fig. 11.8 shows a similar evolution. Even without any active biology, the circulation alone can account for the *K. brevis* bloom demise. As a corollary, the failure of the 2017 bloom to dissipate may also have an ocean circulation explanation. Throughout the 2017 bloom evolution, there were no extended instances when the Loop Current effected an upwelling circulation on the WFS until the end of August 2018. Thus, the lack of substantial flushing by the coastal ocean circulation in 2017 may have enabled the 2017 bloom to persist throughout the year and into 2018. Whereas there remains much to be learned about *K. brevis* ecology, what seems clear is that the ocean circulation is as important for *K. brevis* ecology as is the organism's biology.

One important aspect of the *K. brevis* biology not yet mentioned is the brevetoxin that the organism produces. This is not by accident. By killing fish via toxin production, *K. brevis* is capable of making its own food supply, either indirectly from the nutrients released by the decaying fish or more directly through mixotrophic behavior. In either case, the toxins are what make *K. brevis* bloom so insidious. While it may be difficult for a bloom to initially form by virtue of the organism's slow growth, once a bloom is in place, it is difficult to get rid of it because of the continual food supply via the effects of the breve-toxins. So while the addition of land-derived nutrients via river runoff may further enhance a bloom, such nutrient inputs are not necessary because the organism is capable of producing its own nutrient supply.

Given these findings on how *K. brevis* blooms may be related to the ocean circulation, is it possible to develop a scheme for forecasting whether, or not, a major *K. brevis* bloom will occur in any given year? With the hypothesized determining factor for mid-shelf oligotrophic conditions in spring and early summer months being the Loop Current, it seemed reasonable to relate Loop Current pattern evolution to *K. brevis* bloom evolution.

One of my former students, Prof. Yonggang Liu, was experimenting with a neural network, artificial intelligence, technique known as the self-organizing map (SOM) designed for pattern recognition. The most comprehensive dataset for Loop Current detection is that obtained by satellite altimetry, whereby radio (radar) waves are used to determine the height of the sea surface from which surface currents may be estimated using the geostrophic relationship (e.g., Chap. 3). Satellite altimeters have been aloft as an international effort since 1993, and analyses performed by Archiving,

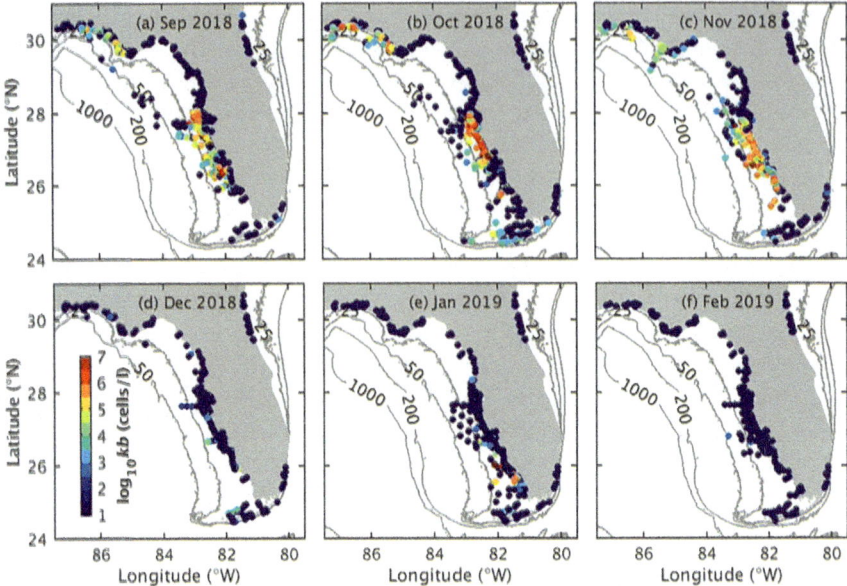

Fig. 11.7 Monthly *K. brevis* cell count distributions for September 2018 through February 2019 color coded on a logarithmic scale, red being 10^6 cells/liter (courtesy of the Florida Wildlife Commission, Florida Wildlife Research Institute) (adapted from Liu, Y., Weisberg, R. H., Zheng, L., Heil, C. A., Hubbard, K. A. (2022), Termination of the 2018 Florida red tide event: A tracer model perspective, Estuarine, Coastal and Shelf Science, 272, 107,901, https://doi.org/10.1016/j.ecss.2022.107901)

Validation and Interpretation of Satellite Oceanographic Data (AVISO) yield daily gridded data over the global ocean with horizontal resolution of 1/3°. Using the SOM technique, we fit the available daily images to 20 patterns that best matched the over 8000 individual images available for analyses. Together with the patterns is a time history of how these patterns evolve, enabling comparisons to be made between the evolution of these Loop Current patterns and the evolution of the *K. brevis* blooms. Being that the time history of pattern occurrence is qualitative, versus quantitative, a correlation analysis is not appropriate. Nonetheless, we can compare when the Loop Current patterns with pressure point (the shelf slope region near the Dry Tortugas) contact in spring through early summer months (the spring bloom period) coincide with a subsequent red tide bloom (Fig. 11.9). The analysis window spans 1993–2022. Magenta lines are the weekly averages of the highest five *K. brevis* cell counts with the ones exceeding 10^6 cells/liter highlighted by red dots. The blue bars indicate the number of days in a given month when the Loop Current was associated with one of the pressure point patterns, and the green bars are a qualitative pressure point forcing index defined as the cumulative number of days for which the relevant pressure point patterns occurred in spring to early summer (4/1 to 6/30) for at least 7 days in each of at least two months.

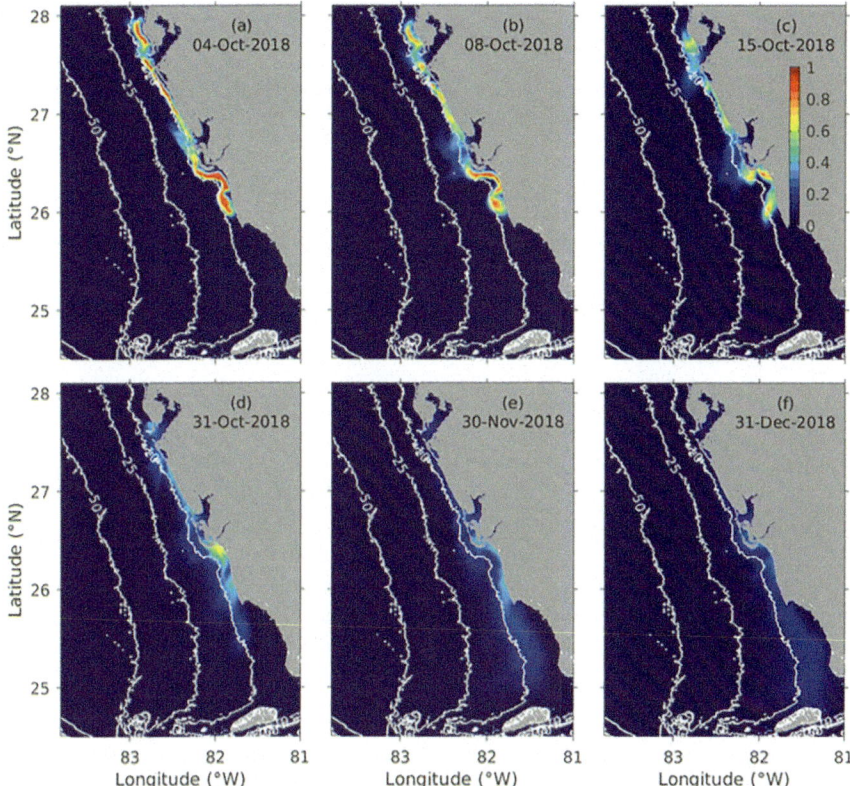

Fig. 11.8 WFCOM simulated tracer concentrations (color coded at values between zero and 1.0) sampled on October 4, 8, 15, 31, November 30 and December 31, 2018, after initialization of a normalized tracer (set at a concentration of 1.0) over a nearshore strip between the shoreline and the 10 m isobaths (adapted from Liu, Y., Weisberg, R. H., Zheng, L., Heil, C. A., Hubbard, K. A. (2022), Termination of the 2018 Florida red tide event: A tracer model perspective, Estuarine, Coastal and Shelf Science, 272, 107,901, https://doi.org/10.1016/j.ecss.2022.107901)

From the red dots alone, it is evident that not all years have major red tides. 1994–1996 were three successive years of major red tides followed by a hiatus with short-lived minor events until 1999. 2005 has a major prolonged red tide, and the one in 2008 just discussed was even worse. In contrast, 2008–2010 was nearly red tide free. Given Fig. 11.9, we can produce a scorecard of how well the qualitative criterion did as a seasonal predictor (Fig. 11.10). Of the 30 years considered, 25 followed the criterion, whereas 5 did not. The suggestion of no major bloom was correct for 7 out of 8 years. The suggestion of a major bloom was correct for 18 out of 22 years.

We originally did this work in 2015, and, on that basis, I went public with an advisory that there would be no major red tide in 2016. As Yogi Berra once said: "It's tough to make predictions, especially about the future." Determining why a

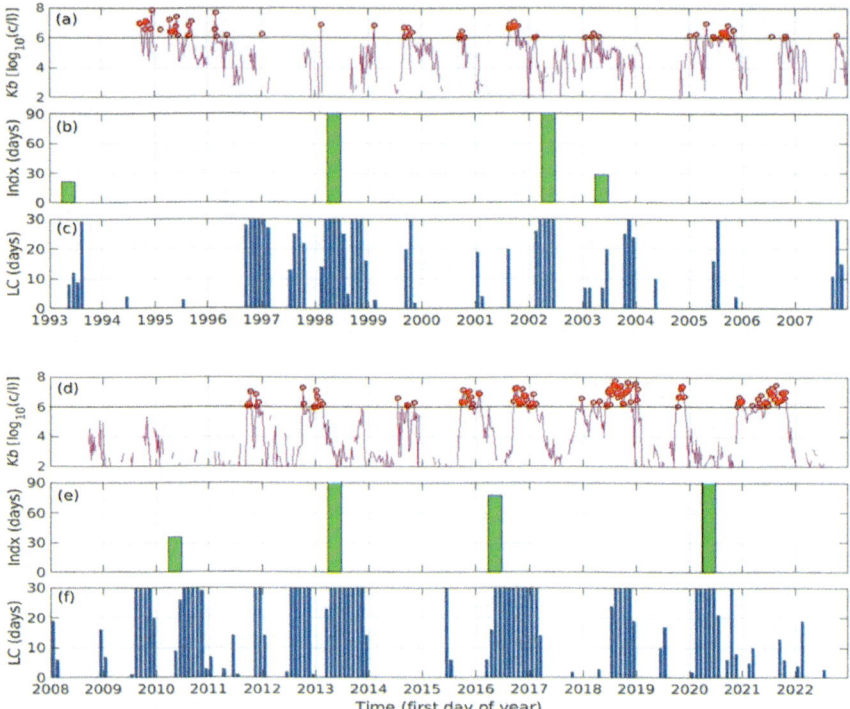

Fig. 11.9 Time history from 1993 through 2022 of *K. brevis* cell counts (magenta with cell counts $> 10^6$ cells/liter as red dots), number of days in a given month of pressure point contact (blue) and a qualitative determination of when a subsequent major red tide bloom should not occur (green bars)

criterion might fail is as useful as the criterion itself. The 2016 red tide event was special in that there were massive sewage spills in Tampa Bay and along the west coast of Florida due to very heavy rainfalls that overloaded sewage treatment plant capacities. *K. brevis* prefers organic nutrients, and because there are always some cells in the environment, it is possible that these *K. brevis* cells responded to the sewage loading because the bloom onset occurred within two weeks of the sewage spills. I got a little shy about making public statements after 2016, but felt confident enough to do so again in 2022, and that time the prediction was borne out by a major red tide.

2021 was another odd year. A red tide materialized as expected, but it became supercharged after there was an intentional drainage of wastewater into Tampa Bay from the defunct Piney Point phosphate stack when its levees were about to rupture. The immediate ecological response to the injection of high nutrient levels (inorganic nutrients this time) was a diatom bloom quite visible to the naked eye. This bloom then died off, resulting in its nutrients being recycled into organic forms. While this was occurring, the *K. brevis* bloom that had been in existence to the south of Tampa Bay was systematically transported northward by persistently anomalous southerly

Year	Persistent offshore forcing in spring & early summer (√)	Major blooms (×)	No major fall bloom (√) or the earlier year bloom reduced intensity (√*)	Outlier years (F)
1993	√		√	
1994		×		
1995		×		
1996		×		
1997			√*	F
1998	√		√*	
1999		×		
2000		×		
2001		×		
2002	√		√	
2003	√		√	
2004			√	F
2005		×		
2006		×		
2007		×		
2008			√	F
2009			√	F
2010	√		√	
2011		×		
2012		×		
2013	√		√*	
2014		×		
2015		×		
2016	√	×		F
2017		×		
2018		×		
2019		×		
2020	√		√	
2021		×		
2022		×		

Fig. 11.10 Pressure point criteria scorecard. Green shading denotes years of no major bloom. Red X's denote years with major blooms. Check marks on the left denote the criteria of persistent pressure point forcing in spring and early summer. The second set of check marks indicates that there was no major bloom in the usual fall onset period and the check marks with a * indicate a bloom of reduced intensity. The F's are those years where the simple qualitative criterion failed

winds of March and April. Once arriving at the mouth of Tampa Bay, these *K. brevis* cells were quickly transported into the bay, and by the end of May 2021, about two months after the start of the Piney Point wastewater release, a full-blown red tide bloom ensued beginning just where the initial diatom bloom had occurred (see Chap. 5 and Fig. 5.13). As this *K. brevis* bloom increased, it killed copious amounts of fish resulting in the recycling of considerable more nutrients than was contained in the Piney Point wastewater, thereby allowing the *K. brevis* bloom to continue to thrive for most of that year.

The lessons from these outlier year experiences are that while *K. brevis* red tide is a natural phenomenon independent of human activities, major perturbations caused by human endeavor can exacerbate or even cause a major red tide bloom when one might otherwise not occur. Three mitigating activities that federal, state and local governments can effect are: (1) to fix the Lake Okeechobee levee system and improve water retention capacities so that high nutrient water releases will no longer be necessary, (2) update aging, overloaded sewage treatment facilities and (3) deal decisively with legacy mining waste heaps to avoid future Piney Point-like incidents.

It is the opinion of this author that all efforts should be made to avoid contaminating the environment; hence, holding commercial and agricultural endeavors to high standards of environmental stewardship is reasonable. However, these will not eliminate *K. brevis* red tide. Instead of the foci of many officials and NGOs being the curtailment of commercial and agricultural activities and the limiting of residential fertilizer usage, a focus on accomplishing the above three objectives would do more to improve coastal ocean water quality while helping to reduce the occurrence of major *K. brevis* red tide outbreaks.

An activity that has worked well is the cleaning up dead fish. Not only does this activity remove odors, if done effectively, it reduces the nutrient supply that supports a bloom. Other forms of mitigation have also been proposed, such as flocculation by adding clay to the ocean or using ozone to kill the *K. brevis* cells. Both of these approaches are demonstrably effective on a laboratory or microcosm scale, but they are impractical on the spatial scales over which *K. brevis* exists. Moreover, except within the nearshore, where *K. brevis* is detectable at the surface via remote sensing by satellites and only on cloudless days, we would not even know the specific location within the much more massive offshore area to be treated.

Even if the large spatial extent of *K. brevis* and the fact that it is hidden beneath the surface when offshore were not mitigation issues, it must be recognized that harmful algae species did evolve as part of the ecosystem. While we may not fully appreciate their ecosystem role, there is a purpose for their existence, just like there are purposes for other perceived nuisance species like ants, termites, bees and so on. The literature is now full of ecological experiments where species were introduced to eliminate a perceived problem. The end result of many of these has been the introduction of invasive species that caused more damage than what they were intended to solve. Kudzu, Brazilian Pepper and Maleluca, Chinaberry are but a few examples.

Despite its description as an oligotrophic environment, the WFS supports abundant fisheries. What role might *K. brevis* play in this abundance, and what might the consequences to the elimination of *K. brevis* be on the rest of the ecosystem. The more that we learn about the ecosystem, the better informed we will be in making decisions regarding mitigation strategies. Given that my biological oceanography training is limited to how to filet a fish, I find the topic of ecological engineering for the purpose of eliminating a certain species to be frightening.

Addendum: Grilled Lamb

It is time for something that is more compatible with a red wine recommendation. Let's discuss lamb and something other than lamb chops, as good as those may be. How about a nice grilled slab of lamb meat.

Let's begin with a boneless leg of lamb. Being that this weighs several lbs, the following dish is for a dinner party (hence a few bottles of red wine). In essence, we will be grilling a large, marinated lamb steak. First, trim away all of the fat that surrounds the meat, especially any membrane-like stuff that would tend to be chewy. Next open up the boneless leg, and carefully think about how to butterfly it so that it ends up being one long piece of lamb steak perhaps about 1½ to 2 in. thick. This is easy to do, but it requires a little thought prior to attacking it.

The marinate is salt, black pepper, garlic, cumin, honey, lemon juice and olive oil. The garlic is to be placed inside the meat so begin by cutting many garlic cloves into long, spear-like pieces and then, using a paring knife, make slits in the meat to insert a spear of garlic, maybe every two inches in width and length. With this done, the rest is quick. Place the garlic laced meat on a tray with sides sufficiently high so that the marinate does not drip off. Liberally salt and pepper the entire slab of lamb, then even more liberally sprinkle on the ground cumin powder, lots of it. Then drizzle honey every couple of inches followed by squeezing fresh lemon over everything and some olive oil. Turn the meat over and repeat on the other side with salt, pepper, cumin, honey, lemon and olive oil. Put the concoction in the fridge and let it sit for at least an hour, preferably a few hours.

Cooking it is like cooking a steak, although being a little thicker, it will take a little longer. A outdoor gas or charcoal grill is what you will use. Go with the steak-grilling rule of say five minutes per side, but with the understanding that you will likely have to repeat this twice. Use a thermometer to check, if you are unsure of your finger touch technique for achieving a medium rare result—don't over-cook. With each turn pour some of marinate over the meat. The caramelization of the honey imparts great flavor to this savory lamb. Like all grilled meat, let it sit for five to ten minutes prior to slicing.

I recommend slicing this into ½ to ¾ in. pieces. Couscous is the natural side, and my preference for this is the Israeli couscous (the coarser pearled variety, versus the more fine stuff). To this side, add some grilled vegetables like zucchini or eggplant slices (1/2 in. slices with salt, pepper and olive oil made on the same grill as the meat).

One last ingredient that I like is a hot mustard. You can buy a jar of Coleman's or better yet make your own by simply adding water to the powdered mustard that you purchase at grocery store spice aisle. You want a mustard paste so add water sparingly and mix until you get a watery, but not too watery consistency. It is rather hot so use it sparingly.

Now for the wine recommendations. A Chateauneuf du Pape is the natural choice, but so is any big red. Being that you will be serving at least two couples open two bottles, the second one being a California Cabernet Sauvignon. These are getting ever more expensive, and frankly, the cheap ones are not very good. If you prefer not to spend 50 dollars or more for decent California Cab, you could try a good Zinfandel because there are many good ones at well under 50 dollars.

There are many choices for California Cabs. For the higher-end ones, my preference is for Hall, and their Hall Napa line, available in local wine shops, presently goes for about the $65. Single vineyard varieties are much more expensive. Other medium-priced California Cabs that tend to be consistently good (to my palate) include Keenan, Honig, Mount Veeder and Sequoia Grove. With so many producers from whom to choose from, it is worth the effort to visit several local wine shops until you find one with both competitive prices and someone whose recommendations you can trust (i.e., who has a palate that is compatible with yours). Sadly, once you start buying more expensive wine, you will realize that there is a difference, but everyone should have at least one vice, and compared with gambling or other vices, wine is relatively inexpensive and may even be healthy.

A difference between California Cabernets and Zinfandels is that you can purchase a higher-end Zinfandel for the price of a lower-end Cab, so these are worth experimenting with. To my palate, Ridge Zinfandels (and there are many different ones to try out) provide a high-quality starting point.

Chapter 12
Natural Climate Variability: What We Are Just Beginning to Learn

As is evident from the instrumental record dating back to the 1880s, the Earth is warming on average. Loosely stated, when averaged globally, the secular rise observed in sea surface temperature (SST) amounts to less than 1 °C per century. Depending upon where the observations may be examined, temperatures on land may be rising more rapidly than SST (or even falling); but, for the reasons stated in Chap. 1, the SST is what really matters with regard to climate. By absorbing most of the incoming solar radiation, thereby determining the interactions that occur between the ocean and atmosphere, which result in both the ocean's currents and the atmosphere's winds, it is the SST, versus the land temperature, that sets the world-wide organization of climate and ecology.

Superimposed on the secular rise in SST are variations of larger magnitude occurring on interannual, decadal and longer time scales. All of these variations have differing origins, so it is important to avoid conflating them with what may be of human (or anthropogenic)-induced origin, as too often is the case.

Published estimates for the secular rise in SST, land temperature or the combined SST and land temperature derive from instrumental temperature records available throughout the globe dating back to the late-nineteenth century. Several compilations exist, each with different usage criteria and adjustments made for perceived observational biases, which may be due either to regional coverage, or to the methods of data collection and the associated sampling techniques. Obviously, with time progressing from the late-nineteenth century though the present, there are ever more observations available with ever more global coverage. In particular, once the orbiting and geostationary satellite era started in the 1970s, with instruments capable of remotely sensing the Earth's surface temperatures from space in addition to the much more limited surface-based instruments, great strides were made in attaining global coverage.

The compilation of global temperature datasets have been the purview of various national agencies. In the USA, NOAA produced an Extended Reconstruction Sea surface Temperature (ERSSTv5) record dating back to 1854. Such reconstruction analyses attempt to correct for biases due to different measurement types (for instance, ship engine room intake readings, bucket sampled surface observations, samples taken during day or night and so on). Controversies may exist because the latest temperature analyses may differ from prior ones, but at least since 1980, both the NOAA and Hadley Center (Great Britain) analyses are comparable. Studies comparing the ERSSTv5 against the previous ERSSTv4 version and those from other agencies show that the trends computed from 1900 to 2015 are 0.70 ± 0.07 °C/century, versus those computed over the more recent 2000–2015 time interval of 1.25 ± 0.77 °C/century, where the error designations are 95% confidence limits. Whereas these different sampling intervals provide different results, the findings for the two different time intervals are not statistically different from one another. Similarly, when comparing trends over different latitude bands, the findings are statistically similar. Arguably, the most important latitude band is that of the tropics because for a given arc of latitude, the tropics contain the largest area for the Earth's oceans, and it is the region whose insolation largely fuels the coupled ocean–atmosphere system. As a corollary, the SST trends, when averaged over the tropics, are nearly identical to those averaged pole to pole.

Examples showing the differences between SST and land temperature anomalies are provided in Fig. 12.1. For SST (upper panel), the anomalies vary from around –0.45 °C in 1903 to around 0.80 °C in 2016, whereas for land temperature (bottom panel), these vary from around − 0.70 °C in 1884 to 1.60 °C in 2020. Thus, the land temperature range is larger than the SST range by about a factor of 2. This is expected because of the heat capacity differences between water and land and also because many land stations tend to be located within highly populated regions, each with their own heat producing infrastructure. Care must be taken not to confuse the two graphs of Fig. 12.1, thereby giving the impression that temperatures are rising much faster now the ever before. For instance, when considering SST, the trend from around 1910 through 1940 (the larger values during WWII are not included as these may be biased due to wartime sampling) is similar in magnitude to that from around 1970 through the present time.

It is also instructive to look at how SST may be varying within certain latitude bands, as shown in Fig. 12.2 for latitude bands segmented into 20° increments from 70 °S to 90 °N, where the dataset used here is from the UK, Hadley Center (HadSST). Note how all of these latitude bands express different results from the overall global average of Fig. 12.1. The closest comparisons with the global average are for the three bands that include the tropics, as suggested earlier. The right-hand panels show the trend lines calculated from 1880 onward at successive yearly increments. The confidence intervals increase with decreasing record length such that trends calculated beginning at 1998 and beyond are not statistically significant.

Superimposed on the long-term trends, whether in Figs. 12.1 or 12.2, are variations occurring at interannual and decadal time scales. The interannual variations are largely attributed to the El Nino-Southern Oscillation (ENSO), and the decadal variations are largely attributed to the Pacific Decadal Oscillation (PDO), both of which are related to the coupling between the ocean and the atmosphere. The ocean and the atmosphere largely derive their properties through the fluxes of heat, mass and momentum across the air–sea interface. These properties are then transported

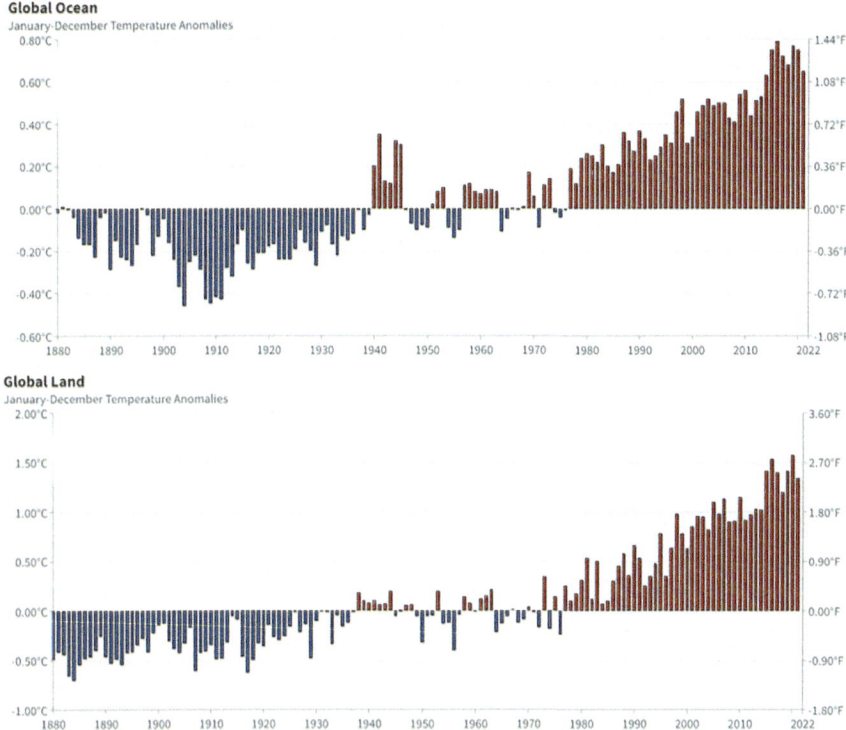

Fig. 12.1 Globally averaged SST (upper panel) and globally averaged land temperature (lower panel) anomalies for the years 1880–2022, where the anomalies are defined as the deviations from the 1901–2000 averages. Note the difference in the temperature scale for each plot (courtesy of NOAA, https://www.ncei.noaa.gov/access/monitoring/climate-at-a-glance/)

elsewhere by the ocean and atmosphere circulations (the currents and the winds). As an example, consider the upper ocean temperature imposed at the surface during winter and mixed downward over the upper ocean mixed layer. When covered over by warmer, less dense water that is heated in spring and summer, these waters with wintertime temperatures may be transported to other regions, where they may re-emerge to the surface, yielding the properties that they obtained elsewhere back to the atmosphere. The time interval between such subduction, transport from elsewhere and eventual re-emergence back to the surface by upwelling and mixing determines the time scales of the coupled, ocean–atmosphere system oscillations. The exchanges relating to ENSO are thought to occur both zonally (east–west) and meridionally (north–south) within a relatively narrow band of latitudes centered on the equator, whereas those for the PDO are thought to occur over a relatively larger latitude band extending to mid-latitudes; hence the shorter, interannual time scales for ENSO, versus the longer, decadal time scales for the PDO.

The use of a decadal descriptor for the PDO may be somewhat misleading because, unlike ENSO, for which many realizations are available, only a few are available for

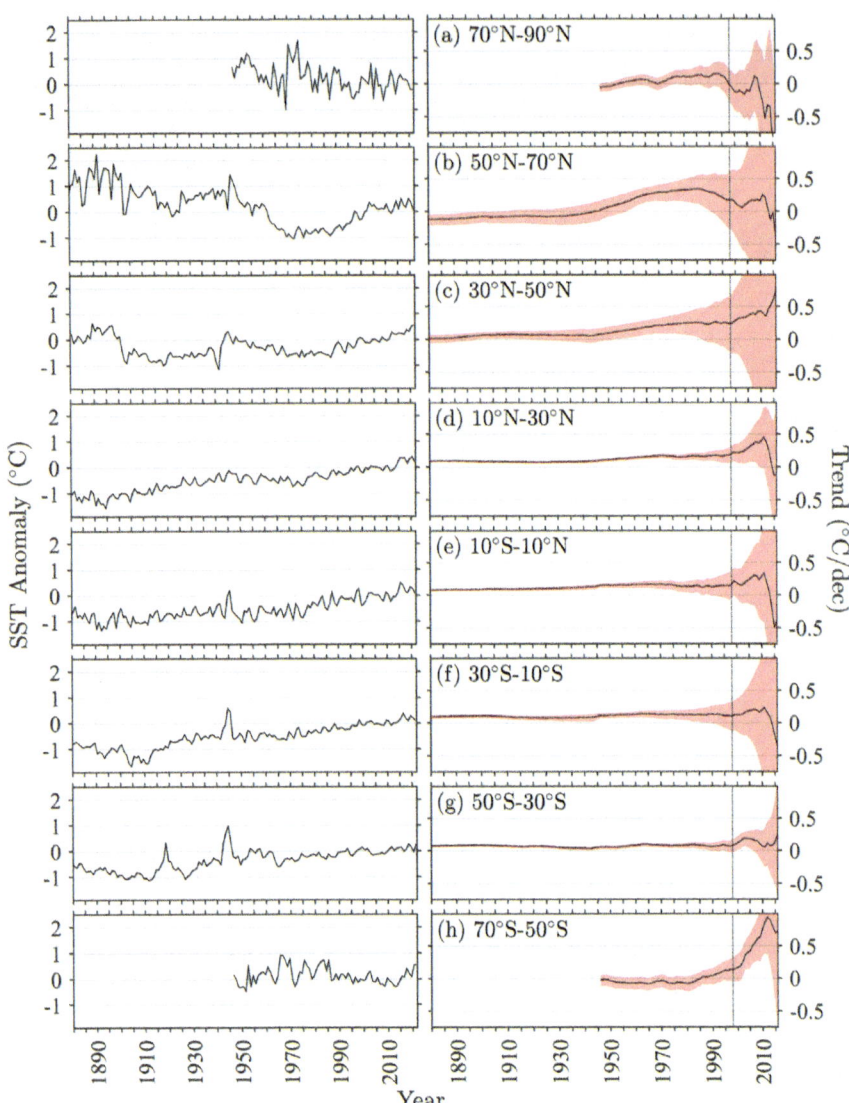

Fig. 12.2 Left: SST anomalies (based on 1991–2020 climatology) from 1946 to 2021 for latitudinal bands **a** 70°S–50°S, **b** 50°S–30°S, **c** 30°S–10°S, **d** 10°S–10°N, **e** 10°N–30°N, **f** 30°N–50°N, **g** 50°N–70°N and **h** 70°N–90°N. Right: SST trends (black line) and 95% confidence interval (red shading) from 1946 to 2021, 1947 to 2021, …, 2016 to 2021 for each latitudinal band. The black dotted line in the right-hand plots identifies 1998 (adapted from Nickerson, A. K., Weisberg, R. H., Zheng, L., Liu, Y. (2023), Sea surface temperature trends for Tampa Bay, West Florida Shelf and the deep Gulf of Mexico, *Deep-Sea Research Part II*, 211, 105,321, https://doi.org/10.1016/j.dsr2.2023.105321)

the PDO. Moreover, the two are not necessarily independent from one another. For instance, water arriving at the equator form higher latitudes as would be associated with the PDO will alter the equatorial thermocline temperature, thereby either adding to or subtracting from any variation due to ENSO alone. Recall from Chap. 3, Fig. 3.2, that most of the ocean is cold, with warm water only occupying a near-surface layer above the thermocline. This surface layer tends to be thinnest at the equator so that the upwelling or mixing of waters arriving from higher latitudes can readily change the SST along the equator and hence the ocean–atmosphere heat flux that occurs there. Consequently, the winds and ocean currents can easily change there, and from Chap. 1, we can appreciate how these changes at the equator can impact climate worldwide. Given this realization, we can imagine the existence of climate variations occurring over a broad range of time scales depending upon the latitude at which a wintertime SST perturbation of large spatial extent may be input to the ocean.

Similarly, there are mechanisms by which equatorial ocean SST perturbations may themselves be transmitted to higher latitudes by ocean planetary wave propagation. Given that the equatorial gyres may interact with the tropical gyres, the subtropical gyres and the subpolar gyres, either directly through ocean planetary wave propagation, or by ocean–atmosphere heat and momentum exchanges, many different potential modes of interannual, decadal and longer time scale oscillations may manifest. This remains a very active area of oceanographic research. For instance, along with ENSO and the PDO there are other identified coupled modes of oscillation such as the Atlantic Multidecadal Oscillation (AMO), the Arctic Oscillation (AO), the North Atlantic Oscillation (NAO) and more, none of which may be thought of as totally independent. That is one reason why not all El Nino and La Nina events are alike. Some may be the result of constructive or destructive interference with the PDO or other modes of oscillation. As recent examples of this, Fig. 12.3 shows a normalized SST anomaly record from 1994 through 2023 averaged between the latitude band of 3.5°N to 3.5°S from 40°E to 80°W, i.e., for the Indian Ocean and the Pacific Ocean. The anomalies are calculated by subtracting the time averaged mean values from each location, and the normalization is performed by dividing the anomalies by their standard deviations. ENSO is normally considered to be a Pacific Ocean process, but with anomaly patterns at times transiting the Indonesian archipelago, Indian Ocean values are also included in this graphic. With red (a warm anomaly) denoting El Nino and blue (a cool anomaly) denoting La Nina, we can discern three major El Nino events over these three decades, plus several minor ones, along with four major La Nina events, plus some minor ones. We also see that some of these events may persist longer than others. The larger, more intense events are likely those for which a constructive interaction occurred between different modes of oscillation (such as ENSO and the PDO), and the lesser events may be the result of destructive interference, not unlike the occurrence of spring and neap tides. Unlike tides however, none of the constituent modes of oscillation adding together are periodic; instead, they may vary over a range of time scales depending on just how any particular cycle gets set up.

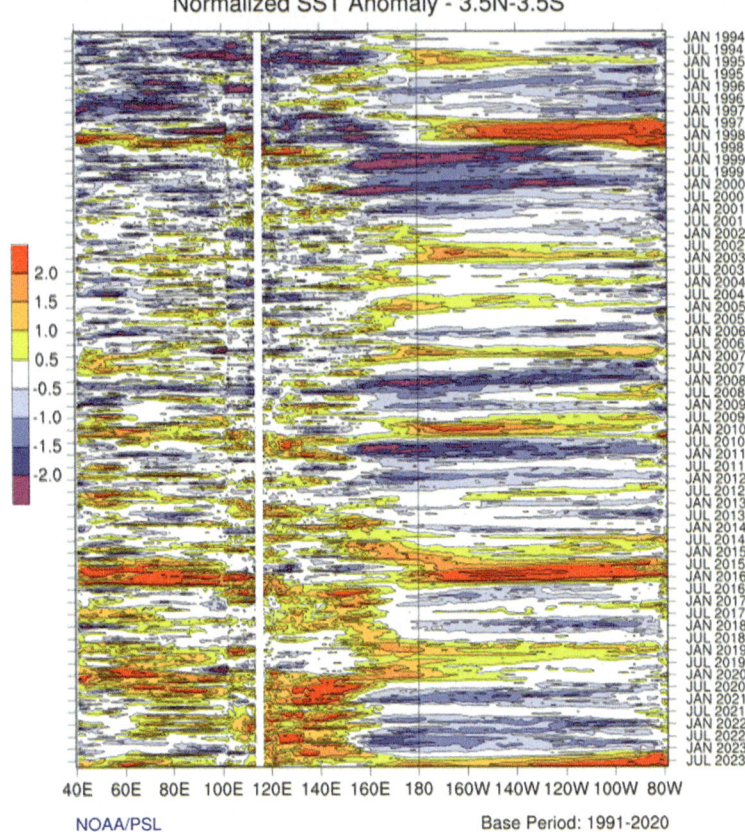

Fig. 12.3 SST anomaly (SST minus the 30 year average of SST from 1991 to 2020) averaged between latitudes of 3.5°N to 3.5°S as a function of longitude from 40°E to 80°W. Normalized by standard deviation, the scale from − 2.0 to + 2.0 means that the anomaly varies between plus and minus two standard deviations (courtesy of the NOAA Physical Sciences Laboratory and available at https://psl.noaa.gov/map/clim/sst.shtml)

Such complexity, and hence an inability to predict what may happen from year to year, is heightened by the fact that an appreciation for such coupled, ocean–atmosphere oscillations is relatively new, and their understandings are just beginning to emerge. For instance, ENSO, while initially recognized by meteorologists more than a century ago, did not achieve a more quantitative, mechanistic appreciation until the late1980s. Similarly, the PDO was not identified in observations until 1997, and it remained without any explanatory hypotheses until 1999. Even the ability to track the movement of the water within the ocean's thermocline itself had to await a theory for the origin and structure of the thermocline that was not introduced until 1983. Hence, while we may have had observations of interannual to decadal variations in climate with global scale implications, it is only recently that we could ascribe these to more specific physical processes linking the ocean and the atmosphere.

Additionally, temperature is but one of the variables determining density. Along with pressure, salinity and water vapor are germane for the ocean and the atmosphere, respectively, and these variables are not independent of one another because ocean salinity derives from evaporation, precipitation, plus both ice formation and melting. For instance, whereas more major rivers empty into the Atlantic Ocean than the Pacific Ocean, the latter has lower salinity because it receives more direct precipitation, which more than compensates for less river inflows.

From Chap. 3, Fig. 3.3, we see that the formation of deep waters at high latitudes is largely due to the contribution that salinity makes to density. Thus, processes that may change global salinity distributions will profoundly affect climate variations, and since deep-water formation occurs at high latitudes, the time scales associated with such variations will necessarily be longer than those for either ENSO or the PDO. Theories have been advanced regarding how such oscillations may occur on century to millennial time scales as are evident within the ice-age periods of the past million years.

All of these factors must be considered when attempting to interpret the rise in SST, part of which may be attributed to human activities and part of which may be occurring quite naturally in ways that we are just beginning to understand. Given the slow maturation of such concepts, it is fair to speculate that the coupled ocean–atmosphere system will reveal new modes of oscillation in the future. Earth systems science is complex. Despite a desire to have simple concrete answers to complex questions, the reality is that these generally do not exist, and there are copious questions and new ones to arise for the next generation of scientists to find answers to.

Addendum: Braised Short Ribs

Here is another red wine worthy meal. Braising meat is simple, but it does take some time and effort. First, you must find a good purveyor of meat because what you may see in the grocery store is generally not worth the effort. You need one big beef short rib (with the bone) per person. By big, I mean about 8 in. long and meaty. Other ingredients are: onions, garlic, carrots, celery (or fennel), a green spice mix and/or a mixture of fresh herbs (thyme, tarragon, rosemary, bay leaf, dill, whatever) placed inside a cheesecloth bag.

Cooking in a Dutch Oven (like Le Creuset) is the way to go. Rinse the short ribs, dry very well, sprinkle on salt and pepper and sear in the Dutch Oven with medium heat using some olive oil. You want a nice brown sear on all sides. Remove and set aside. Add diced onion, carrot and celery (I like to substitute fennel for the celery). For four to six ribs use about one large onion, four cloves of garlic, two carrots and two stalks of celery (or one small fennel bulb). Sweat the vegetables and stir until the brown bits from the meat are all loosened from the bottom of the Dutch Oven and incorporated with the vegetables. Then remove half of the vegetables, lay the meat on what is left in the Dutch Oven and then put the rest of the vegetables atop the meat. Liberally add the green spice mix (Herbs d' Provence, or similar) and/or the cheesecloth bag of fresh herbs. Zest a lemon and squeeze the juice for a little acid, and then pour in an entire bottle of red wine. I generally use a good Zinfandel. Depending on the size of the pot, the wine will reach about half way up the meat, which is about right.

Bring the liquid to a boil, then lower the heat to a minimum level cover and let this slowly cook for about two hours. It is done when the meat will easily separate from the bone. In fact, to serve it, you will remove the meat from the bone.

When the meat is done remove it from the pot and strain all of the liquid into bowl using a spoon to mash some of the vegetable to release their juices. The next step is to remove the fat from the liquid. This can be done using a spoon, but that is difficult. An easier way is to place the bowl in the freezer so that the fat congeals at the surface. Once this occurs you can then simply remove it with a spoon or fork. With the fat removed, return the liquid to a saucepan and reduce by about half by boiling it—the result will be a savory sauce to be spooned over the meat after plating. Make sure to remove the fat, which will ruin the dish. I once ordered short ribs at a restaurant touting a so-called noted chef. Swimming in fat, it was lousy.

To serve, there are several choices. The meat (with the bone removed) needs to be placed atop some base, either polenta, mashed potatoes, couscous or rice. For a vegetable, asparagus seems appropriate, but his can be anything that you like. After plating, put sauce over the meat and sprinkle a little finishing salt.

Polenta is nice either freshly made (soft) or put into a baking dish to chill, cut into strips and then grilled. For mashed, try using 1/4 rutabaga, 3/4 potatoes, but cut the rutabaga in small cubes prior to boiling because it takes longer than the potatoes to get done.

As you may have guessed, Zinfandel provides a good accompaniment, but you might also try an Italian red like a Brunello, Super Tuscan, Barolo or an Amarone. Italian wines can be terrific, but the better ones are expensive and they can also be a little finicky. Again, get to know your wine purveyor and seek some advice. Regardless, Italian reds generally need some time to breathe before they are drinkable. Some take only a half hour or so, others several hours so if you are to serve these, open well in advance, and even think about decanting if they are too astringent upon opening. Amarone is a rather unique form of wine made from desiccated grapes. The flavors can be intense, but quite enjoyable and they are great with braised meat.

Chapter 13
Alternative Energy Generation from the Ocean: What May or May Not Be Feasible and Why

With the Earth's climate and ecological systems essentially fueled by sunlight, including the winds and ocean currents, it seems natural to ask whether, or not these resources are directly accessible for powering machines. The answer dates back to antiquity wherein winds were used to power ships and pump water, and water was used to run mills for grinding grains. More recently, but still more than a century ago, the workings of photovoltaics were described, ushering in the use of solar panels to generate electricity. Thus, solar and mechanical means of power generation for human consumption certainly exist. The remaining question is whether or not such direct harnessing of heat and mechanical energy can replace the use of conventional fuels for powering human endeavor.

Being that everything of economic value to society relies on energy consumption, this is a serious question with equally serious consequences. For instance, food and shelter, the most fundamental of human needs, rely on energy consumption, and a harmonious social order can only exist if these requirements of survival are met. Short of that, we have chaos and war.

Fuels over the millennia have taken many forms, evolving from wood to peat to coal to oil and gas and nuclear fission, with each transition due to some imperative, e.g., the depletion of forests to noxious air pollution to the fear of oil depletion, nuclear disaster and more recently, the warming climate implication of fossil fuel combustion.

With power generation by the utilization of the sun, winds, water and waves being ever more pursued, it is useful to access just what nature may offer and how present-day machinery may convert these natural resources to electricity. Being that Florida is the sunshine state with an adjacent and strong western boundary current, the Gulf Stream, Florida and its surrounding waters provide a useful medium for considering these assessments.

Here we will use observations of surface winds, solar radiation, ocean currents and waves collected by the University of South Florida, Coastal Ocean Monitoring and Prediction System (COMPS), augmented by other data and numerical ocean circulation model simulations, to address Florida's potential for electrical power generation by harnessing the natural energy sources of wind and sunlight, along with ocean currents and waves. We will begin by identifying what nature offers. For wind and solar sources, we will use specifications from existing, commercially available devices to convert nature's bounty to electrical power generation estimates. In the absence of mature, commercially available devices for ocean currents and waves, we will draw upon physical principles to arrive at power generation estimates for these potential sources. On the basis of what nature offers and what machinery may be capable of producing, we will then make reasonable extrapolations on what these estimations may mean in a practical sense for supplying society with its functional power needs.

13.1 Wind Observations

The COMPS moored array (e.g., Fig. 10.4), initiated in 1998, has produced a lengthy set of wind observations that continues through the present time. Hourly wind speed data for the eight-year interval 1999–2007 are shown in Fig. 13.1. Measured using anemometers located at approximately 3 m above the sea surface, these data were first scaled to a standard observing height of 10 m and then further scaled to a hypothesized wind turbine hub height of 100 m using logarithmic boundary layer scaling factor of 1.35. Such a scaling factor is larger than a published empirical value of 1.12, and the speeds are more likely overestimates than underestimates for what the winds aloft may actually be. The speed values range from zero to around 20 ms^{-1}, with the higher values occurring on rare occasions during the passage of either tropical storms, or very pronounced weather fronts. Whereas the subsequent wind analysis is limited to observations at a single point off the west central Florida coast, the winds observed there are representative of winds elsewhere in Florida with some caveats. On both long-term and seasonal averages, the winds tend to increase from north to south by virtue of the trade winds' meridional structure and Florida peninsular land effects, which also results in the trade winds being a little stronger on the east coast. On the diurnal time scale, the east coast sea breeze tends to be more regular than that on the west coast. Nonetheless, for typical synoptic scale weather fronts occurring from fall through spring, the entire State of Florida is similarly influenced.

13.2 The Conversion of Wind Speed Observations to Electrical Power Generation Potential

It is common practice when referring to commercially available wind turbines to quote their nameplate power generation capacity. This can be misleading because the actual power output depends on wind speed so rarely is the nameplate capacity ever

13.2 The Conversion of Wind Speed Observations to Electrical Power ...

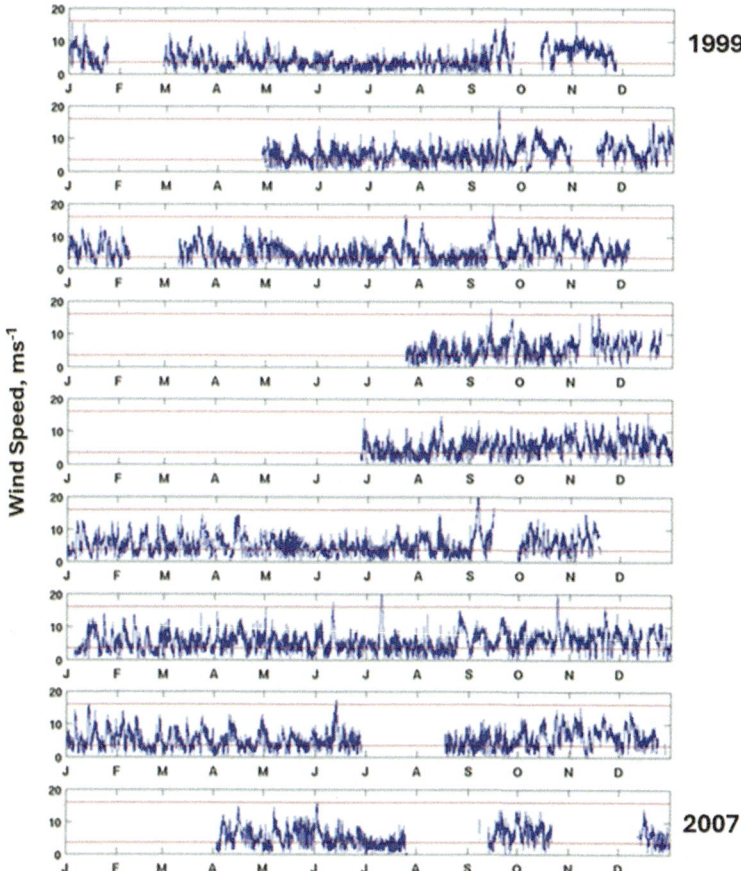

Fig. 13.1 Hourly averaged wind speed observations from COMPS moored buoy C10 located offshore from Sarasota Florida on the 25 m isobath. The observations were made at a distance of about 3 m from the surface, and these were then scaled up using empirical determinations to a hypothesized turbine hub height of 100 m. Each panel represents a successive year beginning with 1999 on the top and 2007 on the bottom (adapted from Weisberg, R. H., Y. Liu, C. R. Merz, J. I. Virmani, and L. Zheng (2012), A critique of alternative power generation for Florida by mechanical and solar means, MTS Journal, 46(5), 12–23, https://doi.org/10.4031/MTSJ.46.5.1)

achieved. As a representative example, here we will consider a General Electric (GE) 3.6 megawatt (MW) Offshore Series Wind Turbine, with specifications available from the manufacturer's internet site. From their wind load to power conversion curve, it is found that the turbine does not begin to produce electrical power until the wind speed exceeds 3.5 ms^{-1} (the cut-in wind speed). Power generation then increases with increasing wind speed, reaching the nameplate rated capacity (3.6 MW) at a wind speed of 14 ms^{-1} (the rated wind speed), and the device ceases power generation and shuts down when the wind speed exceeds 27 ms^{-1} (the cut-out wind speed). The lower and upper horizontal lines on Fig. 13.1 represent the cut-in and the nameplate

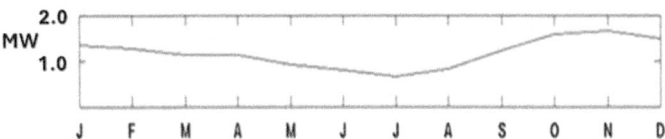

Fig. 13.2 Climatologically averaged, month mean electric power output potential for a GE 3.6 MW wind turbine, as estimated using the manufacturer's specifications and the COMPS mooring C10 (Fig. 13.1) observed winds

rated capacities for the GE 3.6 MW turbine, respectively. From eight years of West Florida Shelf observation we see that the winds at 100 m hub height fail to drive the turbine some 20% of the time and that rarely do the winds reach the nameplate rated capacity.

The power curve (Fig. 13.2) was used to determine what the power output may be when the turbine is running. For wind speeds between 14 m s^{-1} and 27 m s^{-1} the output was held constant at 3.6 MW. The results, when climatologically averaged (i.e., the averages of all Januarys, all Februarys and so on), are shown in Fig. 13.2 from which several points are clear. First, the nameplate capacity is never achieved on climatological average. Second, there are many days (as seen from Fig. 13.1) without any power generation because the cut-in speed of 3.5 ms^{-1} is not exceeded. Third, from the climatological monthly mean results, we see that minimum and maximum power generations occur in summer and fall months, respectively. The monthly mean minimum is about 0.6 MW; the monthly mean maximum is about 1.8 MW and the grand average across all years and months is about 1 MW.

The minimum in summer months is troublesome for Florida because that is when the demand for air conditioning is the largest. The maximum value in fall months is not very beneficial because that is when heating and cooling are least required. In summary, for Florida winds, this analysis suggests that a GE 3.6 MW turbine might provide about 28% of its nameplate capacity, at times more, or less and even zero at some times. Newer turbines are available from GE and other providers, but the finding that the nameplate capacity is a misleading metric remains a fact.

13.3 Ocean Currents

Unlike windmills, where commercial maturity provides known power generation potential, watermills driven by ocean currents remain in development. Estimating the potential for power generation by ocean currents requires that we begin from first principles. Available power, P, for potential extraction from ocean currents is the kinetic energy flux (kinetic energy, $1/2 \, \rho V^2$, times velocity, V) times the cross-sectional area, A, of the device used for extracting this energy flux, or $P = \frac{1}{2} \rho V^3 A$, where the units for P are watts (W). Watermills, like windmills, are subject to the same hydrodynamic limitations (known as Betz's law), which states that the maximum

13.3 Ocean Currents

power that may be extracted is 59% of the kinetic energy flux. Additional loses come from the efficiency of the device itself, such that the expected power generation potential for either a windmill or a watermill is in the approximate range of 40–50%.

Watermills, like windmills, are also expected to have a cut-in speed threshold below which they will not function. In other words, a baseline torque is necessary to drive an electrical generator. For instance, if you try to turn the generator under the hood of your automobile by turning the drive belt, you may be surprised by the resistance encountered. Given that torque equals force times distance, it is proportional to the pressure (force per unit area) on a turbine blade times both the area and the length of the blade. A dimensional analysis, using the cut-in speed for the GE 3.6 MW turbine, its length scale and assuming that a watermill may have a length scale about an order of magnitude smaller than that for the windmill (i.e., instead of 50 m blades, it may have 5 m blades), we arrive at a cut-in speed estimate of around 1 ms^{-1} for the watermill. Granted, this is a very crude estimate, but what it does suggest, even if off by a factor of two or more, is that typical coastal ocean current speeds on the continental shelf (away from tidal inlets), which are of order 0.2–0.5 ms^{-1}, are much too small to drive watermills. Nevertheless, Florida does have a strong western boundary current seaward of its continental shelf. For the west coast of Florida, this western boundary current is the Gulf of Mexico Loop Current with speeds of about 1–2 ms^{-1}, which feeds into the Gulf Stream on the east coast of Florida with similar speeds. Being that the Gulf Stream is tightly constrained to flow between Florida and the Bahamas, the east coast of Florida is the most practical place for considering power generation by watermills.

Despite its constraint, the Gulf Stream is still highly variable in space and in time. Numerical ocean circulation model simulations (e.g., the Naval Research Laboratory Gulf of Mexico HYCOM) demonstrate this, as shown for relatively high- and low-speed realizations at a Palm Beach cross section in Fig. 13.3a, b, respectively. Two generalizations follow. The first is that the Gulf Stream's highest speeds are at the surface and on the western side of the cross section. The second is that rarely do speeds in excess of 100 cm^{-1} (1 ms^{-1}, or about 2 kts), the estimated cut-in speed, extend down below 300 m depth. An additional factor that any machine to be deployed in the Gulf Stream must contend with is the effect of waves.

Thus, for a machine to function, it would have to be deployed deep enough to avoid being adversely impacted by wave motions. Given actual waves observed in the Gulf Stream, especially under northerly winds in fall though winter months, a 50 m deployment depth would be a minimum to avoid compromising the workings of a turbines rotors; 100 m would be a safer working depth, and even at 100 m depth there may be times when wave-induced motions may be detrimental to the turbine machinery.

With highest speeds on the western side of the Gulf Stream and with little potential for power generation below 300 m depth, Fig. 13.4 shows the total kinetic energy flux between depths of 50–200 m and 50–300 m integrated (summed) across the entire Gulf Stream for this Palm Beach cross section. Demonstrations like this are used to argue that the potential for power generation using watermills deployed in

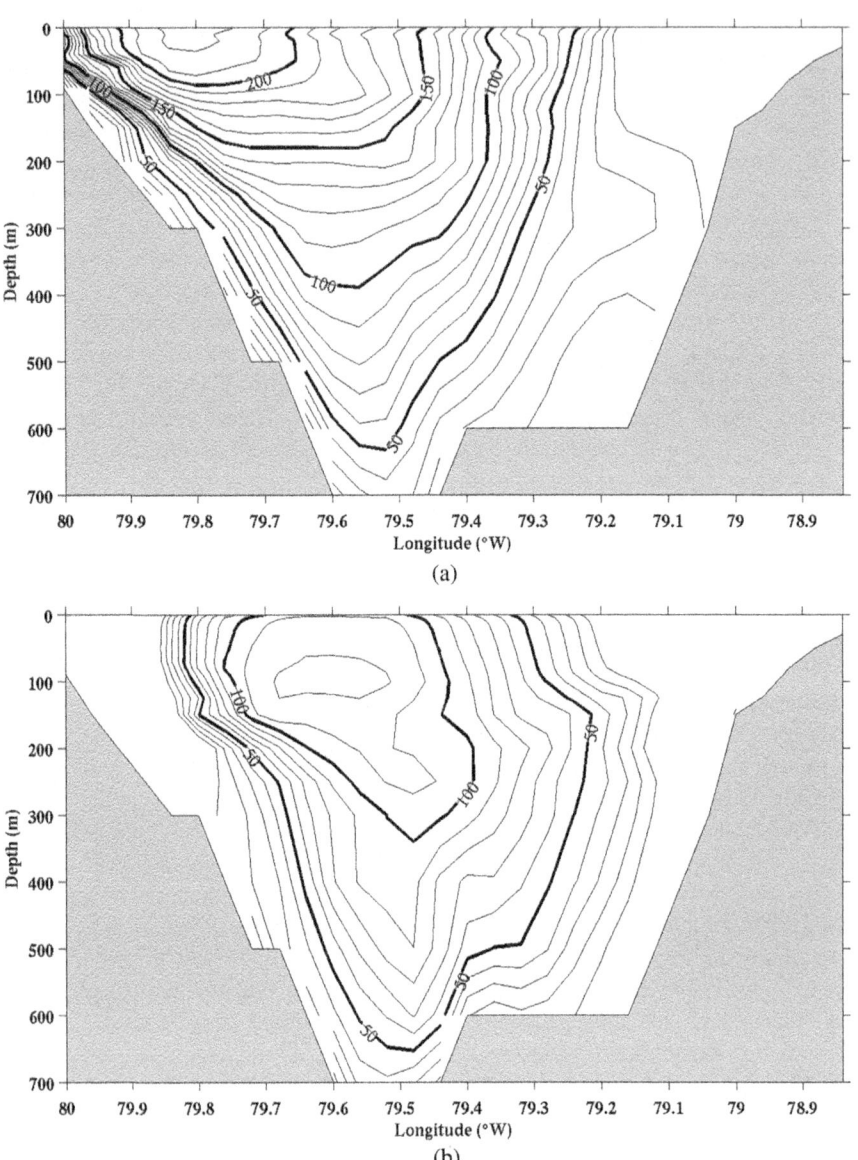

Fig. 13.3 a North–south component of velocity for a Palm Beach to Bahamas cross section on July 17, 2008, sampled from a Gulf of Mexico HYCOM simulation. Contours are in cm/sec. **b** The north–south component of velocity for a Palm Beach to Bahamas cross section on January 4, 2008, sampled from a Gulf of Mexico HYCOM simulation. Contours are in cm/s

13.3 Ocean Currents

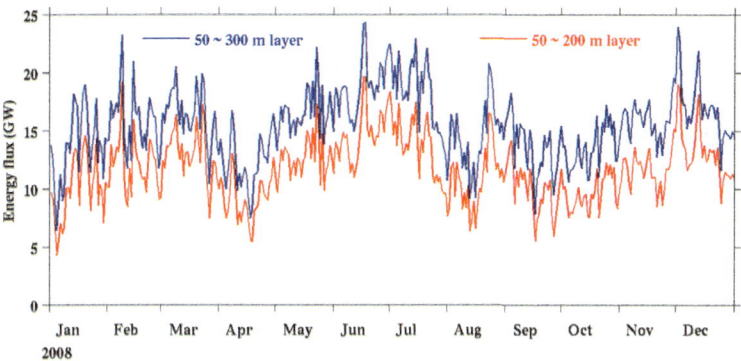

Fig. 13.4 Total kinetic energy flux integrated across the Gulf Stream between depths of either 50–200 m (red) or 50–300 m (blue) for calendar year 2008

the Gulf Stream is large, in this case between 10 and 20 GW. Whereas nature may offer that much, how much of this might actually be harnessed?

If we now just focus on the western side of the Palm Beach cross section from the shoreline to either the 300 m or the 400 m isobaths and further restrict attention to depths between 50 and 200 m where speeds in excess of 1 ms^{-1} may be found, the results reduce to what are shown in Fig. 13.5. As expected, the larger the area summed (or integrated) over, the larger the energy flux, but now we see values ranging from zero to about 6 GW, with averages of about 1–3 GW depending on the area chosen, which is vastly different from Fig. 13.5. To this finding, it must also be reiterated that just as for windmills, there are limitations to how much of the energy flux a machine may extract. Recall that a combination of Betz's Law and mechanical losses reduces the potential by about 50% and the spacing between turbines to avoid entanglements would reduce this by maybe another factor of 10. So instead of GW scale power generation, a more realistic potential might be for a few hundred MW at most, i.e., that of a small conventionally fueled power plant.

Given that windmills installed on either land or offshore and watermills installed within the Gulf Stream (the most powerful of the ocean currents adjacent to the U.S.) tend to have similar promise for power generation, two additional factors should be considered. The first is with regard to the physical size of the required machinery. Being that the power produced by either a windmill or a watermill both entail products of fluid density times fluid velocity cubed times area (i.e., $0.5 \rho V^3 A$), it is apparent that with air being about a thousand times less dense than water while the air velocity is about ten times faster than water, the area necessary to generate comparable amounts of power for either air or water flowing past a machine is about the same. This begs the question: If the size of the machine is about the same for use in air or water, why would we choose to work in a fluid medium (water) that is technologically and practically so much more challenging than in air? The second consideration is even more daunting. Windmills operate within the lower 100 m of the atmosphere, or in the lower part of the frictional boundary layer, which is driven by, and whose

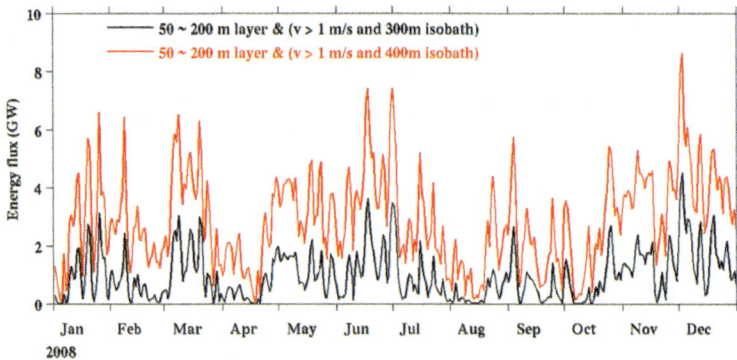

Fig. 13.5 Kinetic energy flux integrated from the western shoreline to either the 300 m isobath (black) or the 400 m isobath (red) within the depth range of 50–200 m for speeds in excess of 1 m/s. Results are for the full calendar year 2008

energy is replenished by, the geostrophic interior that extends up to the tropopause, some 10 km or more aloft. Watermills, in contrast, by operating across a major portion of the water column, are in the geostrophic interior itself, and hence, unlike windmills, there are no natural means for replenishing the energy that a watermill may extract. It is for this reason that windmill farms may have closely spaced units with additional windmills distributed downwind from one another, whereas watermills cannot share this deployment strategy. Once power is removed from a cross section by a watermill, it cannot be readily replenished. For these two reasons, even in the swiftest of currents, like the Gulf Stream, the notion of power generation by watermills tapping the kinetic energy flux is much more costly and limited when compared with what may be achieved by using windmills. It is the opinion of this author that the use of watermills for power generation on any utility scale is simply untenable.

Moreover, if I were to theorize (tongue in cheek, of course) on how to destroy the United States of America, I might suggest deploying lots of watermills between Palm Beach and the Bahamas. By extracting enough of the kinetic energy flux from the Gulf Stream, this might accomplish rerouting the western boundary current (as required to close the North Atlantic Subtropical Gyre, as explained in Chap. 2) to flow to the east of the Bahamas, thereby bypassing the Caribbean Sea and the Gulf of Mexico. As a consequence, the temperature of the Gulf of Mexico would be altered, changing the moisture flux into the atmosphere, which results in much of the rainfall over the nation's heartland, and thereby destroying our agricultural output. We could only imagine how drastically this might impact the nation's economy and our way of life. Fiddling with nature can have consequences, especially when this involves the ocean, the major determinant of climate and ecology.

13.4 Ocean Waves

Waves, like currents, also possess an energy flux that may be tapped by mechanical devices. The available power, P_W, for potential extraction from ocean gravity waves is the total mechanical energy flux per unit wave crest width, $\frac{1}{2}\rho g a^2 C_G$, times the crest width length, L, of the device used for extracting this flux, or $P_W = \frac{1}{2}\rho g a^2 C_G L$, where g is the acceleration of gravity, a is the wave amplitude, C_G is the group velocity, and the units for P_W are watts. An alternative expression for P_W, based on significant wave height and application of the deep water dispersion relation, is: $P_W = \frac{1}{2} H_s^2 T L$, where T is the wave period and significant wave height H_s is defined as the average of the highest one third of the waves.

How power extraction by wave devices may apply to the west coast of Florida may be estimated using numerical wave model simulations. Figure 13.6 shows a monthly mean climatology for the West Florida Shelf wave energy flux per unit crest width with units of KW/m and calculated from NOAA Wave Watch III model results inclusive of 1999–2007. A robust annual cycle is seen with minimum wave energy in summer and maximum wave energy in winter, similar to that of the significant wave height observed at several NOAA buoys. With low wave energy, the West Florida Shelf is not a candidate site alternative power generation by tapping the energy flux of surface gravity waves.

The east coast of Florida does have larger waves, especially north of the Bahamas where the entire North Atlantic fetch comes into play. But even there, the energy flux per unit crest width remains small compared with other higher energy coastlines worldwide, where tens to even a hundred KWm^{-1} are potentially available. Nonetheless, and as with electrical power generation potential using ocean currents, the question for using such a source is one of feasibility. Is there enough energy potentially available to justify the costs for extraction, and can the technical challenges be met?

Commercial advocacy groups suggest that wave energy extraction is economically feasible for regions where the energy flux per unit crest width exceeds 15 KWm^{-1}. Two examples of extraction devices are: (1) a large snake-like set of linked cylinders that extract wave energy via undulations across the linked cylinder length (e.g., 180 m for the 4 m diameter Pelamis WavePower device; http://www.pelamiswave.com) and (2) a 4 m diameter buoy that tracks the vertical motion associated with wave propagation past a fixed point (e.g., Ocean Power Technologies, Inc.; http://www.oceanpowertechnologies.com). The first of these uses the length of the device to extract a major portion of the wave energy flux past the 4 m diameter cross section; the second of these is more limited in the percentage of the flux that can be extracted. Without questioning any of the details (see the industry brochures), it is clear that for either of these devices the total wave energy flux that may potentially be tapped is whatever passes the 4 m device width. So in either case, the question becomes: Can some fraction of 15 KWm^{-1} times 4 m provide an economically feasible source of power (assuming that 15 KWm^{-1}, as promoted, is a viable level)? Assuming complete extraction with no loses (an impossibility), devices such as these can at

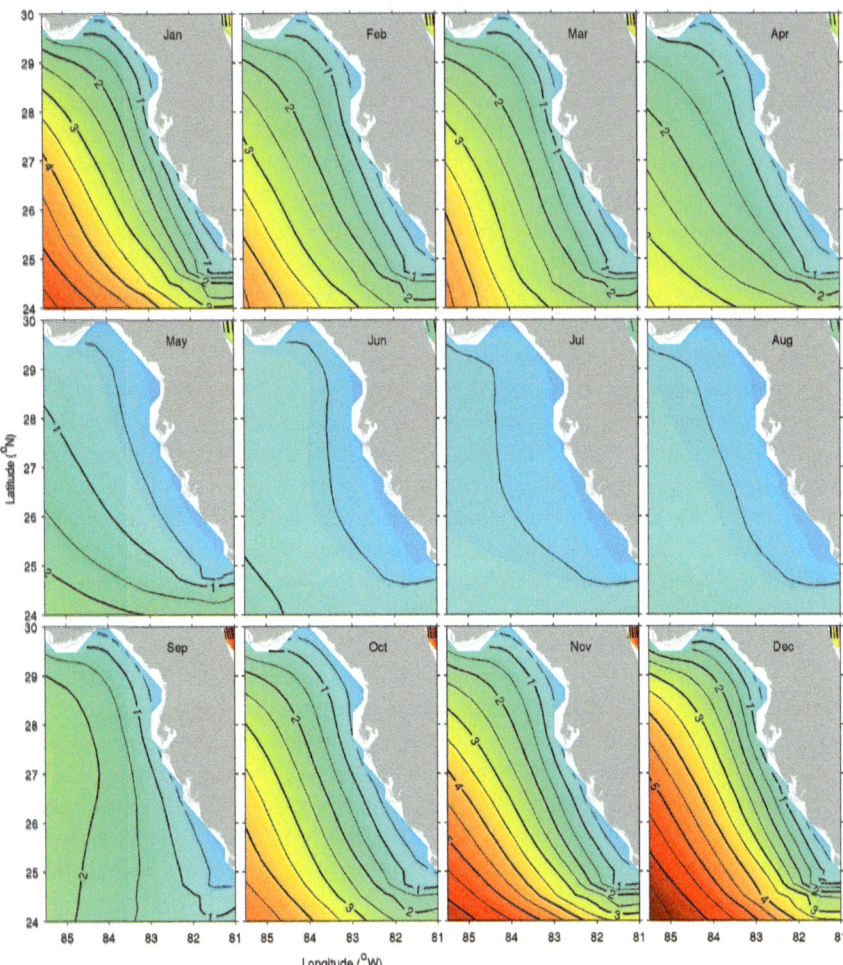

Fig. 13.6 Monthly mean wave climatology for the West Florida Shelf computed using NOAA WaveWatch III numerical wave model simulations for the period 1999–2007 where the contour intervals are KWm^{-1}. The color coding indicates highest (red) to lowest (blue) values (adapted from Weisberg, R. H., Y. Liu, C. R. Merz, J. I. Virmani, and L. Zheng (2012), A critique of alternative power generation for Florida by mechanical and solar means, *MTS Journal*, 46(5), 12–23, https://doi.org/10.4031/MTSJ.46.5.1)

most garner 60 KW, thus lacking any economic feasibility. Just as with watermills, machines for extracting ocean wave energy are not viable tools for electric power generation at a utility scale.

13.5 Solar

Solar panels convert incoming shortwave radiation to electrical power. Direct observations of incoming solar (shortwave) radiation show what nature provides at any given location and the application of specifications from industry rated devices may then be used to determine how much of this energy flux may be tapped for actual use. Observations of incoming shortwave radiation measured on a buoy moored about 25 nm offshore form Sarasota, FL (COMPS station C10), are provided in Fig. 13.7 for the eight-year interval 1999–2007. Despite data gaps due to telemetry failures, these hourly averaged time series show well-defined diurnal (day/night) and seasonal variations. On an hourly basis, we see that the daily maximum insolation varies from as high as 1000 Wm^{-2} in spring and summer to as low as 500 Wm^{-2} in fall and winter. Averaging diurnally to account for the fact that there is no incoming shortwave radiation at night, we find highest daily mean values ranging from about 300 Wm^{-2} in summer to 150 Wm^{-2} in winter. An obvious drawback to the use of solar panels is that they only produce electricity during day light hours and that production varies seasonally.

As with wind power, the maturity of the solar power industry allows for the use of established vendor product specifications to convert what nature offers to what may be available for use as electrical power. As an example, Fig. 13.8 shows the climatological monthly averaged output of electrical power in watts based on a Siemens SP75 solar panel. In accordance with the industry, the method employed to determine the output of such a solar panel is to calculate the equivalent number of hours for which such panel would capture insolation at a 1000 Wm^{-2} level. Thus, for the year with the most data (2001), a yearly averaged hourly insolation curve was constructed and the area under that curve was calculated to arrive at an equivalent area at an insolation level of 1000 Wm^{-2}, yielding 5.39 h or 404 Wh for this 75 W solar panel. Dividing by 24 h then results in 16.8 W, as the annually averaged daily output for this nameplate rated 75 W solar panel, or about 22% of the nameplate rating. Just as with windmills, the nameplate rating may be quite misleading. Dividing by the area of the solar panel (0.63 m^2) gives 26.7 Wm^{-2} as the daily averaged power output per unit area under WFS daily averaged insolation for this manufacturer's particular solar panel. This number may be as high as about 38 Wm^{-2} in May and June to as low as about 18 Wm^{-2} in December.

13.6 Discussion

Having considered the electrical power generation potential in Florida by windmills, solar panels, watermills (ocean current turbines) and wave devices, it is easy to understand why the first two of these have mature industries from which a customer can purchase and install such devices, whereas the other two do not. These facts in and of themselves are supportive of the analyses provided above. If watermills and

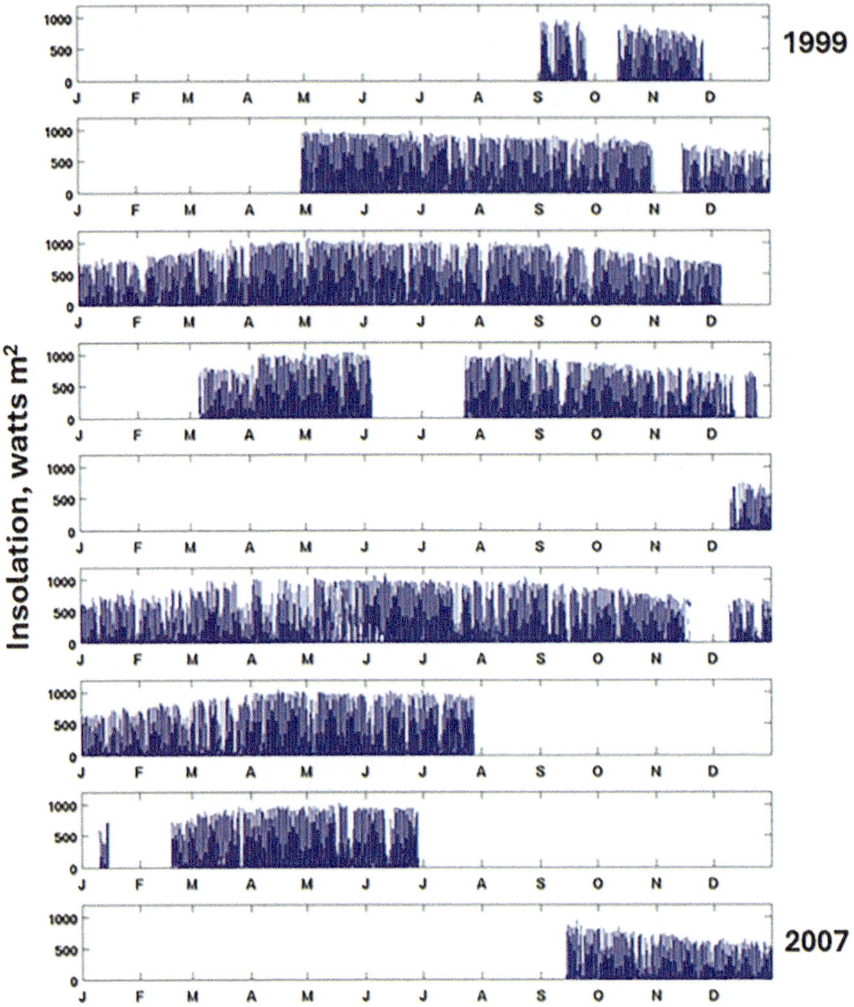

Fig. 13.7 Eight successive years (1999–2007) of hourly averaged, incoming short-wave solar radiation observed on COMPS buoy C10 located about 25 nm offshore from Sarasota FL (adapted from Weisberg, R. H., Y. Liu, C. R. Merz, J. I. Virmani, and L. Zheng (2012), A critique of alternative power generation for Florida by mechanical and solar means, MTS Journal, 46(5), 12–23, https://doi.org/10.4031/MTSJ.46.5.1)

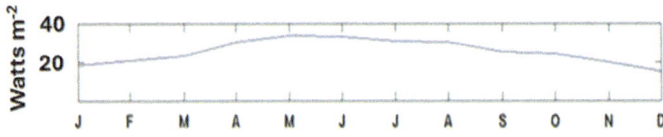

Fig. 13.8 Conversion of observed insolation (Fig. 13.8) to climatological monthly mean electrical power generation using specifications from a commercially available Siemens SP75 solar panel

wave devices were suitable, then mature industries would have come into being by now. As an aside, these concepts are not new. In the early 1970s when my major professor was approached to do such an assessment, he assigned the task to his graduate student. The answer was the same some 50 years ago as it is now (the physics have not changed). Windmills and solar panels are viable alternative power generation devices at some practical scale, whereas watermills and wave devices are not, with the exception of powering a local appliance.

While the economics of wind and solar device implementation are beyond the scope of this chapter, some considerations may be addressed. The first is a matter of size. Solar panels are readily deployable on residential rooftops at costs that may be estimated (e.g., see https://www.nrel.gov/docs/fy22osti/80694.pdf). Given the daily averaged output relative to the nameplate rating for a solar panel, the number of years to recoup the investment likely exceeds the useful lifetime of the installation. Moreover, few homes have enough rooftop surface area to fully support their electrical power needs with solar panels, and even if they did, a backup source would still be required for those inclement weather days of little sunshine.

With windmills, we are faced with a more difficult size issue. These devices are large and not suitable for individual residences. On a utility scale, to generate the equivalent of a municipal power plant, e.g., 1 GW, the requirement would be 1000 of the 3.6 MW turbines considered here. These are enormous devices. With rotor diameters of 111 m, they would have to be deployed away from any residential areas. Like solar panels, there would also be a requirement for a backup source for those days when the winds fail to exceed the cut-in value. Thus, even if affordable, conventionally fueled power plants would still have to be in place as backups. Moreover, it makes little sense to cycle conventionally fueled power plants on and off because not only is there a fuel inefficiency when doing so, the equipment itself can become damaged. In summary, solar and wind devices may supplement conventional electric power generation, but neither of these alternative energy sources can replace the presently required base load power generation capacity, despite what some politicians, driven by enterprising individuals, may want the public to believe.

Addendum: Meatloaf

I am not joking. Who ever said that meatloaf must be lousy? With a little creativity, it can be terrific. I started playing around with various renditions a few years ago when a recalled from a trip in my youth (an obligatory west coast summer trip with a friend during my sophomore year, when we visited and camped at several national parks, gambled in Vegas, roamed the major California cities, failed miserably at surfboard riding, and tried to stay out of trouble in Tijuana). While staying with an Italian family outside of Los Angeles we were served a wonderful Italian meat loaf, which inspired the following.

Sausage and salami tastes good because it has lots of garlic and fennel so why not make meat loaf this way. The main ingredient is ground beef, perhaps of the 85/

15 lean to fat variety. Unlike a basic mirepoix of onions, carrots and celery, for this I use onions, bell peppers and fennel, plus lots of garlic. A loaf is pretty big so let's start with about three lbs of ground chuck, one large sweet onion, a red pepper and a half bulb of fennel, all diced into about ¼ in. pieces. Sweat these in a pan with olive oil, salt and black pepper, and once soft add about six cloves of minced garlic. Set aside and let cool once the garlic is incorporated.

Meatloaf gets stretched by some filler, in this case breadcrumbs, and everything gets glued together by eggs. It is nice to even hide an egg or two in the meatloaf. In a mixing bowl combine the ground beef, about a cup on unseasoned breadcrumbs, two eggs, all of the cooked vegetables, two or three tablespoons of fennel seeds and a good shake of red chili flakes (depending on who else may be sharing this with you— I have to go sparingly with Cindy). Salt and black pepper are also necessary so think about how much you would put on a half-pound burger and increase accordingly (three pounds of meat is about six of those burgers). Be careful; you can always add at the table. Take two eggs that are nearly hard boiled, peel, rinse and place these in the middle of the loaf that you will form on a baking sheet. These will look nice once the meatloaf is finished and sliced for serving. Bake in the oven at about 375° with a thermometer inserted so that you can tell when it is done (medium). Let it set, slice and serve atop garlic mashed potatoes.

Another rendition is an Asian style with rice used as filler and finely cut bok choi replacing the fennel. Instead of salt use soy sauce, and as a glaze, use Thai chili sauce. You could also try a reverse braised meat dish. Instead of braising the meat with a mirepoix, but the mirepoix inside the meat and bake. You can dream up all sorts of ways to make meatloaf so that it is more edible than the blue plate special that you hated in high school cafeteria.

Since we used the word mirepoix, how about a Bordeaux to accompany the meatloaf. I am partial to St. Julien and Pauillac, but the variations by chateau and vintage can be large providing yet again a reason why cultivating relationship with a trusted local wine merchant is important. Attending tastings, when offered, also provides an enjoyable social occasion. For that, an Uber would be a good choice.

Chapter 14
Why Grouper Sandwiches Are Popular on Florida's West Coast

Fish sandwiches are a lunchtime staple, whether Haddock in Boothbay Harbor, ME; Salmon in Seattle, WA; Flying fish in Bridgetown, Barbados; Mahi in Key Largo, FL.; Walleye in Ely, MN; Catfish in New Orleans, LA; or Cod at any pub, anywhere. While all are delicious, a grouper sandwich in St. Petersburg, FL is unsurpassed. Groupers come in several varieties, Black, Red, Yellowedge and Gag being the more popular ones, at least based on what one can find at a local St. Petersburg, FL fish market. Black are the most expensive, but with Black and Gag looking only subtly different, and with Gag landings exceeding Black, my guess is that these may be interchanged, but who knows. If the piece of fish looks thick and cooks up nice and flakey, then it is probably grouper. If it looks thin and at a bargain price, it is likely fish mystere, as unfortunately occurs all too often.

Gag adults, in the eastern Gulf of Mexico, are known to spawn offshore, whereas their juvenile offspring settle near shore. These Gag belong to a single population that spawns from late winter to early spring at hard-bottom habitats between about the middle of the continental shelf and the shelf break, i.e., along the outer portion of the continental shelf. Settlement is estimated to occur some 30–50 days after spawning, and juvenile densities are found to be highest along the west coast of Florida from Tampa Bay to the Charlotte Harbor/Sanibel estuary complex, as compared with locations farther north and south, with a secondary region of juvenile settlement in the vicinity of Apalachicola Bay (see Fig. 14.1 for the overall coastal ocean region in discussion).

Three marine protected areas identified by red shapes provide a general depiction of the vicinity within which spawning is thought to occur, and the three embayments mentioned are also identified, along with other locations to be discussed later.

Being that fish eggs tend to be buoyant, thus floating to the surface upon spawning, it had been thought that the transport of Gag eggs and their eventual larvae to the near shore occurred via the surface currents. However, this hypothesis was found

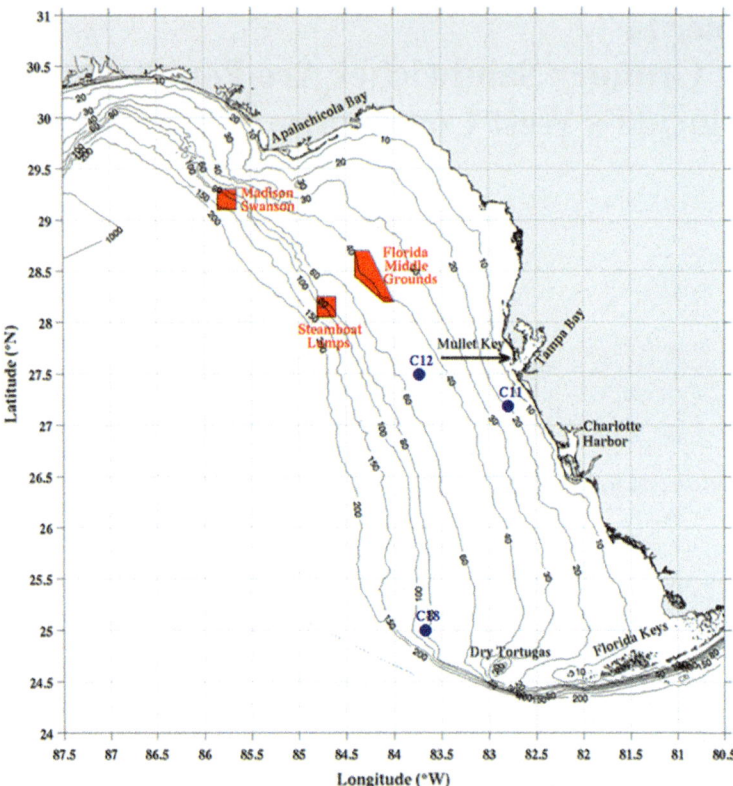

Fig. 14.1 Eastern Gulf of Mexico region where gag are thought to spawn. The general vicinity of the Marine Protected Areas is denoted in red, and the primary and secondary coastal region embayments for juvenile settlement are labeled, along with other locations discussed in text (adapted from Weisberg, R. H., L. Y. Zheng, and E. Peebles (2014), Gag Grouper larvae pathways on the west Florida shelf, *Cont. Shelf Res.*, 88, 11–23, https://doi.org/10.1016/j.csr.2014.06.003)

to be inconsistent with the prevailing wind-driven surface currents. Thus, a conundrum existed: How is it possible for Gag juveniles to arrive at their preferred settlement locations? The answer, of course, lies with the ocean circulation, but demonstrating this necessitated a fully three-dimensional approach, versus just looking at the surface.

Sometime around my birthday in 2007, a colleague (Dr. Ernst Peebles) came into my office carrying a bucket of water. He had something to show me. Reaching into the bucket, he pulled out a little fish, which he identified as a Gag juvenile. Also in his hand was a piece of macro algae known to grow along the bottom in hard, reef-like areas. He had just collected these on the beach at Mullet Key (27°37.761′N, 82°44.210′W), part of Fort DeSoto Park at the mouth of Tampa Bay (Fig. 14.1). This turned out to be a terrific birthday present because it resulted in a collaborative paper solving the Gag grouper recruitment conundrum.

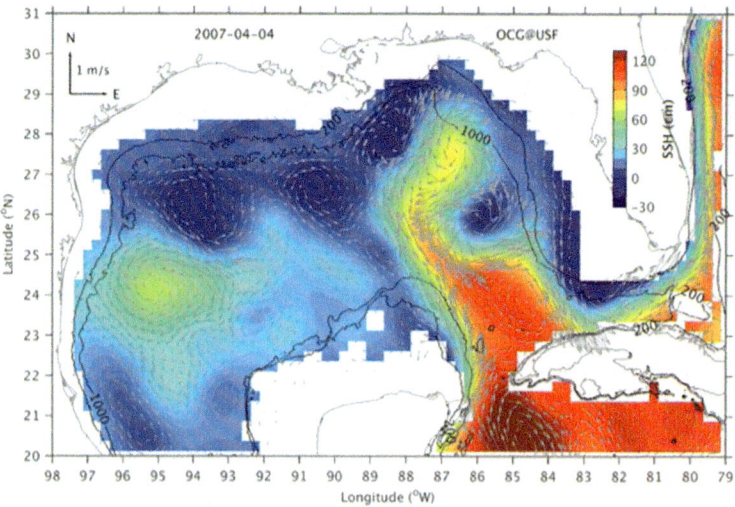

Fig. 14.2 Sea surface height (red being high and blue being low) and surface geostrophic velocity vectors (arrows) estimated from satellite altimetry on April 4, 2007. Note the interaction of the loop current with the shelf slope near the Dry Tortugas, resulting in a southward surface geostrophic current on the WFS

From previous studies, I immediately knew the answer, but to demonstrate this an experiment had to be designed to test the hypothesis that I had in mind. Recall from Chap. 4 that the Gulf of Mexico Loop Current, under certain conditions, can set the entire west Florida shelf in an upwelling favorable motion, wherein near-bottom water is transported across the shelf toward the shore. Thus, the same mechanism that was found to be responsible for the delivery of red tide from its offshore origination zone to the shore (Chap. 11) might also be responsible for delivering gag juveniles from their spawning location to their region of settlement. To see if this was a viable concept in this particular instance, we first needed to know if the Loop Current indeed did satisfy the upwelling circulation inducing conditions in spring 2007 and, if so, did our West Florida Coastal Ocean Model (WFCOM) simulations faithfully mimic the actual circulation as observed during that time. The answers to both of these questions turned out to be yes. This is another example of why the coordination between actual observations and model simulations is necessary. Observations alone are too sparse to fully describe a complex system, so models are required to dynamically interpolate a limited set of observations in order to construct a more complete field of motion. But model simulations are just that—simulations. Observations are required to assess the simulation veracity before it may be used for hypotheses testing. Spring 2007 satisfied both of these needs. The Loop Current did impinge upon the southwest corner of the shelf slope (Fig. 14.2) and we had sufficient observations to demonstrate the effect of the Loop Current on the shelf and for comparison with the model simulations (Fig. 14.3). With these necessities fulfilled, we could construct and test both surface and bottom larvae transport hypotheses.

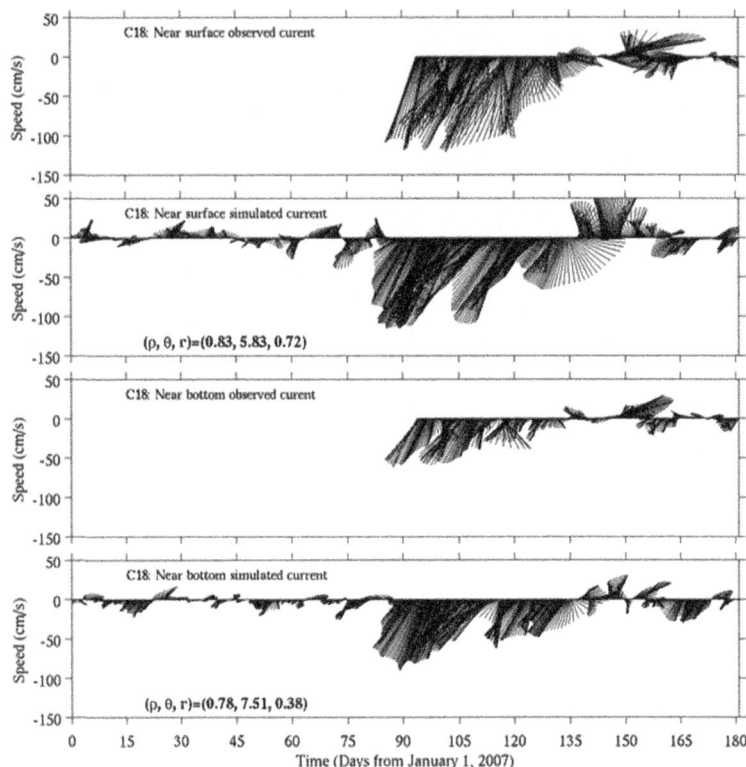

Fig. 14.3 Velocity vector time series comparisons between near-surface and near-bottom currents, either observed at mooring C18 or simulated by the West Florida Coastal Ocean Model at that location (see Fig. 14.1). The times series are all low pass filtered to remove oscillations on time scales shorter than 36 h, and the quantitative metrics of vector correlation coefficient, angular deviation and regression coefficient are provided as the triad of numbers in parentheses. The interval of primary concern here brackets days 90 through 135, corresponding to April through mid-May. Note that each stick represents a vector whose length is proportional to speed as indicated along the ordinate and whose direction of flow is given by the orientation with north directed upward and east to the right (adapted from Weisberg, R. H., L. Y. Zheng, and E. Peebles (2014), Gag Grouper larvae pathways on the west Florida shelf, *Cont. Shelf Res.*, 88, 11–23, https://doi.org/10.1016/j.csr.2014.06.003)

The coincidence of these necessary factors was fortuitous. The Loop Current impingement did not last very long, although it was long enough to be seen in the coincident velocity observations, and it occurred at precisely the right time as evident by the Gag juvenile observations at Mullet Key, which were within the third and fourth weeks of May 2007, peaking on May 23. As a demonstration, consider the sea surface height and geostrophic current findings for April 4, 2007, as shown in Fig. 14.2. We see the Loop Current impinging on the shelf slope near the Dry Tortugas, and accompanying this is a southward directed geostrophic current on the west Florida shelf.

Our mooring C18, which documented the immediate impact of the Loop Current on the shelf, was deployed at the beginning of April, without which we would not have known whether, or not, the Loop Current indeed activated the shelf circulation. Figure 14.3 shows velocity vectors both as observed by mooring C18 and simulated by WFCOM. Each stick is a velocity vector whose length is proportional to speed and whose orientation is the direction of flow (with north oriented up and east to the right). We see that at both the surface and near the bottom, the flows were essentially directed toward the south-southwest from the beginning of April through mid-May, as expected for such a Loop Current impingement (see Chap. 4). Here it is noted that WFCOM derives its deeper Gulf of Mexico forcing from another model run by the Naval Research Laboratory, and that model, the Navy's Hybrid Coordinate Ocean Model (HYCOM) is not always accurate near the Dry Tortugas. Fortunately, it was in April through May of 2007.

Given the accurate forcing of WFCOM by HYCOM near the Dry Tortugas at this time, moorings C12 and C11 further documented the upwelling circulation as occurred at mid-shelf. This is shown in Fig. 14.4 by comparing the near-bottom velocity vectors both observed and simulated by WFCOM, with high fidelity during the salient April through mid-May time period. Combined, these observational and model simulation comparisons provide ample evidence to justify using the WFCOM simulation for hypothesis testing.

Whereas individual moorings provide observations at particular points in space, model simulations fill in the surrounding area. The vertical line of Fig. 14.4 designates April 20, 2007, for which the entirety of the near-bottom flow field (as simulated) is shown in Fig. 14.5.

Thus, along with the individual points for which observations exist, we see that the entire west Florida shelf at this time was undergoing an upwelling circulation for which the near-bottom flow field was directed southeastward and hence toward the near shore. Such shelf-wide upwelling persisted throughout April and into the first two thirds of May, particularly near shore in the latter portion of May.

Given that the peak settlement for juveniles at Mullet Key was on May 23 and that prior evidence suggested a duration of about 45 days from spawning to settlement, tracking experiments were set up to assess the hypotheses of either a near-surface or a near-bottom transport pathway from spawning to settlement. Tracking may be performed in a variety of ways. For this set of experiments, we decided to take the simplest approach of initializing particles in the model either at the surface, or near the bottom and constraining them to stay either at the surface, or near the bottom. Based on what was thought to be the primary spawning site, i.e., the mid-shelf to the shelf slope region of the northern portion of the west Florida shelf, the region including the Madison Swanson, Steamboat Lumps and Middle Ground Marine Protected Areas (Fig. 14.1), particles were initialized along the 40 m, 60 m, 80 m and 100 m isobaths between the latitudes of 28°N–29.5°N at two day intervals on April 7, April 9, April 11 and April 13 and then tracked for 45 days.

The results for the particles released at the surface in the model simulations for any of these days and regardless of the different initial isobaths along which they were deployed were similar in that none of the particles made any substantive headway

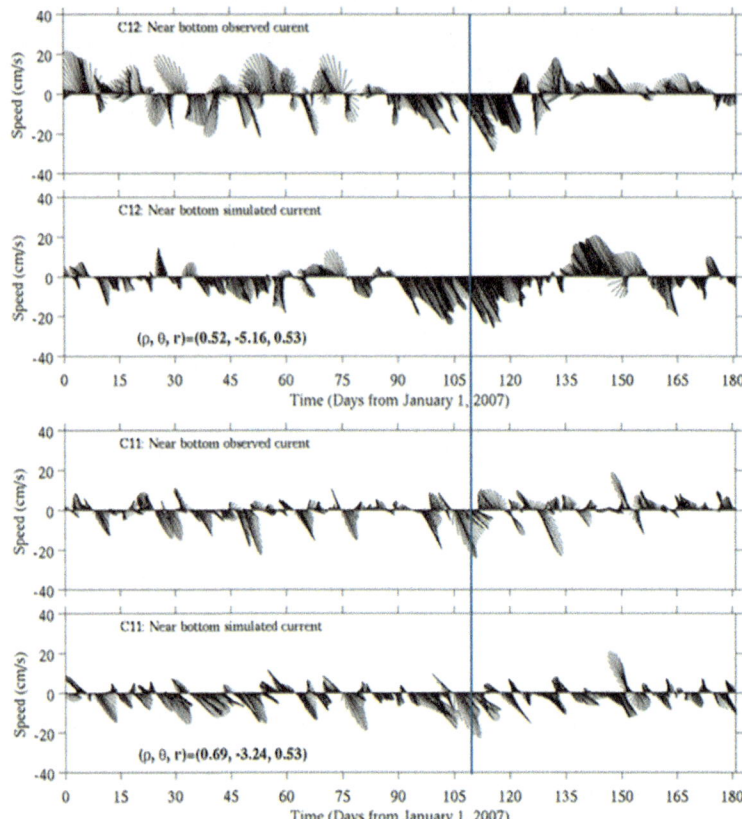

Fig. 14.4 Velocity vector time series comparisons between near-bottom currents, either observed or simulated by the West Florida Coastal Ocean Model at mooring C12 (top two panels) and at mooring C11 (bottom two panels). See Fig. 14.1 for their respective locations. The times series are all low pass filtered to remove oscillations on time scales shorter than 36 h, and the quantitative metrics of vector correlation coefficient, angular deviation and regression coefficient are provided as the triad of numbers in parentheses. The interval of primary concern here brackets days 90 through 135, corresponding to April through mid-May, and the vertical line is day 110 corresponding to April 20, 2007, for which the near-bottom simulated currents are shown in Fig. 14.5

toward the near shore. Instead, they all headed toward the south, primarily along the isobaths, and they all eventually ended up being located farther offshore. Had these particles been actual fish larvae, they would not have made it to settlement. Instead, they would have been food for other zooplankton or fish. Thus, all of the surface-initiated particles would have encountered aberrant drift, versus a trajectory conducive to settlement. As in a previous study by others, the hypothesis of a spawning to settlement pathway being at the surface had to be rejected.

Distinctly different results were found for particles initiated near the bottom. Regardless of the initial time or the initial isobath, all of these particles made headway

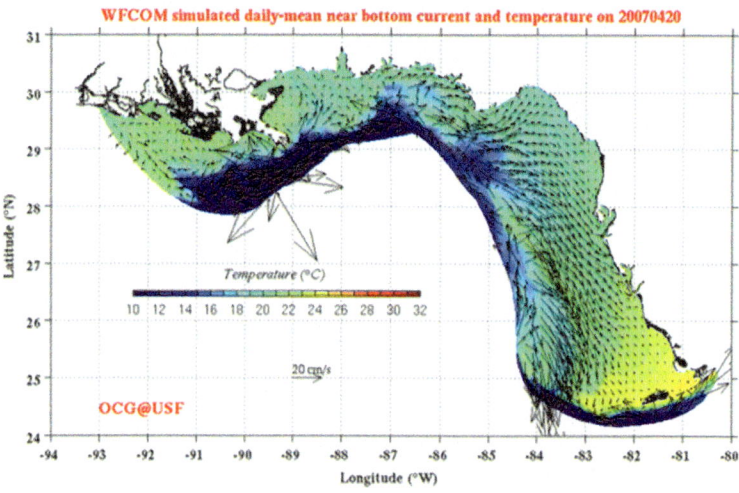

Fig. 14.5 WFCOM simulated near-bottom currents superimposed upon temperature for April 20, 2007. Note the upwelling-oriented near-bottom currents (arrows)

toward the near shore, with some actually landing at Mullet Key. As demonstrated by Fig. 14.6, particles initialized on April 7, 2007. on the 40 m, 60 m and 80 m isobaths actually completed transits to the shoreline, whereas those initialized on the 100 m isobath fell a little short of that mark. Similar results were found for the other initiation times. Given that the Gag juveniles were found at Mullet Key co-located with macro algae of hard-bottom origin, it could be argued that the 100 m isobath is too far offshore owing to the attenuation of light that would preclude macro algae growth at that depth. There is also another reason that we will get too shortly.

From the near-bottom particle pathlines, we can appreciate why the Tampa Bay vicinity and points to its south are the epicenter for gag juvenile settlement and also why Apalachicola Bay is a secondary region of settlement. These locations are where the pathlines intercept the shoreline. The shelf is too wide within the Florida Big Bend (the region to the north of Tampa Bay) for juveniles to arrive at the shore there. However, the Big Bend region is generally quite shallow and with seagrass-beds extending offshore, so settlement may also occur farther offshore in this region.

A further appreciation for the directionality of the pathlines harkens back to Fig. 14.5. The across-shelf, transport pathway is the bottom Ekman layer (see Chap. 4) that was well developed across the entire west Florida shelf in April 2007, whereas it eventually dwindled in the latter part of May 2007 when the Loop Current sidled away from the shelf slope near the Dry Tortugas. Nonetheless, upwelling did continue closer to shore in response to wind forcing.

From these results, we may also surmise that Gag recruitment is not uniformly operant from year to year. Those years for which a Loop Current impingement upon the shelf slope near the Dry Tortugas occurs within the spawning window of winter

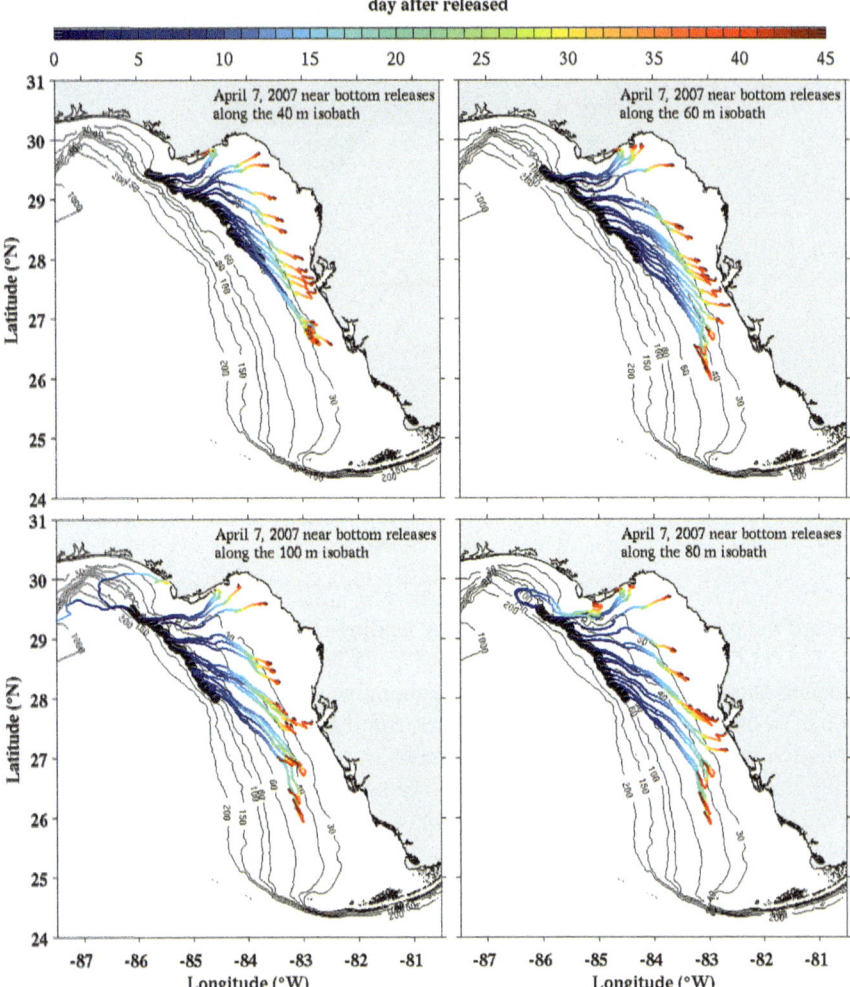

Fig. 14.6 Model simulated near-bottom particle trajectories for the 45-day interval beginning on April 7, 2007. Clockwise from the upper left-hand panel are particles initialized on the 40 m, 60 m, 80 m and 100 m isobaths, respectively (adapted from Weisberg, R. H., L. Y. Zheng, and E. Peebles (2014), Gag Grouper larvae pathways on the west Florida shelf, Cont. Shelf Res., 88, 11–23, https://doi.org/10.1016/j.csr.2014.06.003)

through spring months are likely to be the more prolific recruitment years. So Gag recruitment, just like red tide, should undergo considerable interannual variability, although that remains to be documented for Gag recruitment.

I digress here to offer an opinion on what I sense to be a shortcoming within the field of oceanography, especially as pertains to fisheries. A common thread throughout this book is how the ocean circulation impacts climate and ecology worldwide, i.e., *"From Climate to a Fish Sandwich: Why we Study the Ocean Circulation."*

One of the original oceanography compendiums, *"The Oceans: Their Physics, Chemistry and General Biology"* by Sverdrup, Johnson and Fleming (published in 1942) set the stage for oceanography as an interdisciplinary applied science. Curiously, as the field evolved, fisheries remained relatively aloof. For instance, the National Oceanographic and Atmospheric Administration (NOAA) has among its line offices a National Ocean Service (NOS), a National Weather Service (NWS) and a National Marine Fisheries Service (NMFS), along with Research, Operations and Satellite Offices. By operating independently from NOS, NMFS, at least in my opinion, gets limited inputs and insights from the rest of the oceanographic community. My experience has been that such aloofness also carries into state and local agency levels. Even here at my own institution, excellent agency collaborations extend across most endeavors except for fisheries, despite studies like this one suggesting advances that could occur through improved collaborations. When I look at my citation history, my Gag research paper from which this chapter derives is not well-cited. In other words, it remains largely unknown by the fisheries community. It is fair to ask what other relevant insights may be ignored simply because the fisheries community stands too aloof from mainstream oceanography and might our fisheries management be improved for the benefit of society if barriers to further collaboration were reduced?

I have my own theory on why this aloofness may exist. It dates back to my undergraduate college years, when, as a spring semester junior, I shattered my leg in an accident requiring a 5-month stint in a full leg cast and the need to attend summer school to make up what I had missed that semester. By that summer, I was quite proficient with my crutches, enabling me to hike over considerably rough terrain. I decided to take up trout fishing and went to a local sports shop to get some gear and advice. The proprietor was quick to sell me what was needed, but would not offer any advice on where the fishing might be best. He seemed to care less that I was hobbled. How naïve of me at that time to think that a fisher would share information on any best sites. Even now, if one tries to get location numbers from anyone regarding where to go catch Gag or any other species, they are met with silence. Such information tends to be hoarded by fishers. Might this be a reason for why Fisheries Oceanography tends to be separate from the rest of the oceanographic disciplines—I wonder.

Returning back now to natural science from my drift into social science, given that Gag tend to spawn within the mid-shelf to outer-shelf region during winter to spring months, might there be a reason for this? A likely one is temperature. Aquaculture studies suggest that temperatures above 13 °C are necessary for grouper in general, with most species preferring temperatures higher than this. This raises the question of where spawning may occur with sufficiently warm near-bottom water such that by the time the juveniles arrive at settlement, the temperatures there may be conducive to rapid growth. From Chap. 10 we found that the west Florida coastal ocean begins its seasonal cooling and warming in August and February, respectively. How rapidly either cooling or warming may occur depends upon water depth because the deeper the depth, the more volume there is to disperse the heat energy that is input via the net surface heat flux (in Wattsm^{-2}). From this, one might surmise that shallow water heats and cools most rapidly. Also, seasonal heating and cooling generally penetrate down through what is referred to as the mixed layer, which, for the Gulf of Mexico, is usually limited to about a 100 m in depth. Below 100 m, the water tends to be relatively cold.

These concepts facilitate a straightforward thought experiment. Starting from uniformly warm water that may extend across most of the continental shelf in August, let's begin cooling the surface. The shallowest water will cool most quickly so by the time fall arrives, we see that the near-bottom temperature distribution on October 1, 2021, as simulated using the West Florida Coastal Ocean Model in the upper left-hand panel of Fig. 14.7, shows somewhat cooler waters nearest to the shoreline in the Big Bend region than farther offshore and that the coldest near-bottom temperatures are beyond the shelf break in waters at depths deeper than around 100 m. Continuing through fall months, we arrive two months later at the December 1, 2021, distribution in the upper right-hand panel. Here we see a relatively warm tongue of near-bottom water extending over the mid-shelf to outer-shelf region, with much colder water both beyond the shelf break and in the shallower Big Bend region. Two months later on February 1, 2022 (lower left-hand panel), the near-bottom waters of both the shallow Big Bend region and beyond the shelf break are seen (lower left-hand panel) to be colder than grouper are likely to find tolerable, whereas a relatively warm tongue of near-bottom water remains within the mid-shelf to outer-shelf regions, particularly within the region between the 40 m and the 80 m isobaths, consistent with what is known to be the preferred Gag spawning region. Then, two months later, from April 1, 2002, and beyond into spring, after the seasonal warming sets in, the near-bottom temperatures in the region of settlement are conducive to rapid Gag juvenile growth. Thus, the timing and location of spawning, plus the near-bottom transport mechanism and duration to the location of settlement are all consistent with the seasonal cycle in west Florida continental shelf heating and cooling. None of this is by chance. The Gag species evolved to take advantage of its natural environment, as determined by the coupled, ocean–atmosphere system (Chaps. 1–3).

I have no doubt that the spawning and settlement behaviors by other species follow similar arguments. For instance, Goliath grouper is also thought to spawn at mid-shelf and to settle in the mangrove region to the south of Marco Island. It could be that the same arguments in explanation of Gag may equally apply to Goliath. With Goliath being an endangered species, one would think that by this time fisheries oceanographers, especially those charged with resource management may have sought some assistance from outside of their immediate sphere.

It is my hope that such comments may help to facilitate future productive collaborations on how the coastal ocean, the region where society meets the sea, actually works. For now, it is time to sit back and enjoy a grouper sandwich.

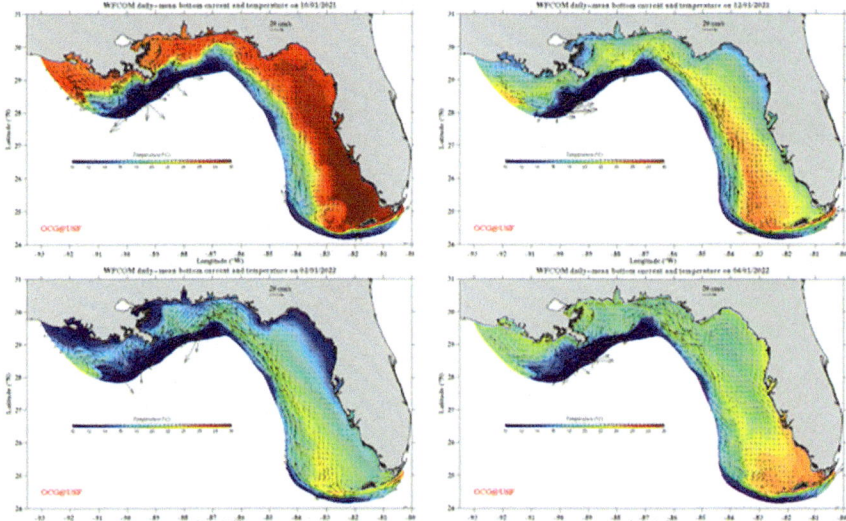

Fig. 14.7 Four snapshots of daily-averaged near-bottom velocity and temperature simulated by the West Florida Coastal Ocean Model (WFCOM). Clockwise from the upper left are results for October 1, 2021, December 1, 2021, February 1, 2022, and April 1, 2022. The choice of year is arbitrary. Each year, while subtly different depending upon Loop Current interactions with the shelf slope near the Dry Tortugas and the vagaries of the weather, behaves in a seasonally similar way

Addendum: A Grouper Sandwich

Grouper sandwiches can take several forms, fried, pan sautéed and grilled. Regardless of the preparation, the key is fresh grouper so find an excellent seafood butcher instead of purchasing fish (other than frozen) at the grocery store.

Not having a deep-fat fryer, I never make fried grouper. Instead, I may pan saute grouper coated in panko crumbs and put this between a soft bun, ciabatta or onion roll with a tartar sauce of some form. Tartar sauce requires a decent mayonnaise like Dukes (never Miracle Whip). Mine is simple, just chopped up brined pickles like those of Ba Tampe half sour, versus most other pickles that are made with vinegar. Mayo, plus these is all that it takes.

Pan searing grouper in butter with a piece of cheese and some avocado or guacamole is also quite good, as is just some lettuce and tomato.

Grilling also works but take a little more effort to light a grill and make sure to oil it so the fish does not stick.

Other renditions on Tartar sauce are worth trying. For instance, chopping up green Spanish olives with or without the pimento provides a different but very nice taste difference from the brine pickle variety.

Grouper sandwiches go well with fried potatoes so to avoid the mess, it might be good idea to just find a restaurant with the best one and have it there. When the kids were little, a Sunday routine was to sail south to Bradenton, have a grouper sandwich for lunch at "The Sandbar" beach restaurant, and then sail home.

Epilog

The preceding text attempted to explain how the oceans affect almost everything on Earth, from climate and ecology to something as simple as a grouper sandwich. It was also intended to inspire high school and college age readers to think about a career in science, the physics of the ocean circulation, in particular. As a science field, oceanography remains relatively new. Whereas tides were studied centuries ago, it was not until the mid-twentieth century that we had any quantitative understanding on the general ocean circulation. As a graduate student, I recall a statement from a noted physical oceanographer to the effect that we once thought we knew the ocean circulation to 10 cms^{-1}, plus or minus 1 cms^{-1}, whereas we now realize that we only know it to 1 cms^{-1}, plus or minus 10 cms^{-1}. In other words, the variability greatly exceeds the mean. As a corollary, whereas we think we know something about climate and how it may vary under the influence of human behavior (such as the rise in CO_2 due to fossil fuel use), in reality, we have just begun to scratch at the surface on what the natural variability in climate may be. There remains much more to learn in distinguishing human from natural influences. Whereas we have now undergone some ten ice-age to interglacial cycles over the past million or so years while the Earth's land masses have been in their present configurations, we do not have a strong basis for understanding exactly how these work, nor how the Earth's climate system worked when its land masses were in prior configurations. The ocean sciences have progressed impressively over the past few generations, but the field is young leaving much more to be discovered!

In choosing a career one should consider more than just earning capacity. When we consider that we sleep, work and recreate, in equal portions (about one third each) of our life, it becomes apparent that half of our waking hours are spent at work. Living an enjoyable life requires sound choices regarding what to pursue and with whom to do so. Science may not be the wealthiest of career paths to follow; but, given the three attributes mentioned in the *Forward*, boredom is never a concern. I can honestly say that while I enjoyed a TGIF, I never needed one, hump day was never something to look forward to and Mondays were never dreaded. I relished my family and personal time, but I also enjoyed my work. As a professor, the student and associate interactions were an added benefit, even to the extent of having what may be considered as a lasting, extended family. The only annoying things were the few acquaintances who may have harbored some disdain for educators. "So you are off all summer," or "What do you do all day if you only teach one course," or "Gosh, I wish I could spend my time going on cruises," are some of the lines that I occasionally suffered. All professions have their detractors. Recall Shakespeare's Henry VI line: "The first thing we do, let's kill all the lawyers." At least no one wanted to kill me, that is as far as I know.

I hope that the reader found this book to be worthwhile. I tried to convey the most general aspects of the ocean's influence on both climate and ecology in Chaps. 1–3. Chapters 4 and 5 then took us into the coastal ocean and the estuaries, where society meets the sea. Sea level (Chap. 6) is a concern for all of us, whether navigating recreationally or commercially, or just wondering if you may someday get flooded. This, of course, led us to the discussion of waves (Chap. 7), where the reader may have found for the first time that these come in all sizes, from the ones we see at the surface to much larger ones occurring sight unseen below. Tsunamis, unto themselves, were then discussed in Chap. 8.

The reader may have wondered why I spent so much time on hurricane storm surge (Chap. 9). Given their destructive potential in regions of high population densities, I deemed this as worthwhile, especially given misconceptions on the potential for destruction by wind, surge and waves. Confusion is fueled, in part, by how insurance is underwritten, wherein flood and wind damage are separated, with flood being underwritten by the federal government (FEMA in the United States) and wind by private carriers.

Hurricane Katrina demonstrated just how misleading this insurance dichotomy could be for the public at large. With most structures destroyed by flood, versus wind, many owners were disappointed with their recoveries, which hardly covered the costs of repair, or replacement. This disparity led to a multitude of legal pursuits, wherein homeowners sued the insurance companies for their damages. In many instances, the insurance companies either: (1) settled to avoid further litigation, (2) were found liable in court, with added punitive damages assessed, even though the damages were by water, versus wind, or (3) won at trial. Regardless of the outcomes, many insurance companies stopped underwriting wind policies in flood-prone areas, leaving the public with fewer carriers to choose from and with greatly increased rates. Thus, the damage caused by Katrina extended far beyond the region that was directly impacted by that storm. We are all paying much higher insurance premiums because if Hurricane Katrina.

As painful as insurance premiums may be, such monetary matters are trivial compared with the risks to life and property. Not only should we be aware of the damage that can result from hurricane storm surge and the accompanying waves, through this awareness we can better appreciate why certain mitigation strategies may be fruitless, why heeding emergency manager calls for evacuation are so important, and also why regional planning, particularly for critical infrastructure placement, requires evaluation. Despite hurricanes being a real threat to coastal communities, land development continues apace—people like to live along the shoreline. It is for that reason why Chap. 9 is particularly long. We should all learn from the lessons of Hurricane Katrina.

Returning to more mundane topics, Chap. 10 explains why sea surface temperatures vary seasonally as they do. Much of this is owing to local ocean–atmosphere interaction, although the underlying coastal ocean circulation may also play an important role in some years. Since the ocean temperature impacts adjacent land temperature, such ocean effect may even help to explain why some summers may seem warmer than others for reasons other than simply global warming.

The next topical application is in explanation of harmful algae blooms, *K. brevis* along Florida's west coast, in particular. This is an excellent example of how ecology is not merely the purview of biology; instead, it entails everything that determines the conditions within which an organism may exist. In other words, ecology is not just biology; it is the whole shebang. By recognizing this fact, we may become better stewards of the environment, instead of just acceding to policy positions advanced by environmental groups or environmental managers that could result in more harm than good. An educated public is essential for helping to guide public policy in the most positive direction.

Behavioral modifications for the purpose of global warming mitigation are at the forefront of discussions regarding climate. Not to downplay the importance of global warming, the evidence for which is compelling, it is equally important to realize that not all climate variation is due to human endeavor, particularly the burning of fossil fuels. Considerable natural variability exists that we are only beginning to comprehend (Chap. 12), and policy decisions based on oversimplifications may be more detrimental than the warming itself.

Humans require power generation for most of what they rely upon, e.g., food, shelter and the economies necessary to provide for these. Power generation reliance has gone from the burning of wood and dung to the succession of peat, coal, oil and natural gas. Added to these fuel uses have been the reliance on wind and water driven machinery, plus solar and nuclear energy conversions. New technologies will undoubtedly continue this evolution of power production, but for now we have what we have, and it is important to be candid and honest regarding what our alternatives may be (Chap. 13). What may be truth; what may be fiction is not always clear when hearing of new, politician-proffered proposals.

Finally, and to end on a positive note, we learned about how the ocean circulation may impact the abundance of grouper and hence why we can enjoy a good grouper sandwich for lunch.

I hope that these topics imparted an appreciation for why we study the ocean circulation and that your time was well-spent in reading this book.

Author Bibliography

1. Duing, W., Hisard, P., Katz, J., Meincke, J., Miller, L., Moroshkin, K. V., Philander, G., Ribnikov, A. A., Voigt, K., & Weisberg, R. (1975). Meander and long waves in the equatorial Atlantic. *Nature, 257*, 380–384.
2. Weisberg, R. H. (1976). The non-tidal flow in the providence river of Narragansett Bay: A stochastic approach to Estuarine circulation. *Journal of Physical Oceanography, 6*, 721–734.
3. Weisberg, R. H. (1976). A note on estuarine mean flow estimation. *Journal of Marine Research, 34*, 387–394.
4. Weisberg, R. H., & Sturges, W. (1976). Velocity observations in the West Passage of Narragansett Bay: A partially mixed estuary. *Journal of Physical Oceanography, 6*, 345–354.
5. Weisberg, R. H. (1979). Equatorial waves during GATE and their relation to the mean zonal circulation, *Deep-Sea Res. Suppl. II to, 26*, 179–198.
6. Weisberg, R. H., Horigan, A., & Colin, C. (1979). Equatorially trapped Rossby-gravity wave propagation in the Gulf of Guinea. *Journal of Marine Research, 37*, 67–86.
7. Weisberg, R. H., Miller, L., Knauss, J., & Horigan, A. (1979). Velocity observations in the equatorial thermocline during GATE, *Deep-Sea Res. Suppl. II to, 26*, 217–248.
8. Horigan, A. M., & Weisberg, R. H. (1981). A systematic search for trapped equatorial waves in the GATE velocity data. *Journal of Physical Oceanography, 11*, 497–509.
9. Weisberg, R. H., & Horigan, A. M. (1981). Low frequency variability in the equatorial Atlantic. *Journal of Physical Oceanography, 11*, 913–920.
10. Weisberg, R. H., & Pietrafesa, L. J. (1983). Kinematics and correlation of the surface Wind field in the South Atlantic Bight. *Journal of Geophysical Research, 88*, 4592–4610.
11. Weisberg, R. H., & Tang, T. Y. (1983). Equatorial ocean response to growing and moving wind systems with application to the Atlantic. *Journal of Marine Research, 41*, 461–486.
12. Tang, T. Y., & Weisberg, R. H. (1984). On the response of the equatorial Pacific Ocean to the 1982/1983 El Nino—Southern Oscillation event. *Journal of Marine Research, 42*, 809–829.
13. Weisberg, R. H. (1984). SEQUAL/FOCAL: First year results on the circulation in the equatorial Atlantic. *Geophysical Research Letters, 11*, 713–714.
14. Weisberg, R. H. (1984). Instability waves observed on the equator in the Atlantic Ocean during 1983. *Geophysical Research Letters, 11*, 753–756.
15. Weisberg, R. H. (1984). Seasonal adjustment in the equatorial Atlantic during 1983 as seen by surface moorings. *Geophysical Research Letters, 11*, 733–735.
16. Weisberg, R. H. (1985). Equatorial Atlantic velocity and temperature observations: February-November 1981. *Journal of Physical Oceanography, 15*, 533–543.
17. Weisberg, R. H., & Tang, T. Y. (1985). On the response of the equatorial thermocline in the Atlantic Ocean to the seasonally varying winds. *Journal of Geophysical Research, 90*, 7117–7128.

18. Philander, G., Halpern, D., Hansen, D., Legeckis, R., Miller, L., Paul, G., Watts, R., Wimbush, & Weisberg, R. (1985). Long waves in the equatorial Pacific Ocean. The Oceanography Report, *EOS*, 4/2/85.
19. Weisberg, R. H., & Colin, C. (1986). Equatorial Atlantic ocean temperature and current variations during 1983–1984. *Nature, 322*, 240–243.
20. Weisberg, R. H., & Weingartner, T. J. (1986). On the baroclinic adjustment of the zonal pressure gradient in the equatorial Atlantic Ocean. *Journal of Geophysical Research, 91*, 11717–11725.
21. Weisberg, R. H., & Tang, T. Y. (1987). Further studies on the response of the equatorial thermocline in the Atlantic Ocean to the seasonally varying trade winds. *Journal of Geophysical Research, 92*, 3709–3727.
22. Weisberg, R. H., Halpern, D., Tang, T. Y., & Hwang, S. M. (1987). M2 tidal currents in the eastern equatorial Pacific Ocean. *Journal of Geophysical Research, 92*, 3821–3826.
23. Weisberg, R. H., Tang, T. Y., Weingartner, T. J., & Hickman, J. H. (1987). Velocity and temperature during the SEQUAL experiment at the equator, 28 °W. *Journal of Geophysical Research, 92*, 5061–5075.
24. Tang, T. Y., Weisberg, R. H., & Halpern, D. (1988). Vertical structure of low frequency variability in the eastern equatorial Pacific Ocean. *Journal of Physical Oceanography, 18*, 1009–1019.
25. Weisberg, R. H., & Weingartner, T. J. (1988). Instability waves in the equatorial Atlantic Ocean. *Journal of Physical Oceanography, 18*, 1641–1657.
26. Halpern, D., & Weisberg, R. H. (1989). Upper ocean thermal and flow fields at 0°, 28°W (Atlantic) and 0°, 140°W (Pacific) during 1983–1985. *Deep-Sea Research, 36*, 407–418.
27. Mayer, D. A., Molinari, R. L., & Weisberg, R. H. (1990). Analysis of volunteer observing ship temperature fields in the tropical Atlantic Ocean. *Oceanologica Acta, 13*, 257–264.
28. Weisberg, R. H., & Tang, T. Y. (1990). A linear analysis of equatorial Atlantic Ocean thermocline variability. *Journal of Physical Oceanography, 20*, 1813–1825.
29. Weingartner, T. J., & Weisberg, R. H. (1991). On the annual cycle of upwelling on the equator in the central Atlantic Ocean. *Journal of Physical Oceanography, 21*, 68–82.
30. Weingartner, T. J., & Weisberg, R. H. (1991). A description of the annual cycle in surface temperature and upper ocean heat in the equatorial Atlantic. *Journal of Physical Oceanography, 21*, 83–96.
31. Galperin, B., Blumberg, A. F., & Weisberg, R. H. (1992). The importance of density driven circulation in well mixed estuaries: The Tampa Bay experience. In *Proceeding Estuarine and Coastal Modelling, 2nd Int'l Conference/WW Div.* ASCE, Tampa, FL 1991.
32. Tang, T. Y., & Weisberg, R. H. (1993). Seasonal variations in equatorial Atlantic Ocean zonal volume transport at 28°W. *Journal of Geophysical Research, 98*, 10145–10153.
33. Mayer, D. A., & Weisberg, R. H. (1993). A description of COADS surface meterological fields and the implied Sverdrup transports for the Atlantic Ocean from 30 S to 60 N. *Journal of Physical Oceanography, 23*, 2201–2221.
34. Wang, C., & Weisberg, R. H. (1994). Equatorially trapped waves of a coupled ocean-atmosphere system. *Journal of Physical Oceanography, 24*, 1978–1998.
35. Wang, C., & Weisberg, R. H. (1994). On the "slow mode" mechanism of coupled ocean atmosphere models of the El Nino-Southern Oscillation (ENSO). *Journal of Climate, 7*, 1657–1667.
36. Jones, W. K., Galperin, B., Weisberg, R.H., & Wu, T.S. (1994). Influence of sikes cut on Apalachicola Bay, FL.; A preliminary analysis from a three-dimensional perspective. In *Proceeding Estuarine and Coastal Modelling, 3rd Int'l Conference/WW Div.* ASCE, 1992.
37. Qiao, L., & Weisberg, R. H. (1995). Tropical instability wave kinematics: Observations from the tropical instability wave experiment (TIWE). *Journal of Geophysical Research, 100*, 8677–8693.
38. Weisberg, R. H., & Hayes, S. P. (1995). Upper ocean variability on the equator in the west central Pacific at 170 W. *Journal of Geophysical Research, 100*, 20485–20498.

39. Squires, A. P., Vargo, G. A., Weisberg, R. H., Fanning, K. A., & Galperin, B. (1995). Review and synthesis of historical Tampa Bay water quality data. *Florida Scientist, 58,* 228–233.
40. Weisberg, R. H., Black, B., & Yang, H. (1996). Seasonal modulation of the west Florida shelf circulation. *Geophysical Research Letters, 23,* 2247–2250.
41. Wang, C., & Weisberg, R. H. (1996). Stability of equatorial modes in a simplified Coupled ocean-atmosphere model. *Journal of Climate, 9,* 3132–3148.
42. Yang, J. Y., Tang, T. Y., & Weisberg, R. H. (1997). Basinwide zonal wind stress and ocean thermal variations in Equatorial Pacific Ocean. *Journal of Geophysical Research, 102,* 911–927.
43. Qiao, L., & Weisberg, R. H. (1997). The zonal momentum balance of the equatorial undercurrent in the central Pacific. *Journal of Physical Oceanography, 27,* 1094–1119.
44. Weisberg, R. H., & Wang, C. (1997). Slow variability in the equatorial west-central Pacific in relation to ENSO. *Journal of Climate, 10,* 1998–2017.
45. Weisberg, R. H., & Wang, C. (1997). A Western Pacific oscillator paradigm for ENSO. *Geophysical Research Letters, 24,* 779–782.
46. Qiao, L., & Weisberg, R. H. (1998). Tropical instability wave energetics: The tropical instability wave experiment. *Journal of Physical Oceanography, 28,* 345–360.
47. Wang, C., & Weisberg, R. H. (1998). Observations of meridional scale frequency dependence in the coupled ocean-atmosphere system. *Journal of Geophysical Research, 103,* 2811–2816.
48. Morris, M., Roemmich, G. M., & Weisberg, R. H. (1998). Upper ocean heat and fresh water advection in the western Pacific. *Journal of Geophysical Research, 103,* 13023–13039.
49. Mayer, D. A., & Weisberg, R. H. (1998). ENSO-related ocean-atmosphere coupling in the western equatorial Pacific. *Journal of Geophysical Research, 103,* 18635–18648.
50. Wang, C., & Weisberg, R. H. (1998). Climate variability in the coupled tropical-extratropical ocean-atmosphere system. *Geophysical Research Letters, 25,* 3979–3982.
51. Mayer, D. A., & Weisberg, R. H. (1999). Correction to: "ENSO-related ocean-atmosphere coupling in the western equatorial Pacific." *Journal of Geophysical Research, 104,* 1579.
52. Wang, C., Weisberg, R. H., & Virmani, J. (1999). Western Pacific interannual variability associated with ENSO. *Journal of Geophysical Research, 104,* 5131–5149.
53. Yang, H., & Weisberg, R. H. (1999). Response of the West Florida continental shelf circulation to climatological wind forcing. *Journal of Geophysical Research, 104,* 5301–5320.
54. Wang, C., Weisberg, R. H., & Yang, H. (1999). Effects of the wind speed-evaporation SST feedback on the El Nino-Southern Oscillation. *Journal of Atmospheric Science, 56,* 1391–1403.
55. Li, Z., & Weisberg, R. H. (1999). West Florida Shelf response to upwelling favorable wind forcing: Kinematics. *Journal of Geophysical Research, 104,* 13507–13527.
56. Yang, H., Weisberg, R. H., Niiler, P. P., Sturges, W., & Johnson, W. (1999). Lagrangian circulation and forbidden zone on the West Florida Shelf. *Continental Shelf Research, 19,* 1221–1245.
57. Li, Z., & Weisberg, R. H. (1999). West Florida continental shelf response to upwelling favorable wind forcing, 2: Dynamics. *Journal of Geophysical Research, 104,* 23427–23442.
58. Wang, C., & Weisberg, R. H. (2000). The 1997–98 El-Nino evolution relative to previous El Nino events. *Journal of Climate, 13,* 488–501.
59. Weisberg, R. H., & Qiao, L. (2000). Equatorial upwelling in the central Pacific estimated from moored velocity profilers. *Journal of Physical Oceanography, 30,* 105–124.
60. Weisberg, R. H., Black, B., & Li, Z. (2000). An upwelling case study on Florida's west coast. *Journal of Geophysical Research, 105,* 11459–11469.
61. Cronin, M. F., McPhaden, M. J., & Weisberg, R. H. (2000). Wind forced reversing jets in the western equatorial Pacific. *Journal of Physical Oceanography, 30,* 657–676.
62. Shay, L. K, Cook, T. M., Haus, B. K., Martinez, J., Peters, H., Mariano, A. J., Van Leer, J., An, P. E., Smith, S., Soloviev, A., Weisberg, R., & Luther, M. (2000). VHF radar detects oceanic submesoscale vortex along Florida coast. *Eos, Transactions American Geophysical Union, 81,* 209 & 213.

63. Harrison, D. E., Vecchi, G. A., & Weisberg, R. H. (2000). Eastward surface jets in the central equatorial Pacific. *Journal of Marine Research, 58*, 735–754.
64. Helber, R. W., & Weisberg, R. H. (2001). Equatorial upwelling in the western Pacific warm pool. *Journal of Geophysical Research, 106*, 8989–9004.
65. Meyers, S. D., Siegel, E. M., & Weisberg, R. H. (2001). Observations of currents on the west Florida shelf break. *Geophysical Research Letters, 28*, 2037–2040.
66. Weisberg, R. H. (2001). An observers view of the equatorial ocean currents. *Oceanography, 14*, 27–33.
67. Wang, C., & Weisberg, R. H. (2001). Ocean circulation influences on sea surface temperature in the equatorial central Pacific. *Journal of Geophysical Research, 106*, 19515–19526.
68. Weisberg, R. H., Li, Z., & Muller-Karger, F. E. (2001). West Florida shelf response to local wind forcing: April 1998. *Journal of Geophysical Research, 106*, 31239–31262.
69. Walsh, J. J. K. D., Haddad, D. A., Dieterle, R. H., Weisberg, Z., Li, H., Yang, F. E., Muller-Karger, C. A. H., & Bissett, W. P. (2002). A numerical analysis of the landfall of 1979 red tide of *Karenia brevis* along the west coast of Florida. *Continental Shelf Research, 22*, 15–38.
70. Vargo, G. A., Heil, C. A., Spence, D., Neely, M. B., Merkt, R., Lester, K., Weisberg, R. H., Walsh, J. J., & Fanning, K. (2001). The hydrographic regime, nutrient requirements, and transport of a gymnodinium breve Davis red tide on the West Florida shelf. In G. M. Hallegraeff, S. I. Blackburn, C. J. Bolch & R. J. Lewis (Eds.), *Proceeding of the IXth International Conference on Harmful Algal Blooms* (pp. 157–160), Feb 7–11, 2000. Hobart, Australia.
71. Shay, L. K., Cook, T. M., Peters, H., Mariano, A. J., Weisberg, R., An, P. E., Soloviev, A., & Luther, M. (2002). Very high frequency radar mapping of surface currents. *IEEE Journal of Oceanic Engineering, 27*, 155–169.
72. He, R., & Weisberg, R. H. (2002). West Florida shelf circulation and temperature budget for the 1999 spring transition. *Continental Shelf Research, 22*, 719–748.
73. Hu, C., et al. (2002). Satellite images track "black water" event off Florida coast. *Eos, Transactions American Geophysical Union, 83*, pp. 281, 285.
74. He, R., & Weisberg, R. H. (2002). Tides on the West Florida Shelf. *Journal of Physical Oceanography, 32*, 3455–3473.
75. Virmani, J. I., & Weisberg, R. H. (2003). Features of the observed Annual Ocean-atmosphere flux variability on the West Florida Shelf. *Journal of Climate, 16*, 734–745.
76. He, R., & Weisberg, R. H. (2003). A loop current intrusion case study on the West Florida Shelf. *Journal of Physical Oceanography, 33*, 465–477.
77. He, R., & Weisberg, R. H. (2003). West Florida shelf circulation and temperature budget for the 1998 fall transition. *Continental Shelf Research, 23*, 777–800.
78. Weisberg, R. H., & He, R. (2003). Local and deep-ocean forcing contributions to anomalous water properties on the West Florida Shelf. *Journal of Geophysical Research, 108*(C6), 15. https://doi.org/10.1029/2002JC001407
79. Walsh, J. J., Weisberg, R. H., Dieterle, D. A., He, R., Darrow, B. P., Jolliff, J. K., Lester, K. M., Vargo, G. A., Kirkpatrick, G. J., Fanning, K. A., Sutton, T. T., Jochens, A. E., Briggs, D. C., Nababan, B., Hu, C., & Muller-Karger, F. (2003). The phytoplankton response to intrusions of slope water on the West Florida Shelf: Models and observations. *Journal of Geophysical Research, 108*(C6), 15. https://doi.org/10.1029/2002JC001406
80. He, R., Weisberg, R. H., Zhang, H., Muller-Karger, F., & Helber, R. W. (2003). A cloud-free, satellite-derived, sea surface temperature analysis for the West Florida Shelf. *Geophysical Research Letters.* https://doi.org/10.1029/2003GL017673
81. Halliwell, G. R., Weisberg, R. H., & Mayer, D. (2003). A synthetic float analysis of upper-limb meridional overturning circulation interior ocean pathways in the tropical/subtropical Atlantic. In G. Goni & P. Malanotte-Rizzoli (Eds.), *Interhemisphere water exchange in the Atlantic Ocean* (pp. 93–136). Elsevier.
82. Soloviev, A. V., Weisberg, R. H., & Luther, M. E. (2003). Energetic Baroclinic super-tidal oscillations on the shelf off Southeast Florida. *Geophysical Research Letters, 30*, 9. https://doi.org/10.1029/2002GL016603

83. Soloviev, A. V., Walker, R. J., Weisberg, R. H., & Luther, M. E. (2003). Coastal observatory investigates energetic current oscillations on the southeast Florida shelf. *Eos, Transactions American Geophysical Union, 84*, 42, 441.
84. Jolliff, J. K., Walsh, J. J., He, R., Weisberg, R. H., Stovall-Leonard, A., Coble, P. G., Comny, R., Heil, C., Nababan, B., Zhang, H., Hu, C., & Muller-Karger, F. (2003). Dispersal of the Suwannee River plume over the West Florida shelf: Simulation and observation of the optical and biochemical consequences of a flushing event. *Geophysical Research Letters, 30*(13), 1709.
85. Weisberg, R. H., & Zheng, L. (2003). How estuaries work: A Charlotte Harbor example. *Journal of Marine Research, 61*, 635–657.
86. Venezia, W., et al. (2003). SFOMC: A successful Navy and academic partnership providing sustained ocean observation capabilities in the Florida Straits. *MTS Journal, 37*, 81–91.
87. Seim, H., Bacon, B., Barans, C., Fletcher, M., Gates, K., Jahnke, R., Kearns, E., Lea, R., Luther, M., Mooers, C., Nelson, J., Porter, D., Shay, L., Spranger, M., Thigpen, J., Weisberg, R., & Werner, F. (2003). SEA-COOS—A model for a multi-state multi-institutional regional observation system. *MTS Journal, 37*(3), 92–101.
88. Zheng, L., & Weisberg, R. H. (2004). Tide, buoyancy, and wind driven circulation of the Charlotte Harbor estuary, a model study. *Journal of Geophysical Research, 109*, C06011. https://doi.org/10.1029/2003JC001996
89. He, R., Liu, Y., & Weisberg, R. H. (2004). Coastal ocean wind fields gauged against the performance of a coastal ocean circulation model. *Geophysical Research Letters, 31*, L14303. https://doi.org/10.1029/2003GL019261
90. Yang, Y. J., Tang, T. Y., & Weisberg, R. H. (2004). Current and thermal variations to westerly wind bursts in the equatorial Pacific Ocean. *Terrestrial, Atmospheric and Oceanic Sciences, 15*, 151–178.
91. Weisberg, R. H., He, R., Kirkpatrick, G., Muller-Karger, F., & Walsh, J. J. (2004). Coastal ocean circulation influences on remotely sensed optical properties: A west Florida shelf case study. *Oceanography, 17*, 68–75.
92. Virmani, J. I., & Weisberg, R. H. (2005). Relative humidity over the west Florida continental shelf. *Monthly Weather Review, 133*, 1671–1686.
93. Liu, Y., & Weisberg, R. H. (2005). Momentum balance diagnoses for the west Florida Shelf. *Continental Shelf Research, 25*, 2054–2074.
94. Katsaros, K. B., Soloviev, A. V., Weisberg, R. H., & Luther, M. E. (2005). Reduced horizontal sea surface temperature gradients under conditions of clear sky and weak winds. *Boundary-Layer Meteorology, 116*, 175–185. https://doi.org/10.1007/s10546-004-2421-4
95. Hu, C., Nelson, J. R., Johns, E., Chen, Z., Weisberg, R. H., & Muller-Karger, F. (2005). Mississippi water in the Florida Straits and in the Gulf Stream off the coast of Georgia in summer 2004. *Geophysical Research Letters, 32*, L14606. https://doi.org/10.1029/2005GL022942
96. Weisberg, R. H., He, R., Liu, Y., & Virmani, J. I. (2005). West Florida shelf circulation on synoptic, seasonal, and inter-annual time scales, in Circulation in the Gulf of Mexico. In W. Sturges & A. Lugo-Fernandez (Eds.), *AGU monograph series, geophysical monograph* (Vol. 161, pp. 325–347).
97. Liu, Y., & Weisberg, R. H. (2005). Patterns of ocean current variability on the West Florida Shelf using the self-organizing map. *Journal of Geophysical Research, 110*(C6), C06003.
98. Weisberg, R. H., & Zheng, L. (2006). Circulation of Tampa Bay driven by buoyancy, tides, and winds, as simulated using a finite volume coastal ocean model. *Journal of Geophysical Research, 111*, C01005. https://doi.org/10.1029/2005JC003067
99. Liu, Y., Weisberg, R. H., & He, R. (2006). Sea surface temperature patterns on the West Florida Shelf using growing hierarchical self-organizing maps. *Journal of Atmospheric and Oceanic Technology, 23*(2), 325–338.
100. Virmani, J. I., & Weisberg, R. H. (2006). The 2005 hurricane season: An echo of the past or a harbinger of the future? *Geophysical Research Letters, 33*, L05707. https://doi.org/10.1029/2005GL025517

101. Liu, Y., Weisberg, R. H., & Mooers, C. N. K. (2006). Performance evaluation of the self organizing map for feature extraction. *Journal of Geophysical Research, 111*, C05018. https://doi.org/10.1029/2005jc003117
102. Aretxabaleta, A., Nelson, J. R., Blanton, J. O., Seim, H. E., Werner, F. E., Bane, J. M., & Weisberg, R. H. (2006). Cold event in the South Atlantic Bight during summer of 2003: Anomalous hydrographic and atmospheric conditions. *Journal of Geophysical Research, 111*, C06007. https://doi.org/10.1029/2005JC003105
103. Walsh, J. J., Jolliff, J. K., Darrow, B. P., Lenes, J. M., Milroy, S. P., Remsen, A., Dieterle, D. A., Carder, K. L., Chen, F. R., Vargo, G. A., Weisberg, R. H., Fanning, K. A., Muller-Karger, F., Steidinger, K. A., Heil, C. A., Tomas, C. R., Prospero, J. S., Lee, T. N., Kirkpatrick, G. J., ... Bontempi, P. S. (2006). Red tides in the Gulf of Mexico: Where, when, and why? *Journal of Geophysical Research, 111*, C11003. https://doi.org/10.1029/2004JC002813
104. Weisberg, R. H., & Zheng, L. (2006). A simulation of the hurricane Charley storm surge and its breach of North Captiva Island. *Florida Scientist, 69*, 152–165.
105. Weisberg, R. H., & Zheng, L. (2006). Hurricane storm surge simulations for Tampa Bay. *Estuaries and Coasts, 29*, 899–913.
106. Shay, L. K., Martinez-Pedrala, J., Cook, T. M., Haus, B. K., & Weisberg, R. H. (2007). High-frequency radar mapping of surface currents using WERA. *Journal of Atmospheric and Oceanic Technology, 24*, 484–503.
107. Helber, R. W., Weisberg, R. H., Bonjean, F., & Lagerloef, G. S. E. (2007). Satellite derived surface current divergence in relation to tropical Atlantic SST and wind. *Journal of Physical Oceanography, 37*, 1357–1375.
108. Liu, Y., & Weisberg, R. H. (2007). Ocean currents and sea surface heights estimated across the West Florida Shelf. *Journal of Physical Oceanography, 37*, 1697–1713.
109. Liu, Y., Weisberg, R. H., & Shay, L. K. (2007). Current patterns on the West Florida shelf from joint self-organizing map analyses of HF radar and ADCP data. *Journal of the Seismological Society of Japan, 24*, 702–712.
110. Mayer, D. A., Virmani, J. I., & Weisberg, R. H. (2007). Velocity comparisons from upward and downward acoustic Doppler current profilers on the West Florida Shelf. *Journal of Atmospheric and Oceanic Technology, 24*, 1950–1960.
111. Barth, A., Beckers, J.-M., Alvera-Azcárate, A., & Weisberg, R. H. (2007). Filtering inertia-gravity waves from the initial conditions of the linear shallow water equations. *Ocean Modelling, 19*, 204–218.
112. Liu, Y., Liang, X. S., & Weisberg, R. H. (2007). A note on the wavelet power spectrum. *Journal of the Seismological Society of Japan, 24*, 2093–2102.
113. Alvera-Azcárate, A., Barth, A., Beckers, J. M., & Weisberg, R. H. (2007). Multivariate reconstruction of missing data in sea surface temperature, chlorophyll and wind satellite fields. *Journal of Geophysical Research, 112*, C03008. https://doi.org/10.1029/2006JC003660
114. Liu, Y., Weisberg, R. H., & Yuan, Y. (2008). Patterns of upper layer circulation variability in the South China Sea from satellite altimetry using the self-organizing map. *Acta Oceanologica Sinica, 27*(Supp.), 129–144.
115. Milroy, S. P., Dieterle, D. A., He, R., Kirkpatrick, G. J., Lester, K. M., Steidinger, K. A., Vargo, G. A., Walsh, J. J., & Weisberg, R. H. (2008). A three-dimensional biophysical model of Karenia brevis dynamics on the west Florida shelf: A look at physical transport and zooplankton grazing controls. *Continental Shelf Research, 28*, 112–136.
116. Barth, A., Alvera-Azcárate, A., & Weisberg, R. H. (2008). Benefit of nesting a regional model into a large-scale ocean model instead of climatology. Application to the West Florida Shelf. *Continental Shelf Research, 28*, 561–573.
117. Barth, A., Alvera-Azcárate, A., & Weisberg, R. H. (2008). A nested model study of the loop current generated variability and its impact on the West Florida Shelf. *Journal of Geophysical Research, 113*, C05009. https://doi.org/10.1029/2007JC004492
118. Lenes, J. M., Darrow, B. A., Walsh, J. J., Prospero, J. M., He, R., Weisberg, R. H., Vargo, G. A., & Heil, C. A. (2008). Saharan dust and phosphatic fidelity: A three-dimensional biogeochemical model of *Trichodesmium* as a nutrient source for red tides on the West Florida Shelf. *Continental Shelf Research, 28*, 1091–1115.

119. Barth, A., Alvera-Azcárate, A., & Weisberg, R. H. (2008). Assimilation of high-frequency radar currents in a nested model of the West Florida Shelf. *Journal of Geophysical Research, 113*, C08033. https://doi.org/10.1029/2007JC004585
120. Seim, H. E., Nelson, J., Fletcher, M., Mooers, C. N. K., Spence, L., Weisberg, R. H., Werner, C., Smith, S., & Lea, R. (2008). SEACOOS program management. *MTS Journal, 42*(3), 17–27.
121. Nelson, J., & Weisberg, R. H. (2008). In situ observations and satellite remote sensing in SEACOOS: Program development and lessons learned. *MTS Journal, 42*(3), 41–54.
122. Shay, L. K., Seim, H. E., Savidge, D., Styles, R., & Weisberg, R. H. (2008). High frequency radar observing systems in SEACOOS. *MTS Journal, 42*(3), 55–67.
123. Voulgaris, G., Haus, B. K., Work, P., Shay, L. K., Seim, H. E., Nelson, J. R., & Weisberg, R. H. (2008). Waves initiative within SEACOOS. *MTS Journal, 42*(3), 58–80.
124. Weisberg, R. H. (2008). Epilogue to SEACOOS. *MTS Journal, 42*(3), 21–23.
125. Weisberg, R. H., & Zheng, L. (2008). Hurricane storm surge simulations comparing three-dimensional with two-dimensional formulations based on an Ivan-like storm over the Tampa Bay, Florida region. *Journal of Geophysical Research, 113*, C12001. https://doi.org/10.1029/2008JC005115
126. Alvera-Azcárate, A., Barth, A., & Weisberg, R. H. (2009). A nested model of the Cariaco Basin (Venezuela): Description of the basin's interior hydrography and interactions with the open ocean. *Ocean Dynamics (special issue GODAE Coastal and Shelf Seas Working Group), 59*, 97–120. https://doi.org/10.1007/s10236-008-0169-y.
127. Kourafalou, V. H., Peng, G., Kang, H., Hogan, P. J., Smedstad, O. M., & Weisberg, R. H. (2009). Evaluation of global ocean data assimilation experiment products on South Florida nested simulations with the hybrid coordinate ocean model. *Ocean Dynamics (special issue GODAE Coastal and Shelf Seas Working Group), 59*(1), 47–66. https://doi.org/10.1007/s10236-008-0160-7
128. Weisberg, R. H., Barth, A., Alvera-Azcárate, A., & Zheng, L. (2009). A coordinated coastal ocean observing and modeling system for the West Florida Shelf. *Harmful Algae, 8*, 585–598.
129. Walsh, J. J., Weisberg, R. H., Lenes, J. M., Chen, F. R., Dieterle, D. A., Zheng, L., Carder, K. L., Vargo, G. A., Havens, J. A., Peebles, E., Hollander, D. J., He, R., Heil, C. A., Mahmoudi, B., & Landsberg, J. H. (2009). Isotopic evidence for dead fish maintenance of Florida red tides, with implications for coastal fisheries over both source regions of the West Florida Shelf and within downstream waters of the South Atlantic Bight. *Program in Oceanography, 80*, 51–73.
130. Halliwell, G. R., Barth, A., Weisberg, R. H., Hogan, P., Smedstad, O. M., & Cummings, J. (2009). Impact of GODAE products on nested HYCOM simulations of the West Florida Shelf. *Ocean Dynamics (special issue GODAE Coastal and Shelf Seas Working Group), 59*(1). https://doi.org/10.1007/s10236-008-0173-2
131. Virmani, J. I., & Weisberg, R. H. (2009). Fish effects on ocean current observations in the Cariaco basin. *Journal of Geophysical Research, 114*, C03028. https://doi.org/10.1029/2008JC004889
132. Seim, H. E., Fletcher, M., Mooers, C. N. K., Nelson, J., & Weisberg, R. H. (2009). Towards a regional coastal ocean observing system: An initial design for the Southeast Coastal Ocean observing. *Journal of Marine Systems, 77*, 261–277. https://doi.org/10.1016/j.jmarsys.2007.12.016
133. Alvera-Azcárate, A., Barth, A., & Weisberg, R. H. (2009). The surface circulation of the Caribbean Sea and the Gulf of Mexico as inferred from satellite altimetry. *Journal of Physical Oceanography, 39*, 640–657.
134. Chassignet, E. P., Hurlburt, H. E., Metzger, E. J., Smedstad, O. M., Cummings, J., Halliwell, G. R., Bleck, R., Baraille, R., Wallcraft, A. J., Lozano, C., Tolman, H., Srinivasan, A., Hankin, S., Cornillon, P., Weisberg, R., Barth, A., He, R., Werner, C., & Wilkin, J. (2009). U.S. GODAE: Global ocean prediction with the hybrid coordinate ocean model (HYCOM). *Oceanography, 22*, 48–59.

135. Barth, A., Alvera-Azcárate, A., Beckers, J. M., Weisberg, R. H., Vandenbulcke, L., Lenartz, F., & Rixen, M. (2009). Dynamically constrained ensemble perturbations—Applications to tides on the West Florida Shelf. *Ocean Science, 5*, 259–270.
136. Zheng, L., & Weisberg, R. H. (2009). Rookery Bay and Naples Bay circulation simulations: Applications to tides and fresh water inflow regulation. *Ecological Modelling, 221*, 986–996. https://doi.org/10.1016/j.ecolmodel.2009.01.024
137. Weisberg, R. H., Liu, Y., & Mayer, D. (2009). West Florida shelf mean circulation observed with long-term moorings. *Geophysical Research Letters, 36*, L19610. https://doi.org/10.1029/2009GL040028
138. Liu, Y., Weisberg, R. H., Merz, C. R., Lichtenwalner, S., & Kirkpatrick, G. J. (2010). HF radar performance in a low energy environment: CODAR SeaSonde experience on the West Florida Shelf. *Journal of Atmospheric and Oceanic Technology, 27*(10), 1689–1710.
139. Huang, Y., Weisberg, R. H., & Zheng, L. (2010). The coupling of surge and waves for an Ivan-like hurricane impacting the Tampa Bay, Florida region. *Journal of Geophysical Research, 115*, C12009. https://doi.org/10.1029/2009JC006090
140. Liu, Y., & Weisberg, R. H. (2011). A review of Self-Organizing Map applications in meteorology and oceanography. In J. I. Mwasiagi (Ed.), *Self-organizing maps—Applications and novel algorithm design* (pp. 253–272). InTech, Rijeka, Croatia. ISBN 978-953-307-546-4.
141. Alvera-Azcárate, A. A., Barth, R. H., Weisberg, A. J. J., Casteneda, L. V., & Beckers, J. M. (2011). Thermocline characterization in the Cariaco basin: A modelling study of the thermocline annual variation and its relation with winds and chlorophyll-a concentration. *Continental Shelf Research, 31*, 73–84.
142. Liu, Y., Weisberg, R. H., Hu, C., & Zheng, L. (2011). Tracking the Deepwater Horizon oil spill: A modeling perspective. *EOS Transactions, American Geophysical Union, 92*(6), 45–46. https://doi.org/10.1029/2010ES003187
143. Weisberg, R. H. (2011). Coastal ocean pollution, water quality and ecology: A commentary. *MTS Journal, 45*(2), 35–42.
144. Hu, C., Weisberg, R. H., Liu, Y., Zheng, L., Daly, K. L., English, D. C., Zhao, J., & Vargo, G. A. (2011). Did the Northeastern Gulf of Mexico become greener after the Deepwater Horizon oil spill? *Geophysical Research Letters, 38*, L09601. https://doi.org/10.1029/2011GL047184
145. Walsh, J. J., Tomas, C. R., Steidinger, K. A., Lenes, J. M., Chen, F. R., Weisberg, R. H., Zheng, L., Landsberg, J. H., Vargo, G. A., & Heil, C. A. (2011). Imprudent fishing harvests and consequent trophic cascades on the West Florida Shelf over the last half century: A harbinger of increased human deaths from paralytic shellfish poisoning along the Southeastern United States in response to oligotrophication. *Continental Shelf Research, 31*, 891–911. https://doi.org/10.1016/j.csr.2011.02.007
146. Liu, Y., Weisberg, R. H., Hu, C., & Zheng, L. (2011). Satellites, models combine to track Deepwater Horizon oil spill. *SPIE Newsroom*. https://doi.org/10.1117/2.1201104.003575
147. Liu, Y., & Weisberg, R. H. (2011). Evaluation of trajectory modeling in different dynamic regions using normalized cumulative Lagrangian separation. *Journal of Geophysical Research, 116*, C09013. https://doi.org/10.1029/2010JC006837
148. Liu, Y., MacFadyen, A., Ji, Z.-G., & Weisberg, R. H. (Eds.). (2011). *Monitoring and modeling the Deepwater horizon oil spill: A record-breaking enterprise, geophysical monograph series* (Vol. 195, 271 pp). ISSN: 0065-8448, ISBN 978-0-87590-485-6. AGU/Geopress, Washington D.C.
149. Liu, Y., MacFadyen, A., Ji, Z.-G., & Weisberg, R. H. (2011). Preface. In *Monitoring and modeling the Deepwater horizon oil spill: A record-breaking enterprise, geophysical monograph series, 195*.https://doi.org/10.1029/2011GM001146
150. Weisberg, R. H., Zheng, L., & Liu, Y. (2011). Tracking subsurface oil in the aftermath of the Deepwater Horizon well blowout. *Monitoring and Modeling the Deepwater Horizon Oil Spill: A Record-Breaking Enterprise, Geophysical Monograph Series, 195*, 205–215. https://doi.org/10.1029/2011GM001131
151. Liu, Y., MacFadyen, A., Ji, Z.-G., & Weisberg, R. H. (2011). Introduction to monitoring and modeling the Deepwater horizon oil spill. *Monitoring and Modeling the Deepwater Horizon*

Oil Spill: A Record-Breaking Enterprise, Geophysical Monograph Series, 195, 1–7. https://doi.org/10.1029/2011GM001147

152. Liu, Y., Weisberg, R. H., Hu, C., & Zheng, L. (2011). Trajectory forecast as a rapid response to the Deepwater Horizon oil spill. *Monitoring and Modeling the Deepwater Horizon Oil Spill: A Record-Breaking Enterprise, Geophysical Monograph Series, 195*, 153–165. https://doi.org/10.1029/2011GM001121

153. Liu, Y., Weisberg, R. H., Hu, C., Kovach, C., & Riethmüller, R. (2011). Evolution of the Loop Current system during the Deepwater Horizon oil spill event as observed with drifters and satellites. *Monitoring and Modeling the Deepwater Horizon Oil Spill: A Record-Breaking Enterprise, Geophysical Monograph Series, 195*, 91–101. https://doi.org/10.1029/2011GM001127

154. Zheng, L., & Weisberg, R. H. (2012). Modeling the West Florida Coastal ocean by downscaling from the deep ocean, across the continental shelf and into the Estuaries. *Ocean Modeling, 48*(2012), 10–29. https://doi.org/10.1016/j.ocemod.2012.02.002

155. Lenes, J. M., Darrow, B. P., Walsh, J. J., Jolliff, J. K., Chen, F. R., Weisberg, R. H., & Zheng, L. (2012). A 1-D simulation analysis of the development and maintenance of the 2001 red tide of the ichthyotoxic dinoflagellate *Karenia brevis* on the West Florida shelf. *Continental Shelf Research, 41*, 92–100. https://doi.org/10.1016/j.csr.2012.04.007

156. Liu, Y., Weisberg, R. H., Vignudelli, S., Roblou, L., & Merz, C. R. (2012). Comparison of the X-TRACK altimetry estimated currents with moored ADCP and HF radar observations on the West Florida Shelf. *Advances in Space Research, 50*, 1085–1098. https://doi.org/10.1016/j.asr.2011.09.012

157. Liu, Y., & Weisberg, R. H. (2012). Seasonal variability on the West Florida Shelf. *Progress in Oceanography, 104*, 80–98. https://doi.org/10.1016/j.pocean.2012.06.001

158. Weisberg, R. H., Liu, Y., Merz, C. R., Virmani, J. I., & Zheng, L. (2012). A critique of alternative power generation for Florida by mechanical and solar means. *Marine Technology Society Journal, 46*(5), 12–23. https://doi.org/10.4031/MTSJ.46.5.1

159. Merz, C. R., Weisberg, R. H., & Liu, Y. (2012). Evolution of the USF/CMS CODAR and WERA HF Radar Network. *IEEE Oceans.* https://doi.org/10.1109/OCEANS.2012.6404947

160. Huang, Y., Weisberg, R. H., & Zheng, L. (2013). Gulf of Mexico hurricane wave simulations using SWAN: Bulk formula based drag coefficient sensitivity for Hurricane Ike. *Journal of Geophysical Research: Oceans, 118*, 1–23. https://doi.org/10.1002/jgrc.20283

161. Zheng, L., Weisberg, R. H., Huang, Y., et al. (2013). Implication from the comparisons between two- and three-dimensional model simulations of the Hurricane Ike storm surge. *Journal of Geophysical Research: Oceans, 118*, 3350–3369, https://doi.org/10.1002/jgrc.20248

162. Kerr, P. C., Donahue, A. S., Westerink, J. J., Luettich Jr., R. A., Zheng, L. Y., Weisberg, R. H., Huang, Y., Wang, H. V., Teng, Y., Forrest, D. R., Roland, A., Haase, A. T., Kramer, A. W., Taylor, A. A., Rhome, J. R., Feyen, J. C., Signell, R. P., Hanson, J. L., Hope, M. E., Estes, R. M., Dominguez, R. A., et al. (2013). U.S. IOOS coastal and ocean modeling testbed: Inter-model evaluation of tides, waves, and hurricane surge in the Gulf of Mexico. *Journal of Geophysical Research: Oceans, 118*, 5129–5172. https://doi.org/10.1002/jgrc.20376

163. Zhao, J., Hu, C., Lenes, J. M., Weisberg, R. H., Lembke, C., et al. (2013). Three-dimensional structure of a Karenia brevis bloom: Observations from gliders, satellites, field measurements, and numerical models. *Harmful Algae, 29*, 22–30.

164. Weisberg, R. H., Zheng, L., Liu, Y., Lembke, C., Lenes, J. M., & Walsh, J. J. (2014). Why a red tide was not observed on the West Florida continental shelf in 2010. *Harmful Algae, 38*, 119–126. https://doi.org/10.1016/j.hal.2014.04.010

165. Heil, C. A., Bronk, D. A., Dixon, L. K., Hitchcock, G. L., Kirkpatrick, G. J., Mulholland, M. R., O'Neil, J. M., Walsh, J. J., Weisberg, R. H., & Garrett, M. (2014). The Gulf of Mexico ECOHAB: Karenia program 2006–2012. *Harmful Algae, 38*, 3–7.https://doi.org/10.1016/j.hal.2014.07.015

166. Heil, C. A., Dixon, L. K., Hall, E., Garrett, M., Lenes, J. M., O'Neil, J. M., Walsh, B. M., Bronk, D. A., Killberg-Thoreson, L., Hichcock, G. L., Meyer, K. A., Mulholland, M. R., Procise, L., Kirkpatrick, G. J., Walsh. J. J., & Weisberg, R. H. (2014). Blooms of *Karenia*

brevis (Davis) G. Hansen & O. Moestrup on the West Florida Shelf: Nutrient sources and potential management strategies based on a multi-year regional study. *Harmful Algae, 38*, 127–140.https://doi.org/10.1016/j.hal.2014.07.016

167. Liu, Y., Weisberg, R. H., & Merz, C. R. (2014). Assessment of CODAR and WERA HF radars in mapping currents on the West Florida Shelf. *Journal of the Seismological Society of Japan, 31*(6), 1363–1382. https://doi.org/10.1175/JTECH-D-13-00107.1

168. Bert, T. M., Arnold, W. S., Wilbur, A. E., Seyoum, S., McMillen-Jackson, A. L., Stephenson, S. P., Weisberg, R. H., & Yarbro, L. A. (2014). Florida Gulf Bay Scallop (*Argopecten Irradians Concentricus*) population genetic structure: Form, variation, and influential factors. *Journal of Shellfish Research, 33*, 99–136. https://doi.org/10.2983/035.033.0112

169. Liu, Y., Weisberg, R. H., Vignudelli, S., & Mitchum, G. T. (2014). Evaluation of altimetry-derived surface current products using Lagrangian drifter trajectories in the eastern Gulf of Mexico. *Journal of Geophysical Research, 119*, 2827–2842. https://doi.org/10.1002/2013JC009710

170. Weisberg, R. H., Zheng, L., & Peebles, E. (2014). Gag grouper larvae pathways on the West Florida Shelf. *Continental Shelf Research, 88*, 11–23. https://doi.org/10.1016/j.csr.2014.06.003

171. Pan, C., Zheng, L., Weisberg, R. H., Liu, Y., & Lembke, C. (2014). Comparisons of different ensemble schemes for glider data assimilation on West Florida Shelf. *Ocean Modelling, 81*, 13–24. https://doi.org/10.1016/j.ocemod.2014.06.005

172. Zhu, J., Weisberg, R. H., Zheng, L., & Han, S. (2015). On the flushing of Tampa Bay. *Estuaries and Coasts, 38*, 118–131. https://doi.org/10.1007/s12237-014-9793-6

173. Zhu, J., Weisberg, R. H., Zheng, L., & Han, S. (2015). Influences of channel deepening and widening on the tidal and non-tidal circulation of Tampa Bay. *Estuaries and Coasts, 38*, 132–150. https://doi.org/10.1007/s12237-014-9815-4

174. Kourafalou, V. H., De Mey, P., Staneva, J., Ayoub, N., Barth, A., Chao, Y., Cirano, M., Fiechter, J., Herzfeld, M., Kurapov, A., Moore, A. M., Oddo, P., Pullen, J., Van Der Westhuysen, A., & Weisberg, R. (2015) Coastal ocean forecasting: Science foundation and user benefits. *Journal of Operational Oceanography, 7*, 3, 149–159.https://doi.org/10.1080/1755876X.2015.1022348

175. Weisberg, R. H., Zheng, L., & Liu, Y. (2015). Basic tenets for coastal ocean ecosystems monitoring. In Y. Liu, H. Kerkering & R. H. Weisberg (Eds.), *Coastal ocean observing systems* (461pp.). Elsevier, London. ISBN: 978-0-12-802022-7.

176. Liu, Y., Kerkering, H., & Weisberg, R. H. (2015). Introduction to coastal ocean observing systems. In Y. Liu, H. Kerkering & R. H. Weisberg (Eds.), *Coastal ocean observing systems* (461pp.). Elsevier, London. ISBN: 978-0-12-802022-7.

177. Liu, Y., Weisberg, R. H., & Lembke, C. (2015). Glider salinity correction for unpumped CTD sensors across a sharp thermocline. In Y. Liu, H. Kerkering & R. H. Weisberg (Eds.), *Coastal ocean observing systems* (461pp.). Elsevier, London. ISBN: 978-0-12-802022-7.

178. Merz, C., Liu, Y., Gurgel, K.-W., Pedersen, L., & Weisberg, R. H. (2015). Effect of radio frequency interference (RFI) noise energy on WERA performance using "Listen Before Talk" adaptive noise procedure on the west Florida shelf. In Y. Liu, H. Kerkering & R. H. Weisberg (Eds.), *Coastal ocean observing systems* (461pp.). Elsevier, London, ISBN: 978-0-12-802022-7.

179. Zhu, J., Weisberg, R. H., Zheng, L., & Han, S. (2015). On the salt balance of Tampa Bay. *Continental Shelf Research, 107*, 115–131. https://doi.org/10.1016/j.csr2015.07.001

180. Walsh, J. J., Lenes, J. M., Darrow, B. P., Parks, A. A., Weisberg, R. H., Zheng, L., Hu, C., Barnes, B. B., Daly, K. L., Shin, S.-I., Brooks, G. R., Jeffrey, W. H., Snyder, R. A., & Hollander, D. (2015). A simulation analysis of the plankton fate of the *Deepwater Horizon* oil spills. *Continental Shelf Research, 107*, 50–68. https://doi.org/10.1016/j.csr.2015.07.002

181. Abascal, A. J., Castanedo, S., Mínguez, R., Medina, R., Liu, Y., & Weisberg, R. H. (2015). Stochastic Lagrangian trajectory modeling of surface drifters deployed during the Deepwater Horizon oil spill. In *Proceedings of the 38th AMOP Technical Seminar on Environmental Contamination and Response, Environment and Climate Change* (pp. 71–99). Canada, Ottawa, ON.

182. Hu, C., Murch, B., Corcoran, A. A., Zheng, L., Barnes, B. B., Weisberg, R. H., Atwood, K., & Lenes, J. M. (2016). Developing a smart semantic Web with linked data and models for near-real-time monitoring of red tides in the eastern Gulf of Mexico. *IEEE Systems Journal.* https://doi.org/10.1109/JSYST.2015.2440782
183. Walsh, J. J., Lenes, J. M., Darrow, B. P., Parks, A. A., & Weisberg, R. H. (2016). Impacts of combined overfishing and oil spills on the plankton trophodynamics of the West Florida shelf over the last half century of 1957–2011: A two-dimensional simulation analysis. *Continental Shelf Research, 116*, 54–73.
184. Weisberg, R. H., Zheng, L. Y., Liu, Y., Corcoran, A., Lembke, C., Hu, C., Lenes, J., & Walsh, J. (2016). *Kerenia brevis* blooms on the west Florida shelf: A comparative study of the robust 2012 bloom and the nearly null 2013 event. *Continental Shelf Research, 120*, 106–121. https://doi.org/10.1016/j.csr.2016.03.011
185. Weisberg, R. H., Zheng, L., Liu, Y., Murawski, S., Hu, C., & Paul, J. (2016). Did Deepwater horizon hydrocarbons transit to the West Florida Continental Shelf? *Deep-Sea Res., Part II, 129*, 259–272. https://doi.org/10.1016/j.dsr2.2014.02.002
186. Liu, Y., Weisberg, R. H., Vignudelli, S., & Mitchum, G. T. (2016). Patterns of the loop current system and regions of sea surface height variability in the Eastern Gulf of Mexico revealed by the self-organizing maps. *Journal of Geophysical Research, 121*, 2347–2366. https://doi.org/10.1002/2015JC011493
187. Rubec, P. J., Lewis, J., Reed, D., Santi, C., Weisberg, R. H., Zheng, L., Jenkins, C., Ashbaugh, C. F., Lashley, C., & Versaggi, S. (2016). Linking oceanographic modeling and benthic mapping with habitat suitability models for Pink Shrimp on the West Florida shelf. *Marine and Coastal Fisheries: Dynamics, Management, and Ecosystem Science, 8*, 160–176. https://doi.org/10.1080/19425120.2015.1082519
188. Liu, Y., Weisberg, R. H., Lenes, J. M., Zheng, L., Hubbard, K., & Walsh, J. J. (2016). Offshore forcing on the "pressure point" of the West Florida Shelf: Anomalous upwelling and its influence on harmful algal blooms. *Journal of Geophysical Research: Oceans, 121*, 5501–5515. https://doi.org/10.1002/2016JC011938
189. Weisberg, R. H., Zheng, L., & Liu, Y. (2016). West Florida shelf upwelling: Origins and pathways. *Journal of Geophysical Research: Oceans, 121*, 5672–5681. https://doi.org/10.1002/2015JC011384
190. Walsh, J. J., Lenes, J. M., Weisberg, R. H., Zheng, L., Hu, C., Fanning, K. A., Snyder, R., & Smith, J. (2017). More surprises in the global greenhouse: Human health impacts from recent toxic marine aerosol formations, due to centennial alterations of world-wide coastal food webs. *Marine Pollution Bulletin.* https://doi.org/10.1016/j.marpolbul.2016.12.053
191. Mayer, D. A., Weisberg, R. H., Zheng, L., & Liu, Y. (2017). Winds on the West Florida shelf: Regional comparisons between observations and model estimates. *Journal of Geophysical Research: Oceans, 122*, 834–846, https://doi.org/10.1002/2016JC012112.
192. Weisberg, R. H., Zheng, L., & Liu, Y. (2017). On the movement of Deepwater Horizon oil to Northern Gulf beaches. *Ocean Modelling, 111*, 81–97. https://doi.org/10.1016/j.ocemod.2017.02.002
193. Wang, C., Wang, X., Weisberg, R. H., & Black, M. L. (2017). Variability of tropical cyclone rapid intensification in the North Atlantic and its relationship with climate variations. *Climate Dynamics.* https://doi.org/10.1007/s00382-017-3537-9
194. Weisberg, R. H., & Liu, Y. (2017). On the Loop Current penetration into the Gulf of Mexico. *Journal of Geophysical Research: Oceans, 122*, 9679–9694. https://doi.org/10.1002/2017JC013330, https://doi.org/10.1002/2017JC013330
195. Soloviev, A. V., Hirons, A., Maingot, C., Dodge, R. E., Yankovsky, A. E., Wood, J., Weisberg, R. H., Luther, M. E., & McCreary, J. P. (2017). Southward flow on the coastal flank of the Florida Current. *Deep Sea Research, 125*, 94–105. https://doi.org/10.1016/j.dsr.2017.05.002
196. Liu, Y., Weisberg, R. H., Law, J., & Huang, B. (2018). Evaluation of satellite-derived SST products in identifying the rapid temperature drop on the West Florida Shelf associated with hurricane Irma. *MTS Journal, 52*(3), 43–50. https://doi.org/10.4031/MTSJ.52.3.7

197. Chen, J., Weisberg, R. H., Liu, Y., & Zheng, L. (2018). The Tampa Bay Coastal Ocean circulation model performance for Hurricane Irma. *MTS Journal, 52*(3), 33–42. https://doi.org/10.4031/MTSJ.52.3.6
198. Chiri, H., Liu, Y., Castanedo, S., Medina, R., Weisberg, R., Abascal Santillana, A. J., & Antolínez, J. A. (2019). Statistical simulation of ocean current patterns using autoregressive logistic regression models: A case study in the Gulf of Mexico. *Ocean Modelling, 136*, 1–12. https://doi.org/10.1016/j.ocemod.2019.02.010
199. Mayer, D. A., Zhang, J. A., & Weisberg, R. H. (2019). Surface layer turbulence parameters derived from 1s wind observations on the West Florida Shelf. *Journal of Geophysical Research: Atmospheres, 124.* https://doi.org/10.1029/2018JD029392
200. Weisberg, R. H. (2019). Chapter 3: Material transports. In J. J. Walsh, R. H. Weisberg & J. M. Lenes (Eds.), *Wind-borne illness from coastal seas: Present and future consequences of toxic marine aerosols*. Elsevier. https://doi.org/10.1016/B978-0-12-812131-3.00003-3
201. Weisberg, R. H., Liu, Y., Lembke, C., Hu, C., Hubbard, K., Garrett, M. (2019). The coastal ocean circulation influence on the 2018 West Florida Shelf *K. brevis* Red Tide Bloom. *Journal of Geophysical Research: Oceans, 124.* https://doi.org/10.1029/2018JC014887
202. Chen, J., Weisberg, R. H., Liu, Y., Zheng, L., & Zhu, J. (2019). On the momentum balance of Tampa Bay. *Journal of Geophysical Research: Oceans.* https://doi.org/10.1029/2018JC014890
203. Zhang, Y., Hu, C., Liu, Y., Weisberg, R. H., & Kourafalou, V. H. (2019). Submesoscale and mesoscale eddies in the Florida straits: Observations from satellite ocean color measurements. *Geophysical Research Letters, 46*, 13262. https://doi.org/10.1029/2019GL083999
204. Liu, Y., Weisberg, R. H., & Zheng, L. (2020). Impacts of hurricane Irma on the circulation and transport in Florida Bay and the Charlotte Harbor estuary. *Estuaries and Coasts, 42*, 1194. https://doi.org/10.1007/s12237-019-00647-6
205. Yang, Y., Weisberg, R. H., Liu, Y., & Liang, X. S. (2020). Instabilities and multiscale interactions underlying the Loop Current eddy shedding in the Gulf of Mexico. *Journal of Physical Oceanography, 50*(5), 1289–1317. https://doi.org/10.1175/JPO-D-19-0202.1
206. Yang, Y., McWilliams, J. C., Liang, X. S., Zhang, H., Weisberg, R. H., Liu, Y., & Menemenlis, D. (2020). Spatial and temporal characteristics of the submesoscale energetics in the Gulf of Mexico. *Journal of Physical Oceanography, 51*, 475–488. https://doi.org/10.1175/JPO-D-20-0247.1
207. Huang, M. X., Liang, Y., Zhu, Y., & Liu, R. H. W. (2021). Eddies connect the tropical Atlantic Ocean and the Gulf of Mexico. *Geophysical Research Letters, 48*, e2020GL091277. https://doi.org/10.1029/2020GL091277
208. Liu, Y., Merz, C. R., Weisberg, R. H., Shay, L. K., Glenn, S., & Smith, M. (2021). An initial evaluation of high-frequency radar radial currents in the Straits of Florida with altimetry and model products. In E. Gill & W. Huang (Eds.), *Ocean remote sensing technologies: HF, marine and GNSS-based radar*. IET (pp. 117–144). https://doi.org/10.1049/SBRA537E_ch5
209. Merz, C. R., Liu, Y., & Weisberg, R. H. (2021). Sea surface current mapping with HF radar: A premier. In W. Huang & E. Gill (Eds.), *Ocean remote sensing technologies: HF, marine and GNSS-based radar* (pp. 95–116). IET. https://doi.org/10.1049/SBRA537E_ch4
210. Justic, D., et al. (22 co-authors including R.H. Weisberg and Y. Liu). (2021). A review of transport processes in the Gulf of Mexico along the river-estuary-shelf-ocean continuum: a synthesis of research from the Gulf of Mexico Research Initiative. *Estuaries and Coasts.*https://doi.org/10.1007/s12237-021-01005-1
211. Solo-Gabriele, H., et al. (43 co-authors including R.H. Weisberg and Y. Liu). (2021). Towards integrated modeling of the long-term impacts of oil spills. *Marine Policy, 131*, 104554. https://doi.org/10.1016/j.marpol.2021.104554
212. Allen, T., et al. (32 co-authors, including R.H. Weisberg). (2021). Anticipating and adapting to the future impacts of climate change on the health, security and welfare of low elevation coastal zone (LECZ) communities in Southeastern USA. *Journal of Marine Science and Engineering, 9*, 1196.https://doi.org/10.3390/jmse9111196

213. Ainsworth, C. H., Chassignet, E. P., French-McCay, D., Beegle-Krause, C. J., Berenshtein, I., Englehardt, J., Fiddaman, T., Huang, H., Huettel, M., Justic, D., Kourafalou, V. H., Liu, Y., Mauritzen, C., Murawski, S., Morey, S., Ozgokmen, T., Paris, C. B., Ruzicka, J., Saul, S., ... Zheng, Y. (2021). Ten years of modeling the Deepwater Horizon oil spill. *Environmental Modelling and Software, 142*, 105070. https://doi.org/10.1016/j.envsoft.2021.105070

214. Xie, S., Chen, J., Dixon, T. H., Weisberg, R. H., & Zumberge, M. A. (2021). Offshore sea levels measured with an anchored spar-buoy system using GPS interferometric reflectometry. *Journal of Geophysical Research: Oceans, 126*, e2021JC017734. https://doi.org/10.1029/2021JC017734

215. Beck, M., Altieri, A., Angelini, C., Burke, M. C., Chen, J., Chin, D. W., Gardiner, J., Hu, C., Hubbard, K. A., Liu, Y., Lopez, C., Medina, M., Morrison, E., Phlips, E. J., Raulerson, G. E., Scolaro, S., Sherwood, E. T., Tomasko, D., Weisberg, R. H., & Whalen, J. (2022). Initial estuarine response to inorganic nutrient inputs from a legacy mining facility adjacent to Tampa Bay, Florida. *Marine Pollution Bulletin, 178*, 113598. https://doi.org/10.1016/j.marpolbul.2022.113598

216. Nickerson, A., Weisberg, R. H., & Liu, Y. (2022). On the evolution of the Gulf of Mexico Loop Current through its penetrative, ring shedding and retracted states. *Advances in Space Research, 69*, 4058–4077. https://doi.org/10.1016/j.asr.2022.03.039

217. Weisberg, R. H., & Liu, Y. (2022). Local and deep-ocean forcing effects on the West Florida continental shelf circulation and ecology. *Frontiers in Marine Science, 9*, 863227. https://doi.org/10.3389/fmars.2022.863227

218. Liu, Y., Weisberg, R. H., Zheng, L., Heil, C., & Hubbard, K. (2022). Termination of the 2018 Florida red tide event: A tracer model perspective. *Estuarine, Coastal and Shelf Science* (in press).

219. Vasbinder, K., Ainsworth, C. H., Liu, Y., & Weisberg, R. H. (2023). Gulf of Mexico Larval dispersal: Combining concurrent sampling, behavioral, and hydrodynamic data to inform end-to-end modeling efforts through a Lagrangian dispersal model. *Deep-Sea Research Part II, 211*, 105323. https://doi.org/10.1016/j.dsr2.2023.105323

220. Chen, J., Weisberg, R. H., Liu, Y., Zheng, L., Law, J., Gilbert, S., & Murawski, S. A. (2023). Tampa Bay coastal ocean model (TBCOM) nowcast/forecast system. *Deep-Sea Research Part II, 211*, 105322. https://doi.org/10.1016/j.dsr2.2023.105322

221. Nickerson, A. K., Weisberg, R. H., Zheng, L., & Liu, Y. (2023). Sea surface temperature trends for Tampa Bay, West Florida Shelf and the deep Gulf of Mexico. *Deep-Sea Research Part II, 211*, 105321. https://doi.org/10.1016/j.dsr2.2023.105321

222. Sorinas, L., Weisberg, R. H., Liu, Y., & Law, J. (2023). Ocean-atmosphere heat exchange seasonal cycle on the West Florida Shelf derived from long term moored data. *Deep-Sea Research Part I, 1*, 212. https://doi.org/10.1016/j.dsr2.2023.105341

223. Liu, Y., Weisberg, R. H., Zheng, L., Hubbard, K. A., Muhlbach, E. G., Garrett, M. J., Hu, C., Cannizzaro, J. P., Xie, Y., Chen, J., John, S., & Liu, L. Y. (2023). Short-term forecast of *Karenia brevis* trajectory on the West Florida shelf. *Deep-Sea Research Part II, 211*, 105335. https://doi.org/10.1016/j.dsr2.2023.105335

224. Chen, J., Liu, Y., Weisberg, R. H., Murawski, S. A., Gilbert, S., Naar, D. F., Zheng, L., Hommeyer, M., Dietrick, C., Luther, M. E., Hapke, C., Myers, E., Moghimi, S., Allen, C., Tang, L., Khazaei, B., Peeri, S., & Wang, P. (2023). Hydrodynamic response to bathymetric changes in Tampa Bay, Florida. *Deep-Sea Research Part II, 212*, 105344. https://doi.org/10.1016/j.dsr2.2023.105344

225. Yang, Y., Fu, G., Liang, X. S., Weisberg, R. H., & Liu, Y. (2023). Causal relations between the Loop Current penetration and the inflow/outflow conditions inferred with a rigorous quantitative causality analysis. *Deep-Sea Research Part II, 209*, 105298. https://doi.org/10.1016/j.dsr2.2023.105298

226. Nguyen, B. V. V., Liu, Y., Stallings, C. D., Breitbart, M., Murawski, S. A., Weisberg, R. H., Kerr, M., Bonnelycke, E.-M.S., & Peebles, E. B. (2024). Retention and export of planktonic fish eggs in the Northeastern Gulf of Mexico. *Fisheries Oceanography, 33*, e12655. https://doi.org/10.1111/fog.12655

227. Law, J., Weisberg, R. H., Liu, Y., Mayer, D. A., & Donovan, J. C. (2024). Mean circulation and its seasonal cycle on the West Florida shelf as evidenced by multi-decadal time series. *Deep-Sea Research Part II, 213*, 105346. https://doi.org/10.1016/j.dsr2.2023.105346

228. Liu, Y., Weisberg, R. H., Zheng, L., Sun, Y., Chen, J., Law, J. A., Hu, C., Cannizzaro, J. P., & Frazer, T. K. (2024). A tracer model nowcast/forecast study of the Tampa Bay, Piney point effluent plume: Rapid response to an environmental hazard. *Marine Pollution Bulletin, 198*, 115840. https://doi.org/10.1016/j.marpolbul.2023.115840

The manufacturer's authorised representative in the EU is Springer Nature Customer Service Centre GmbH, Europaplatz 3, 69115 Heidelberg, Germany. If you have any concerns regarding our products, please contact ProductSafety@springernature.com

Printed and bound by CPI Group (UK) Ltd, Croydon, CR0 4YY

26/03/2026

02078941-0005